Atoms and Molecules in Strong External Fields

Atoms and Molecules in Strong External Fields

Edited by

P. Schmelcher
University of Heidelberg
Heidelberg, Germany

and

W. Schweizer
University of Tübingen
Tübingen, Germany

Plenum Press • New York and London

Library of Congress Cataloging-in-Publication Data

Atoms and molecules in strong external fields / edited by P. Schmelcher and W. Schweizer.
p. cm.
"Proceedings of the 172nd WE-Heraeus-Seminar on Atoms and Molecules in Strong External Fields, held April 7-11, 1997 at Bad Honnef, Germany"-- T.p. verso.
Includes bibliographical references and index.
ISBN 0-306-45811-X
1. Electronic structure--Congresses. 2. Rydberg states--Congresses. 3. Electric fields--Congresses. 4. Magnetic fields--Congresses. 5. White dwarfs--Congresses. 6. Neutron stars--Congresses. 7. Astrophysics--Congresses. I. Schmelcher, P. II. Schweizer, W. III. W.E. Heraeus Seminar (172nd : 1997 : Bad Honnef, Germany)
QC176.8.E4A86 1998
530.4'16--dc21 97-52806
CIP

Proceedings of the 172nd WE-Heraeus-Seminar on Atoms and Molecules in Strong External Fields, held April 7 – 11, 1997, at Bad Honnef, Germany

ISBN 0-306-45811-X

A Division of Plenum Publishing Corporation
233 Spring Street, New York, N.Y. 10013

http://www.plenum.com

10 9 8 7 6 5 4 3 2 1

Printed in the United States of America

PREFACE

This book contains contributions to the 172. WE-Heraeus-Seminar "Atoms and Molecules in Strong External Fields," which took place April 7–11 1997 at the Physikzentrum Bad Honnef (Germany).

The designation "strong fields" applies to external static magnetic and/or electric fields that are sufficiently intense to cause alterations in the atomic or molecular structure and dynamics. The specific topics treated are the behavior and properties of atoms in strong static fields, the fundamental aspects and electronic structure of molecules in strong magnetic fields, the dynamics and aspects of chaos in highly excited Rydberg atoms in external fields, matter in the atmosphere of astrophysical objects (white dwarfs, neutron stars), and quantum nanostructures in strong magnetic fields. It is obvious that the elaboration of the corresponding properties in these regimes causes the greatest difficulties, and is incomplete even today.

Present-day technology has made it possible for many research groups to study the behavior of matter in strong external fields, both experimentally and theoretically, where the phrase "experimentally" includes the astronomical observations. Understanding these systems requires the development of modern theories and powerful computational techniques. Interdisciplinary collaborations will be helpful and useful in developing more efficient methods to understand these important systems. Hence the idea was to bring together people from different fields like atomic and molecular physics, theoretical chemistry, astrophysics and all those colleagues interested in aspects of few-body systems in external fields.

In combination or individually, the articles present a broad and timely review of the recent progress and the current state of the art in the theoretical, computational, and experimental studies of atoms and molecules in strong external fields. Astrophysical aspects related to magnetic white dwarfs and neutron stars are discussed. The computational problems in the strong field regime where the valence electrons experience electric and magnetic forces of comparable strength are discussed, and some new and effective methods based on discretization and finite element methods as well as novel basis set approaches are presented.

New experiments of Rydberg states in strong external fields are reported and related theoretical and computational aspects as well as the quest of quantum chaos are discussed. Attention is drawn to the non-separability of the center-of-mass for atomic and molecular systems in strong magnetic fields. This non-separability gives rise to effects important in the Rydberg as well as in the astrophysical region. But not only atoms and molecules in strong magnetic fields are reviewed; this book is rounded off by the discussion of quantum dots and shallow donor states in strong magnetic fields.

Due to the scientific importance of the subject we hope that the articles presented in this book will prove valuable to a wide scientific audience, ranging from the experienced researcher to the newcomer. The 172. WE-Heraeus-Seminar brought together about 50 scientists from many countries. As scientific organizers, we wish to thank them for their participation, their presentation, and their enthusiasm, which created a very stimulating and scientifically fruitful atmosphere. We would like to express our thanks to Jutta Hartmann and Dr. Volker Schafer from the WE-Heraeus-Stiftung for the unbureaucratic procedure of funding, general organization and realization, and, of course, to the founders Dr. Wilhelm Heinrich Heraeus and Else Heraeus. We thank the Deutsche Forschungsgemeinschaft for their financial support for the East-European participants.

Tübingen and Heidelberg

W. Schweizer
P. Schmelcher

CONTENTS

Atoms and Molecules in Strong External Fields

WHITE DWARFS FOR PHYSICISTS

Detlev Koester

Institut für Astronomie und Astrophysik, Universität Kiel
D-24098 Kiel, Germany

INTRODUCTION

A small number of white dwarf stars show extremely high magnetic fields, of the order of 10^8 G. This is the only possibility to observe the behavior of the hydrogen atom in such fields, and to compare energy shifts and transition probabilities with the predictions of theory. These strange objects clearly deserve to be a topic at this meeting, and observations of magnetic white dwarfs as well as theoretical interpretations will be presented in a later talk by S. Jordan. This paper is meant as an introduction for the non-specialist. Using extremely simplified models and avoiding astronomical terminology as far as possible, I will attempt to describe what are white dwarfs, where do they come from, and what are the physical conditions we find in them.

These questions are answered by the theory of stellar structure and stellar evolution, and we understand already the most important facts about stellar evolution, if we realize the overwhelming importance of gravitational forces. The life of a star is dominated by a battle between the gravitational attraction of matter, which attempts to compress the stellar matter to higher and higher densities, and the pressure of the gas, which tries to resist this compression. Since stars are losing energy from the surface into interstellar space, an internal energy source is necessary to maintain the pressure, at least as long as the equation of state is given by the ideal gas law, where pressure depends on density *and* temperature. As we know today, these energy sources are nuclear fusion reactions, and a critical phase in the life of a star comes, when the nuclear fuel is exhausted and stellar evolution reaches the final stages. According to theory there are three different possibilities for these end-products: a black hole, which means the ultimate victory of gravitation, a neutron star, where the pressure of degenerate neutrons (modified by nuclear interactions) supplies the pressure independent of temperature, and, finally, white dwarfs, where the pressure is supplied by the degenerate electron gas.

EXTREMELY SIMPLIFIED OVERVIEW OF STELLAR EVOLUTION

Let us start from the beginning, the formation of stars, and a little more quantitatively. We consider a spherical mass of gas, with radial coordinate r measured from

Atoms and Molecules in Strong External Fields
Edited by Schmelcher and Schweizer, Plenum Press, New York, 1998

the center, and m the mass inside a sphere of radius r, dm the mass of a shell between r and $r + dr$. The gravitational force between the sphere and the overlying shell is then

$$F = \frac{G\,m\,dm}{r^2}$$

with gravitational constant G. This force creates an increase of pressure, going inward over a shell dm of

$$dP = \frac{G\,m\,dm}{r^2\,4\pi r^2}.$$

In order to integrate this equation exactly, we would have to know the distribution of matter density $\rho(r)$ inside the sphere. But on dimensional grounds as well as from integrations with simple assumptions (e.g. a homogeneous sphere, $\rho = const$) it is clear that the "gravitational pressure" at the center of the sphere, caused by the "weight" of the matter in the gravitational field, has to be

$$P_{grav} \approx \frac{GM^2}{R^4} \qquad \text{or} \qquad P_{grav} \propto GM^{2/3}\rho_c^{4/3},$$

where M and R are the total mass and radius of the sphere, and for the second form we have used the fact that $\rho_c \propto M/R^3$. The constant of proportionality in the second expression above is 0.81 for a homogeneous sphere, 0.59 for a quadratic increase of density inward, and always of the order of 1. In our future estimates we will just use 1.

Star formation and early evolution

We can apply this result to study the conditions for the formation of stars out of thin interstellar matter. Considering a spherical cloud of density ρ and temperature T, we estimate that the cloud will start to contract under its own gravity, if at the center the gravitational pressure is larger than the gas pressure

$$GM^{2/3}\rho^{4/3} > \frac{\mathcal{R}}{\mu}\rho T$$

with the gas constant $\mathcal{R}$ and molecular weight μ. A simple calculation determines the minimum mass necessary for this to occur as

$$M > \left[\frac{\mathcal{R}}{G\mu}\right]^{3/2} \rho^{-1/2}\,T^{3/2},$$

which in astronomy is called the Jeans criterium for star formation. Under typical conditions of the interstellar matter ($\rho = 10^{-24} g\,cm^{-3}, T = 100\,K, \mu = 1$) this corresponds to about 22000 $M_\odot$ (solar masses). Stars are formed in larger groups (clusters) — only when the density gets higher, smaller masses of the order of a solar mass become unstable and the fractionation of the interstellar cloud continues. It should be emphasized again, that this description is extremely simplistic, and that in fact the star formation is rather poorly understood, even by the experts.

What happens next, after the cloud has started to contract, decreasing the radius and increasing the density? That depends on how the two pressures in the balance react to increasing density

$$P_{grav} \propto \rho^{4/3} \qquad \text{and} \qquad P_{gas} \propto \rho T$$

using again the equation of state for an ideal gas. In the beginning the matter is opti-

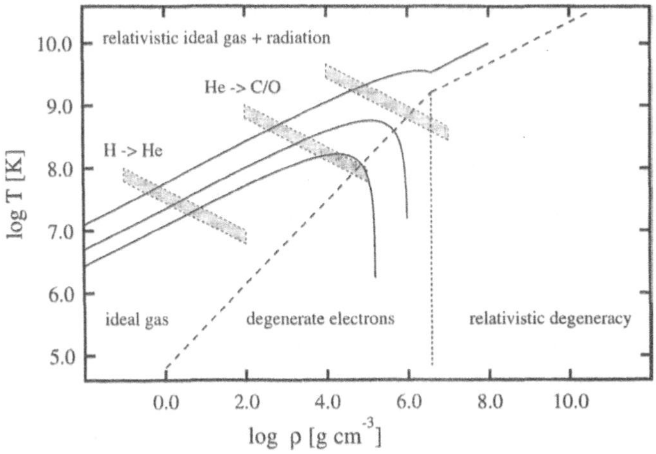

Figure 1. Very schematic evolution of stars with different initial masses in the $T - \rho$ diagram. The path followed by the central parts of the stars is indicated by the continuous lines, from the lower left to upper right. The lower curves correspond to stars with small masses, which, after going through the H- and He-burning stages (shaded areas) reach the domain of electron degeneracy. Here the temperature decreases during the further contraction and the star ends its nuclear life as a white dwarf.

cally thin, meaning that photons can freely escape and carry away the heat produced by contraction and release of gravitational binding energy. The temperature remains approximately constant, and therefore $P_{gas} \propto \rho$. Gravitational forces increase steeper with density and very soon dominate completely over the gas pressure. This leads to a free-fall collapse of the cloud. The timescale for this collapse is the dynamical timescale, which can be estimated in several different ways (for example from the time a sound wave needs to travel the radius of the cloud R). The typical result is always

$$\tau_{dyn} = \frac{1}{\sqrt{G\rho}},$$

which in the case considered means a few million years.

When the density becomes high enough, photons can no longer escape freely and a better model is the opposite extreme of adiabatic changes (no exchange of heat with the outside world). For a monatomic gas (e.g. neutral hydrogen), we then get $P_{gas} \propto \rho^{5/3}$. This is a steeper increase than for the gravitational pressure, and the protostar can find a new hydrostatic equilibrium, where both pressures are in complete balance, $P_{grav} = P_{gas}$.

As the energy loss from the surface continues (called L, the luminosity, by astronomers), the protostar continues to contract, transforming gravitational binding energy into heat, but the evolution is slow and the object always remains extremely close to mechanical equilibrium. Such a phase is called gravitational contraction. The gravitational binding energy of a protostar or star is approximately

$$\Omega \approx -\frac{GM^2}{R} \approx P_{grav} R^3.$$

The release of this energy could supply the luminosity L of a star for a time called the thermal or Kelvin-Helmholtz timescale

$$\tau_{th} \approx \left|\frac{\Omega}{L}\right|$$

which is about 10^7 years for our sun.

Evolution in the density-temperature plane

The key point to understanding the essentials of stellar evolution, and especially the formation of white dwarfs, is the study of the behavior of the central parts in a density-temperature diagram (Fig. 1). Using the hydrostatic equilibrium condition $P_{grav} = P_{gas}$, we find

$$GM^{2/3}\rho^{4/3} = \frac{\mathcal{R}}{\mu}\rho T \longrightarrow T = \frac{G\mu}{\mathcal{R}}M^{2/3}\rho^{1/3}$$

or

$$\log T = \frac{1}{3}\log\rho + \frac{2}{3}\log M + const.$$

The central parts move on a straight line with slope 1/3 in the double-logarithmic diagram, and therefore the temperature increases, until the conditions necessary for "hydrogen burning", the fusion of hydrogen to helium, are reached. This marks the change from protostar to star; nuclear fusion provides so much energy that the star changes very little for several billion years (nuclear timescale). For a star like our sun this is the longest phase in its life.

When finally the hydrogen in the central parts is transformed to helium, the energy generation moves farther out, to a shell around the helium core. This core again starts gravitational contraction, until conditions for He burning are reached. For a massive star, e.g. 15 $M_\odot$, this pattern of nuclear burning and gravitational contraction continues until the central parts consist of the most tightly bound element iron, and no further energy source is available. The interior then collapses to a neutron star or black hole, releasing so much energy in one second that we observe it as a very spectacular event, a supernova.

What is different for less massive stars? According to our condition for gravitational contraction less massive stars evolve at lower temperature and higher density. They eventually reach regions in the $T-\rho$ diagram, where the assumption of a classical ideal gas for the equation of state is no longer valid. The matter in the interior is completely ionized, consisting of the heavy nuclei and electrons. When the electrons are squeezed into a smaller and smaller volume by the overall gravitational forces, they start to feel the effect of the quantum mechanical Pauli principle. Because all low lying states for the momenta are occupied, they are forced into higher and higher states, increasing the pressure (= transfer of momentum) provided by the electron gas. In the extreme case of complete degeneracy, the pressure does not depend anymore on temperature, but only on density as

$$P \propto \rho^{5/3} \qquad \text{or} \qquad P \propto \rho^{4/3}$$

depending on whether the velocities of the electrons are non-relativistic (5/3) or relativistic (4/3). We can estimate the location of the transition region by equating the pressure of the limiting expressions ideal gas, and completely degenerate, non-relativistic electron gas

$$\rho T \propto \rho^{5/3} \longrightarrow T \propto \rho^{1/3}.$$

The slope of this line marking the transition is obviously steeper than the slope of the path during gravitational contraction (2/3), so sooner or later a low-mass star will reach this region.

Once the central parts reach the region of degeneracy, this results in a profound change of evolution. We can understand this qualitatively with a simple approximation to the equation of state in the transition region by taking the sum of both contributions. In the limiting cases this is correct, while in the transition region the error may be a factor of 2, but that is good enough to understand the basic principle. The equilibrium condition becomes

$$GM^{2/3}\rho^{4/3} = \frac{\mathcal{R}}{\mu_0}\rho T + \frac{K_1}{\mu_e^{5/3}}\rho^{5/3},$$

where some new symbols are constants from the exact formulation of the equation of state, but not important for our argument here. The evolution in the $T - \rho$ plane is given by

$$T = \frac{GM^{2/3}\mu_0}{\mathcal{R}}\rho^{1/3} - \frac{K_1\,\mu_0}{\mathcal{R}\,\mu_e^{5/3}}\rho^{2/3}.$$

The first term is the well known result for the ideal gas, with the temperature increasing with contraction. However, when the region of electron degeneracy is reached, the second term will gradually become more and more important, the central temperature will go through a maximum and then start to *decrease* steeply upon further contraction. This is still a gravitational contraction with some release of gravitational binding energy, but since the star cools down internally, no new nuclear energy source will be reached and this is a final state of evolution. Our current theory predicts that most stars, including our own sun, will reach this stage after the He burning phase. Their interior will then be composed of the ashes of this process, that is carbon and oxygen.

WHITE DWARFS — COOLING HIGHLY DEGENERATE CONFIGURATIONS

The astronomical objects called "white dwarfs" are identified with these theoretical configurations, which do not reach iron in the sequence of nuclear burning phases, but enter the regime of electron degeneracy (in most cases after the He burning) and then quietly cool down into invisibility. Observationally they were recognized about 90 years ago as stars with normal surface temperatures, but much lower total energy output (luminosity). The only explanation was a small radius, of the order of 1/100 of the solar value. In the case of binary stars, e.g. the famous example of Sirius A and its companion Sirius B the mass was known to be about one solar mass, which meant extremely high densities. This puzzle was only solved in 1926, after the discovery of quantum mechanics and the degenerate electron gas.

Masses, radii, cooling times

Typical parameters of these stars are masses around 0.6 $M_\odot$ sun, with a rather narrow distribution, although a few stars are known below 0.4 and above 1.0 $M_\odot$. Average densities are then 10^5 to 10^6 g cm^{-3}, and typical luminosities around 0.01 $L_\odot$. This luminosity ultimately comes from the change of gravitational binding energy

$$\Delta\Omega \approx \frac{GM^2}{R}\frac{\Delta R}{R}$$

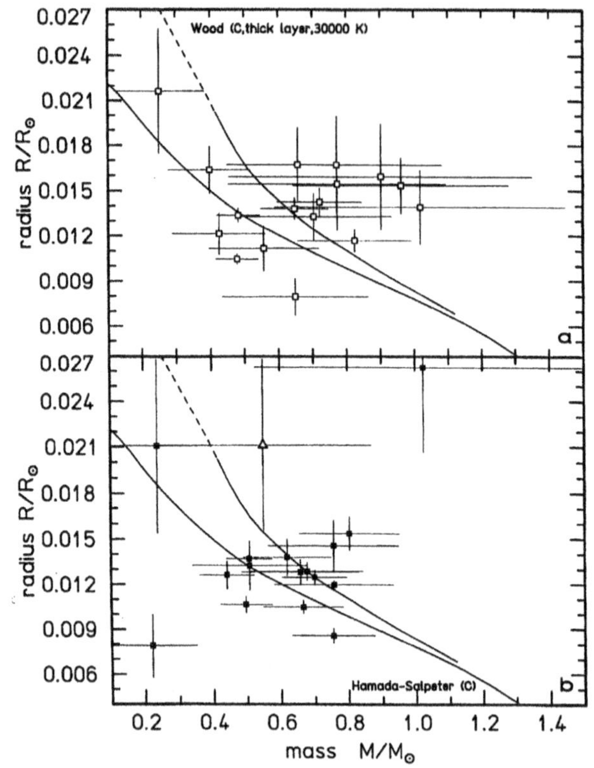

Figure 2. Masses and radii obtained using ground-based distances (top) compared to data derived from the recent HIPPARCOS space mission. The continuous lines are more sophisticated versions of the MRR, including non-ideal effects like Coulomb interactions in the equation of state, and the effect of finite temperatures (non-complete degeneracy). The results generally agree with the theoretical relations (with a few exceptions, which can be understood as perturbation of the measurement by a very close companion star), but due to the clustering of points around 0.6 $M_\odot$ they cannot confirm the detailed shape. From Vauclair et al. (1997)

leading to a cooling timescale of

$$\tau_{cool} \approx \frac{\Delta\Omega}{L} \approx \frac{GM^2}{RL}\frac{\Delta R}{R} \approx 10^{11}\frac{\Delta R}{R} \qquad \text{years.}$$

Even a very small change in R is sufficient to supply the luminosity of a white dwarf for billions of years; this is another long-lived phase for low-mass stars.

The best-known fact about the physics of white dwarfs is probably the existence of a mass-radius relation (MRR) and of a limiting mass. We can understand this qualitatively using the same argumentation as before for the mechanical equilibrium, but now using the equation of state for the degenerate electron gas and the form including the radius instead of density

$$\frac{GM^2}{R^4} \propto \rho_c^{5/3} \propto \left(\frac{M}{R^3}\right)^{5/3} \propto \frac{M^{5/3}}{R^5}.$$

This leads to a relation $R \propto M^{-1/3}$, valid for non-relativistic electrons, that is low-mass white dwarfs. The radius decreases with increasing mass and increasing central density.

When the electrons become relativistic, we have

$$\frac{GM^2}{R^4} \propto \rho_c^{4/3} \propto \frac{M^{4/3}}{R^4},$$

and equilibrium is now possible for one single mass only, but arbitrary radius. This is, however, not a stable equilibrium; a small perturbation would either lead to a collapse to infinite density at radius zero, or to an expansion. In such an expansion the electrons in the outer parts will become non-relativistic and a stable equilibrium is possible. The single solution for the mass in the ultra-relativistic case is the critical, or Chandrasekhar mass. It is the upper limit for white dwarf masses, and for an interior composition of carbon or oxygen its value is 1.4 $M_\odot$.

Although this MRR and the limiting mass are firmly established theoretically, the empirical evidence is still not very convincing. The most important reasons are that the observed white dwarfs seem to cluster around 0.6 $M_\odot$, making it difficult to establish the relation for small and large masses, and the difficulty to measure distances to these objects, which are necessary for the determination of masses and radii. In recent years the European Space Agency ESA has used the satellite HIPPARCOS, to measure accurate distances to a large number of stars, including about 20 white dwarfs. Fig. 2 shows the results for the MRR obtained with these new data, compared to the use of ground-based measurements only. Because the white dwarfs are very faint, the improvement is not as obvious as for other, brighter stars. The general agreement with the theoretical calculations is considered satisfactory, although the observations certainly do not prove the detailed shape of the relation, nor distinguish between different versions for slight differences in the internal structure of white dwarfs.

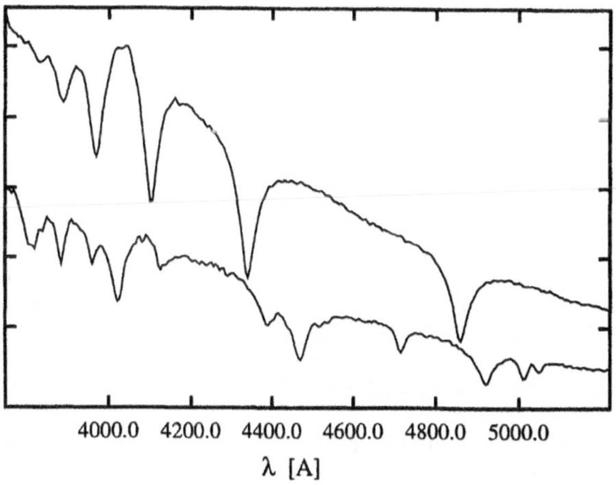

Figure 3. Optical spectra of typical DA (upper curve, Balmer lines of hydrogen only) and DB (lower curve, HeI lines only). Vertical axis is intensity in arbitrary units.

Observable Atmospheres

Directly observable are only the atmospheres, the outermost layers of white dwarfs, which are accessible to photometry (measuring brightness through different filters) and spectroscopy. From the observed spectra we distinguish two main spectral groups of white dwarfs. By far the largest subgroup ($\approx 80\%$) shows only spectral lines of hydrogen; this is the type DA, and the surface layers consist indeed of extremely pure hydrogen. On the other hand, in the remaining 20 %, the atmospheres are almost pure helium and show only spectral lines of neutral or ionized helium (spectral types DB, DO, + some smaller groups). Fig. 3 shows typical representatives of these two spectral groups; the most apparent features are extremely broad lines (broadened by pressure broadening) due to either hydrogen (DA) or helium (DB). These mono-elemental compositions are unknown in any other object in the universe; the basic explanation for this is "gravitational separation", an effect known since almost 50 years. In the strong gravitational fields on the surfaces of these stars the heavy elements sink down, leaving the lightest element present floating on top. The physical process is element diffusion, and it seems to work efficiently in white dwarfs, because there are no other velocity fields (due to convection, circulation, stellar winds) to disturb it.

Of the few white dwarfs with very strong magnetic fields, all objects with identified features belong to the DA class. Whether this is a selection effect due to small numbers, or whether helium is responsible for some objects with unidentified features, is currently unknown, and will probably only be understood, when calculations for He in extreme fields become available.

This concludes our journey from interstellar matter to the surfaces of magnetic white dwarfs. White dwarfs are very interesting objects from an astronomical point of view, since they are the most common end-product of stellar evolution, and since they offer the opportunity to study important astrophysical processes as convection, diffusion, pulsation, accretion. But they are also fascinating for a physicist, because they offer conditions that cannot, or not easily be achieved in terrestrial laboratories. We can study macroscopic effects of quantum mechanics with the equation of state, various aspects of line broadening theories, and, finally, the effect of extremely strong magnetic fields on atoms, which is the topic of this meeting. In the spirit of this very elementary physical discussion I have given almost no references in the text; however, for the reader interested in more of the physical or astronomical details I include below a few review papers and the most relevant recent conference proceedings.

REFERENCES

Barstow, M.A. (ed.), 1993, White Dwarfs: Advances in Observation and Theory, Kluwer (Dordrecht)

Chanmugam, G., 1992, Magnetic fields of degenerate stars, Ann. Rev. Astr. Ap. 30:143

Koester, D., Chanmugam, G., 1990, The physics of white dwarf stars, Rep. Prog. Phys. 53:837

Koester D., Werner, K. (eds.), 1995, White Dwarfs, Lecture Notes in Physics, Vol. 443, Springer-Verlag (Heidelberg)

Shapiro, S.L., Teukolsky, S.A., 1983, Black Holes, White Dwarfs, and Neutron Stars, Wiley & Sons (New York)

Vauclair, G., Schmidt, H., Koester, D., Allard, N., 1997, White dwarfs observed with the HIPPARCOS satellite, A&A, in press

MAGNETIC WHITE DWARFS: OBSERVATIONS IN COSMIC LABORATORIES

Stefan Jordan

Institut für Astronomie und Astrophysik, Universität Kiel
D-24098 Kiel, Germany

INTRODUCTION

Magnetic white dwarfs are the only known physical system in which the behaviour of spectral lines, especially of hydrogen, in the presence of very strong magnetic fields (up to 10^9 G) can directly be studied. Presently, the analysis of the radiation from neutron stars is much more complicated and less unique. As discussed in the paper by Detlev Koester (this conference) the atmospheres of white dwarfs (i.e. the layers in which the observed radiation originates) are often of very simple chemical composition (almost pure hydrogen or helium); the reason is element separation due to the strong gravitational accelaration of about $g = 10^8\,\mathrm{cm\,sec^{-2}}$. Therefore the shifted line components of hydrogen and helium can be observed, often without taking into account a complicated mixture of different elements.

MAGNETIC FIELD ON STARS

Magnetic fields have been measured in many different types of stars. For obvious reasons the first star on which magnetic fields could be detected was the sun on which Hale (1908) observed the magnetic splitting of spectral lines in sunspots. The solar magnetic field is quite complex and mostly concentrated in magnetic flux tubes with field strengths of a few kG. Babcock (1947) discovered a large (≈ 1500 G) and variable magnetic field on 78 Vir. With spectral type A1 p this star belongs to the peculiar A and B main sequence stars (hot stars, burning hydrogen to helium in their center) on which magnetic fields up to 16 kG have been found (Landstreet 1992). It was not until 1980 when Robinson et al. discovered magnetic fields of about 2000 G on limited parts ($\approx 10\%$) of the stellar surface of cooler main sequence stars (spectral type G and K).

MAGNETIC FIELD ON WHITE DWARFS

Blackett (1947) predicted that much stronger magnetic fields (≈ 1 MG) could exist in white dwarfs if the magnetic moment of a star is proportional to its angular momentum,

Atoms and Molecules in Strong External Fields
Edited by Schmelcher and Schweizer, Plenum Press, New York, 1998

which he assumed to be conserved during the stellar evolution and the collapse. This is, however, probably not the case since most isolated white dwarfs seem to be relatively slow rotators ($v \lesssim 40$ km/sec, e.g. Koester & Herrero 1988, Heber et al. 1997), although a few exceptions from this rule exist (e.g. RE J 0317-853, see below). The fact that white dwarfs are typically slow rotators is rather surprising since most of the known white dwarfs stem from progenitors with masses $> 1.5\ M_\odot$ which had typical rotational velocities of $v_{rot} \approx 100$ km/sec; if angular momentum is completely conserved during the evolution we would expect the white dwarf remnant to have $v_{rot} \approx 10,000$ km/sec.

Another possibility was proposed by Ginzburg (1964) and Woltjer (1964). They argued that if the magnetic flux, which is proportional to BR^2, is conserved during evolution and collapse, very strong magnetic fields can be reached in degenerate stars. A main sequence star with a radius $R \approx 10^{11}$ cm and a surface magnetic field of 1-10 kG can therefore become a white dwarf ($R \approx 10^9$ cm) with a magnetic field strength of $10^7 - 10^8$ G.

The search for magnetic white dwarfs began in 1970 when Preston looked for quadratic Zeeman shifts in the spectra of DA white dwarfs. Due to the extremely strongly Stark broadned Balmer lines and the limited spectral resolution he was only able to place upper limits of about 0.5 MG for the magnetic fields in several white dwarfs.

A rather sensitive method to detect magnetic fields in white dwarfs is the measurement of circular polarization. Kemp (1970) proposed that a field of $10^6 - 10^7$ G would produce detectable circular polarization due to circular dichroism, caused by different free-free opacities for the ordinary and extraordinary mode of radiative propagation. After his failure to find polarization in DA white dwarfs he applied his method to several of the almost featureless white dwarfs (classified as DC). In Grw +70°8247, an object that was known for its rather shallow and unidentified "Minkowski bands" (Minkowski 1938, Greenstein 1956, Wegner 1971), he detected circular polarization of several percent. With the help of a magnetoemission model he derived a magnetic field strength of 10 MG, although the circular polarization was not proportional to the wavelength as predicted by Kemp's model. Later his value for the magnetic field strength turned out to be much too low (due to the fact that the free-free opacity is not the dominating absorption process in Grw +70°8247); his idea that the strange spectrum of Grw +70°8247 can be explained by a strong magnetic field was, however, correct. Nevertheless, all attempts to identify the Minkowski bands with various atoms or molecules in magnetic fields of a few MG failed.

Even for the simplest atoms, hydrogen and helium, accurate calculations for the line components did not exist at that time for field strengths above 20-100 MG (depending on the line transitions, Kemic 1974a, 1974b); only for extremely intense fields ($10^9 - 10^{10}$ G) data were available again (Garstang 1977), but none of the predicted line positions were in agreement with the wavelengths of the Grw +70°8247 features. For this reason Angel (1979) proposed that the star must possess a field strength above 100 MG (but below the intense-field regime).

For hydrogen the intermediate-field gap has been closed partly during the last twelve years with numerical calculations of energy level shifts and transition probabilities for bound-bound transitions by groups in Tübingen and Baton Rouge (Forster et al. 1984; Rösner et al. 1984; Henry and O'Connell 1984, 1985).

Since the magnetic field on the surface of a white dwarf normally is not homogeneous but often better described by a magnetic dipole, the variation of the field strengths from the pole to the equator (a factor of two in the case of a pure dipole field) smears out most of the absorption lines; this explains why the spectral features

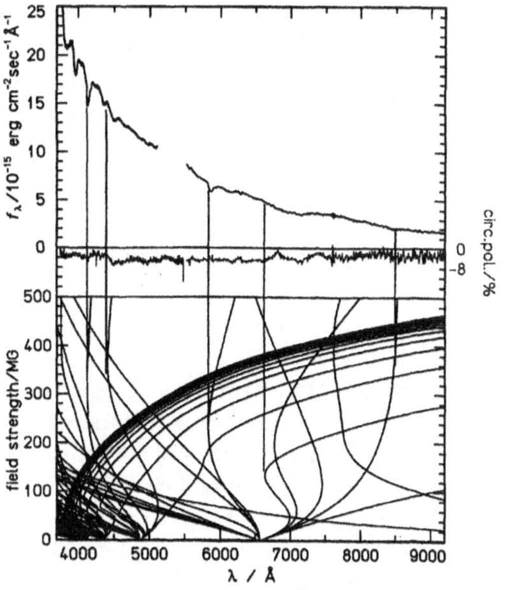

Figure 1. Spectrum and circular polarization of Grw +70°8247 (Obs. A. Putney). The star has a magnetic dipole with a polar field strength of about 320 MG (Wickramasinghe & Ferrario 1988, Jordan 1988). The variation of the wavelengths of the line components of hydrogen with magnetic field strength is indicated below. Some of the stronger features can be identified with stationary line components of hydrogen, others are a superposition of components which vary only moderately between about 160 and 320 MG

on Grw +70°8247 are so shallow for strong magnetic fields. However, a few of the line components become stationary, i.e. their wavelengths go through maxima or minima as functions of the magnetic field strength. These stationary components are visible in the spectra of magnetic white dwarfs despite a considerable variation of the field strengths.

It was a great confirmation for the correctness of the theoretical calculations that indeed the unidentified features in the optical and UV spectrum of Grw +70°8247 could be attributed to stationary components of hydrogen in fields between about 150 and 500 MG (Greenstein 1984, Greenstein et al. 1985, Angel et al. 1985, Wunner et al. 1985, cf. Fig. 1).

These identifications allowed an estimation of the approximate range of field strengths covering the stellar surface. However, the detailed field structure could not be inferred. This was only possible by simulating the radiative transfer through magnetized stellar atmospheres using the line opacities published by the groups in Baton Rouge and Tübingen. Wickramasinghe & Ferrario (1988) have obtained a good fit to most of the Minkowski bands by assuming a pure dipole model with a polar field strength of 320 MG. This result was confirmed by Jordan (1988; 1989) who used more recent atomic data and made improvements to the treatment of the bound-free opacities.

Up to now on about 50 (2%) of the 2100 known white dwarfs (McCook & Sion 1996) magnetic fields have been detected with fields ranging from about 40 kG up to 1 GG. A list of all currently known magnetic white dwarfs is found in Jordan (1997). Although some selection effects may exist (e.g. shallow features are not easily recognized in faint stars) we believe that the number statistics is consistent with the assumption that Ap stars are the progenitors of magnetic white dwarfs, in which the field strengths are enhanced by magnetic flux conservation during the evolution.

The goal of magnetic white dwarf spectroscopy is to determine the field strength, the detailed geometry of the magnetic field, and the rotational period of the star (which is very difficult to measure in non-magnetic white dwarfs). The results provide important constraints for the theory of the origin of magnetic white dwarfs.

MODELS FOR THE RADIATIVE TRANSFER

The spectrum and polarization of a magnetic white dwarf is the superposition of the radiation originating from all different parts of the visible hemisphere of a white dwarf (which may vary due to rotation). Observations of the spectra and wavelength dependent polarization can be analyzed by simulating the transport of polarized radiation through a magnetized stellar atmosphere. The methods for the calculations of synthetic spectra and the wavelength dependent linear and circular polarization are described by Jordan (1988, 1992). The basis is the solution of the four coupled radiative transfer equations (Beckers 1969) for the four Stokes parameters which describe the intensity and polarization of the radiation. With the help of the atomic data the absorption coefficients for $\Delta m = -1$, 0, and +1 and the magneto-optical parameters for Faraday rotation and Voigt effect are calculated for a given magnetic field strength and orientation. With these values the radiative transfer equations are solved for the temperature and pressure structure of a (currently zero-field) white dwarf model atmosphere.

For the line data of hydrogen we use the data from the Tübingen group (Forster et al. 1984, Rösner et al. 1984, Wunner et al. 1985). For the bound-free opacities either a simple and probably unrealistic approximation (Lamb & Sutherland 1974) with some improvements by Jordan (1988, 1992) is used or complex energy eigenvalues and dipole matrix elements calculated by Merani et al. (1995) were utilized in order to study the influence of the bound-free opacities on the polarization (Jordan & Merani 1995).

The magnetic field configuration cannot be derrived from the observed flux and polarization in a unique way by a simple inversion process, since different magnetic geometries can in principle lead to the same observational data. The current strategy is to assume that the global field can be described by a magnetic dipole, which does not necessarily need to be located in the center of the star, or by a dipole+quadrupole combination. In principle higher order multipoles could be included, but this would increase the number of fit parameters. After the magnetic geometry has been fixed, the stellar surface is divided into a large number (typically 1000-10 000) of surface elements on which the radiative transfer equation are calculated. Finally, the Stokes parameters are added up according to the projected size of the surface elements.

RESULTS OF THE ANALYSES

The main result of the analyses of magnetic white dwarfs is that many spectra and polarization measurements can be sucessfully reproduced with our models. In order to do so it is, however, often necessary to assume off-centered dipoles or dipole+quadrupole configurations for the magnetic field geometry (e.g. Putney & Jordan 1995, see Fig. 2). One important questions is, how the higher order multipoles of the magnetic field can survive during the cooling time of a white dwarf.

Chanmugam & Gabriel (1972) and Fontaine et al. (1973) have calculated the time scale for the decay of magnetic fields of white dwarfs. They showed that the decay

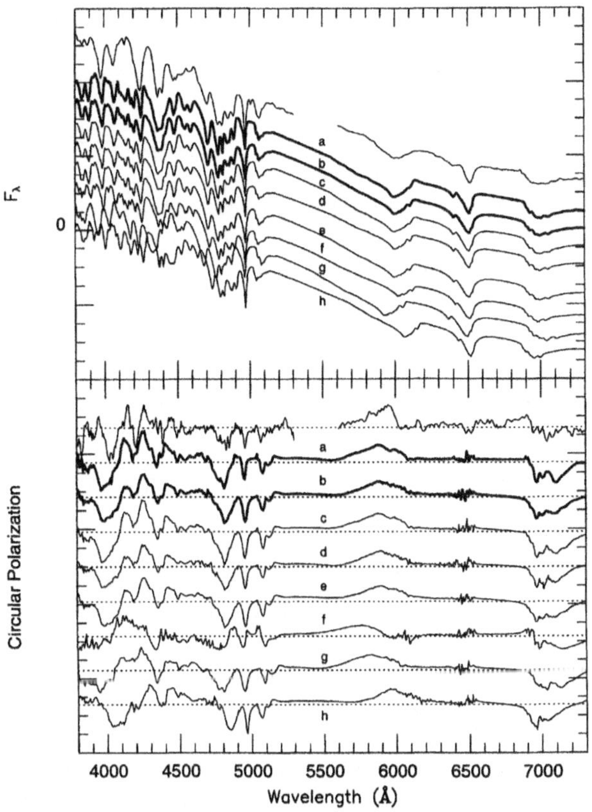

Figure 2. Upper panel: the observed spectrum of the magnetic white dwarf KUV 813-14 (upper curve) is compared to several very similar theoretical models (a-h). A good agreement can be reached with a dipole (43 MG) +quadrupole (8 MG) combination (model a). Lower panel: observed and calculated circular polarization for the same models. The dottet lines correspond to 0% polarization; the maximum amplitude is about 10%

times are $\gtrsim 10^{10}$ yr with the higher modes decaying more rapidly than the fundamental. This could lead to the assumption that the magnetic field becomes more dipolar during evolution. However, Muslimov et al. (1995) have shown that a weak quadrupole (or octupole, etc.) component on the surface magnetic field of a white dwarf may survive the dipole component under specific initial conditions: Particularly the evolution of the quadrupole mode is very sensitive (via Hall effect) to the presence of internal toroidal field. For a 0.6 solar masses white dwarf with a toroidal fossil magnetic field of strength $< 10^9$ G the dipole component declines by a factor of three in 10^9 yr, while the quadrupole component is practically unaffected. Without an internal toroidal field the dipole component still declines by a factor of three but the quadrupole component is a factor of six smaller after 10 Gyr.

This shows that the detection of higher-order multipoles provides us with information about internal magnetization of white dwarfs and the initial conditions from the pre-white dwarf evolution. Therefore, further investigations of the complex magnetic fields of white dwarfs remain important.

With the exception of narrow NLTE cores sometimes present in the H α profiles of some white dwarfs, the spectral lines in white dwarfs are strongly Stark broadened so that it is difficult to measure the rotational period of these stars via Doppler broadening. Spectropolarimatric data from magnetic white dwarfs provide a possibilty to measure the rotational velocity of these stars, due to the strong dependence of the absorption coefficients on the local magnetic field. If the rotational axis is not perfectly alligned with a symmetry axis of the field, variation of both the spectra and the polarization should be detectable.

As an example, we have taken phase resolved spectra of the star HE 1211-1707 with an exposure time of five minutes each during one night and found that the period of spectral variation is about 110 minutes (Jordan 1997). The fastest rotation of a white dwarf has been measured by Barstow et al. (1995) who found that the magnetic white dwarf RE J 0317-853 is rotating with a period of only 725 seconds.

On the other hand there are several objects like Grw +70°8247 in which the observed features in the spectra look constant with time, so that rotationalal periods longer than about a hundred years can be inferred. Why these stars have lost almost all of there angular momentum while others have not remains a mystery.

HELIUM AND CARBON IN MAGNETIC WHITE DWARFS

In some cases both hydrogen and helium are present in the atmosphere as in the case of Feige 7 (Liebert et al. 1977, Martin & Wickramasinghe 1986). Achilleos et al. (1992) could show that a rather complicated model with a displaced magnetic dipole ($z_{\rm off} = 0.15$) having a polar field strength of 35 MG and variable surface abundances of H and He can reproduce the spectra observed during different rotational phases. For such a moderate magnetic field it was, however, already necessary to extrapolate the atomic data for He II calculated by Kemic (1974b), which exist only up to 20 MG.

A mixture of hydrogen and helium is most likely also present in the spectrum of LB 11146B, which probably possesses a polar magnetic field strength of about 670 MG (Liebert et al. 1993, Glenn et al. 1994). However, no detailed modelling was possible due to the lack of atomic data for helium in a strong magnetic field.

The most famous magnetic white dwarf whose spectrum and polarization is still unexplained is GD 229. Angel (1979) proposed that the absorption features in this star are due to He I. Östreicher et al. (1987) have proposed that some of the absorption bands may be due to stationary lines of hydrogen in a field as low as 25-26 MG, but this idea has never been confirmed by model atmosphere analyses. Engelhardt & Bues (1995) have tried to explain the regular almost periodical structure of the GD 229 spectrum by quasi-Landau resonances (O'Connell 1974) of hydrogen in a magnetic field of 2.5 GG, but it is not clear at the moment, whether their approximations are valid. A strong indication that no hydrogen is present in this star comes from the fact that no components of Lyman α could be identified in the GD 229 spectrum (Schmidt et al. 1996); since Lyman α originates from transitions between rather strongly bound states one would expect to see such an absorption even in rather strong fields.

Recently, the first approximate data of some He I line components have become available (Thurner et al. 1993). However, none of the unknown features in the high S/N UV and optical spectra of GD 229 taken by Schmidt et al. (1996) could be explained by these data.

As announced at this conference, several groups are presently calculating atomic data for He I or have data ready for publication. Ceperly et al. (this conference) have found some agreement between the position of some spectral features with He I (calculated with Monte-Carlo calculations) and He II lines at fields between 352 and 590 MG. However, at the effective temperature of GD 229 ($\lesssim$24000 K) we do not expect He II lines to be present in the spectrum, although at present we cannot fully exclude that the ionization equilibrium of helium is strongly modified by the strong magnetic field:

Firstly, the ionization energies calculated as the energy difference between the ground state and the lowest Landau threshold differ from the zero-field situation. Secondly, the Saha equation is modified since the motion (transverse to the magnetic field) of the electrons in phase space is restricted by the magnetic field (see e.g. Ventura et al. 1992 for a discussion of the Saha equation for hydrogen).

In two helium rich magnetic white dwarfs with temperatures below 9000 K carbon molecules (C_2, CH) are responsible for the absorption. Bues & Pragal (1989) have derived a magnetic field strength of about 10-20 MG on the surface of G 99-37. Bues (1993) found an even stronger field of about 150 MG on LP 790-29; it is, however, not quite clear how accurate these values are in detail, since no reliable theory for the Swan bands of C_2 exist at these strong fields.

IMPROVEMENTS NEEDED

While the flux spectrum of the prototype Grw +70°8247 can be well reproduced by the models its polarization shows still strong deviations from the predictions. This must be due to the shortcomings of the present models.

Presently the influence of the magnetic field on the temperature and pressure structure is neglected. A modification of the zero-field stratification is possible via magnetic pressure terms from field configurations which are not force-free (i.e. cannot be described by a scalar potential $\vec{B} = \vec{\nabla}\psi$). Moreover, the polarization of the radiation also slightly modifies the hydrostatic structure of the outer layers.

Another difficulty arrises from the fact that at the effective temperature of Grw +70°8247 convection is present in non-magnetic white dwarfs. Currently it is not clear whether convection is fully suppressed or whether some of the energy is still transported by convection depending on the field strength and the angle between the stellar surface and the magnetic field.

As far as atomic data or molecular data are concerned there is still a strong need for further calculations: At the temperature of Grw +70°8247 H^- is one of the major opacity sources in the non-magnetic case. Since it is reasonable to assume that H^- also plays an important role in the presence of magnetic fields, a grid of absorption coefficients for H^- would be needed for realistic radiative transfer calculations.

Schmelcher (this conference) has presented numrical calculations for the chemical bond and electronic structure of the H_2 and H_2^+ molecules. Such data are very important for the analysis of UV spectra of white dwarfs in the range of effective temperatures between 9000 and 19000 K, where quasimolecular satellite features are observerved due to interactions of H atoms with H and H II perturbers. In the absence of a magnetic field absorption features occur in the wings of Lyman α at 1400 and 1600 Å (Koester et al. 1985, Nelan & Wegner 1985, Allard et al. 1994). In a HST spectrum of Grw +70°8247 small bumps are visible at 1420 and 1620 Å (Allen & Jordan 1994). Since the (non-magnetic) 1600 Å feature is relatively weak at 15 000 K, we may speculate that the bump at 1620 Å is a shifted "1400 Å feature". With the molecular data for H_2 and H_2^+ in a magnetic field it would be possible to calculate the full Lyman α profile including the satellite features.

There are several papers at this conference in which the calculation of He I in the presence of strong magnetic fields is discussed. We can hope that the spectrum and polarization of GD 229 can be explained when enough accurate energy levels and oscillator strengths become available.

Finally, consistently calculated molecular data for C_2 are needed in order to perform a reliable analysis of magnetic white dwarfs showing polarized Swan bands in there spectra.

The line components in magnetic white dwarfs are not only shifted by the magnetic field but also by the electric field at the location of the absorbing atoms. The Stark effect may be important for the line profiles at low magnetic fields ($\lesssim$20 MG), where the electric field is not small compared to the magnetic field and for slight shifts in the line positions of stationary line components. Moreover, line transitions forbidden by the selection rules for dipole radiation may occur if both electric and magnetic fields are present (see Friedrich, this conference).

Therefore, we can conclude that magnetic white dwarfs are still important and interesting laboratories in which present day calculations of atomic data for atoms and molecules in strong magnetic fields can be tested.

Acknowledgements. We thank the DARA (project 50 OR 9409 1) and DFG (KO 738/7/1) for financial support.

REFERENCES

Allard N.F., Koester D., Feautrier N., Spielfiedel A.,1994, A&AS 108, 417
Allen R.G., Jordan S., 1994, Bull.American Astron.Soc., 26, 1383
Angel J.R.P., 1979, in White Dwarfs and Variable Degenerate Stars, Univ. of Rochester Press, p. 313
Angel J.R.P. Liebert J., Stockman H.S., 1985, ApJ 292, 260
Babcock H.W., 1947, ApJ 105, 105
Barstow M.A, Jordan S., O'Donoghue D., Burleigh M.R., Napiwotzki R., Harrop-Allin M.K., 1995, MNRAS 277, 971
Beckers, J.M. 1969, Sol.Phys. 9, 372
Blackett P.M.S., 1947, Nature 159, 658
Bues I, 1993, in White Dwarfs: Advances in Observation and Theory, ed. M.A. Barstow, Kluwer, p. 213
Bues I., Pragal M., 1989, in White Dwarfs, Lecture Notes in Physics 328, ed. G. Wegner, Springer-Verlag, New York, p. 329
Chanmugam G., Gabriel M., 1972, A&A 16, 149
Engelhardt D., Bues I., 1995, in White Dwarfs, eds. D. Koester & K. Werner, Lecture Notes in Physics, Springer, Berlin, p. 123
Fontaine G., Thomas J.H., Van Horn H.M., 1973, ApJ 184, 911
Forster H., Strupat W., Rösner W., Wunner G., Ruder H., Herold H., 1984, J.Phys., V 17, 1301
Garstang R.H., 1977, Rep.Prog.Phys. 40, 105
Ginzburg V.L., 1964, Sov. Phys. Dokl. 9, 329
Glenn J., Liebert J., Schmidt G.D., 1994, PASP, 106, 722
Greenstein J.L., 1956, in Proc. 3rd Berkeley Symp. on Math. Probab. and Stat., ed. J. Neyman, Berkeley, Univ.Cal.Press, p. 11
Greenstein J.L., 1984, ApJL 281, L47
Greenstein J.L., Henry R.J.W., O'Connell R.F., 1985, ApJL 289, L25
Heber U., Napiwotzki R., Reid I.N., A&A, in press
Henry R.J.W., O'Connell R.F., 1984, ApJ, 282, L97
Henry R.J.W., O'Connell R.F., 1985, PASP, 97, 333
Jordan S., 1988, PhD thesis, University of Kiel
Jordan S., 1989, in White Dwarfs, Lecture Notes in Physics 328, Ed. G. Wegner, Springer, p. 333
Jordan S., 1992, A&A 265, 570
Jordan S., 1997, in White Dwarfs, eds. J. Isern, M. Hernanz, E. Garciía-Berro, Kluwer, p. 397
Jordan S., Merani N., 1995, in White Dwarfs, eds. D. Koester and K. Werner, Lecture Notes in Physics, Springer, Berlin, p. 134
Kemp J.C., 1970, ApJ 162, 169
Kemic S.B., 1974a, ApJ 193, 213
Kemic S.B., 1974b, JILA Rep. 133
Koester D., Weidemann V., Zeidler-K.T. E.-M., Vauclair G., 1985, A&A 142, 5
Koester D., Herrero A., 1988, A&A 332, 910

Lamb F.K., Sutherland P.G., 1974, Physics of Dense Matter, ed. C.J. Hansen, Dordrecht: Reidel, p. 265
Landstreet J.D., 1992, A&AR 4, 35
Liebert J., Angel J.R.P., Stockman H.S., Spinrad H., Beaver E.A., 1977, ApJ 214, 457
Liebert, J., Bergeron P., Schmidt G.D., Saffer R.A., 1993, ApJ 418, 426
Martin B., Wickramasinghe D.T., 1986, ApJ 301, 177
Merani N., Main J., Wunner G., 1995, A&A 298, 193
Minkowski R., 1938, Annu.Rep.Mt.Wilson Obs., p.38
Muslimov A.G., Van Horn H.M., Wood M.A., 1995, ApJ 442, 758
Östreicher R., Seifert W., Ruder H., Wunner G., 1987, A&A 173, L15
Nelan E.P., Wegner G., 1985, ApJ 289, L31
O'Connell R.F., 1974, ApJ 187, 275
Preston G.W., 1970, ApJL 160, L143
Rösner W., Wunner G., Herold H., Ruder H., 1984, J.Phys., V 17, 29
Robinson R.D., Worden S.P., Harvey J.W., 1980, ApJL 236, L155
Schmidt G.D., Allen R.G., Smith P.S., Liebert J., 1996, ApJ 463, 320
Thurner G., Körbel H., Braun M., Herold H., Wunner G., 1993, J.Phys.B 26, 4719
Wegner G., 1971, PASP 83, 205
Wickramasinghe D.T., Ferrario L., 1988, ApJ, 327, 222
Woltjer L.,1964, ApJ 140, 1309
Wunner G., Rösner W., Herold H., Ruder H., 1985, A&A 149, 102

HYDROGEN IN STRONG ELECTRIC AND MAGNETIC FIELDS AND ITS APPLICATION TO MAGNETIC WHITE DWARFS

Susanne Friedrich,[1,2*] Peter Faßbinder,[2] Ingo Seipp,[2] and Wolfgang Schweizer[2*]

[1]Institut für Astronomie und Astrophysik der Universität Kiel
D-24098 Kiel, Germany
[2]Institut für Astronomie und Astrophysik der Universität Tübingen
D-72076 Tübingen, Germany

INTRODUCTION

The wavelengths of hydrogen transitions in the strong magnetic fields of white dwarfs are strongly dependent on the magnetic field strength. Taking into account that the magnetic field strength varies at least by a factor of two across the surface of a magnetic white dwarf, it is obvious that most lines are broadened to invisibility in the spectra. A few lines, however, exhibit maxima or minima as functions of the magnetic field strength and thus can produce sharp absorption features. Atomic data for all magnetic field strengths relevant for DA white dwarfs are available (Ruder et al. 1994) and a substantial number of magnetic white dwarfs has been successfully analyzed.

One outstanding problem is the calculation of bound-free opacities of the hydrogen atom in magnetic fields. First results for some magnetic field strengths have recently become available (Merani et al. 1995; Seipp et al. 1996). As it is known from the Stark broadened lines in the spectra of DA white dwarfs without magnetic fields there are strong electric fields present caused by free electrons and ions in the stellar atmospheres. In addition to their influence on the wavelengths and oscillator strengths these electric fields might be responsible for further absorption features, since some transitions forbidden in the diamagnetic case are turned into allowed transitions.

*Visiting Astronomer, German-Spanish Astronomical Center, Calar Alto, operated by the Max-Planck-Institut für Astronomie Heidelberg jointly with the Spanish National Commission for Astronomy

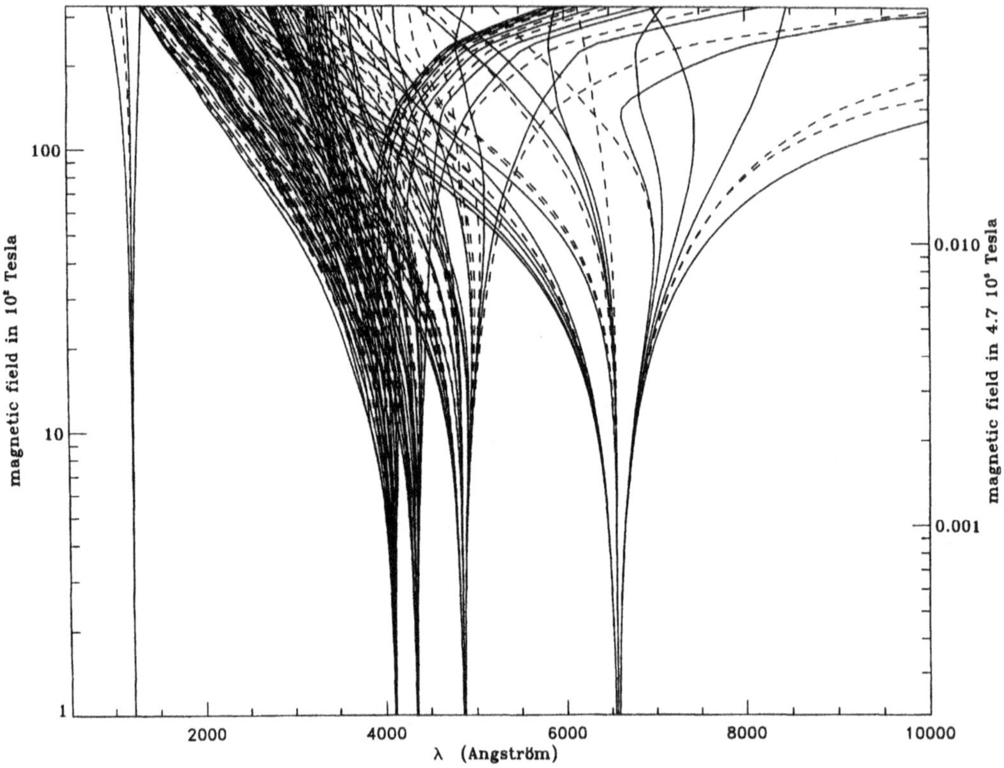

Figure 1. Wavelengths of all Lyman-α and Balmer-α to Balmer-δ transitions of the hydrogen atom as a function of magnetic field strength between 100 T and 33000 T. Solid lines correspond to transitions already present without electric fields and dashed lines are dipole-forbidden without an additional parallel electric field.

QUANTUM MECHANICAL CALCULATIONS

Bound-bound Transitions

The non-relativistic single particle Hamiltonian of a hydrogen atom in an external magnetic field B and electric field F (in units of $F_0 = 5.12 \cdot 10^{11}$V/m) reads:

$$H = \underbrace{\frac{|p|^2}{2m_e}}_{kin.} - \underbrace{\frac{e^2}{|r|}}_{Coul.} + \underbrace{\frac{e}{2m_e} B \cdot L}_{paramagn.} + \underbrace{\frac{e^2}{8m_e}(B \times r)^2}_{diamagn.} + \underbrace{eF \cdot r}_{electr.}$$

For an effective numerical treatment we reduced the three-dimensional Schrödinger equation via the discrete variable technique to a system of unidimensional differential equations. This system of differential equations is solved by the Finite-Element method. For details see Faßbinder & Schweizer and Schweizer et al. (this volume) and references therein.

As expected, an additional electric field can have drastic effects on the structure of the calculated spectra. On the other hand, it could be shown (Faßbinder & Schweizer, 1996) that for most magnetic and electric field strengths relevant for white dwarfs the influence of the electric field component perpendicular to the magnetic field is negligible compared to the influence of its parallel component. Therefore we can restrict ourselves to parallel magnetic and electric fields.

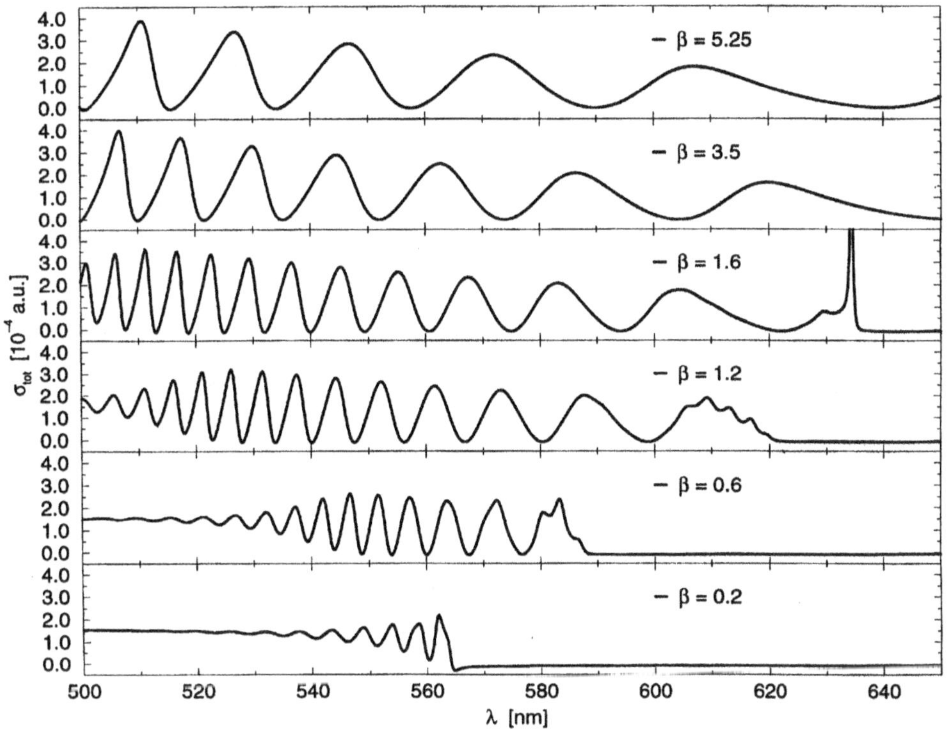

Figure 2. Photoionization cross-sections for a magnetic field strength of 28200 T and parallel electric fields with field strengths in units of 10^7 V/m.

Bound-free transitions

The Hamiltonian for hydrogen in parallel magnetic and electric fields in atomic units and spherical coordinates (r, ϑ, ϕ) reads:

$$H = \underbrace{-\Delta}_{kin.} - \underbrace{\frac{2}{|r|}}_{Coul.} + \underbrace{2\beta(l_z + 2s_z)}_{paramagn.} + \underbrace{\beta^2 r^2 \sin^2\vartheta}_{diamagn.} + \underbrace{2Fr\cos\vartheta}_{electr.},$$

with the magnetic field β in units of $B_0 = 4.7 \cdot 10^5$T and the electric field strength F in units of F_0. In recent years the complex coordinate method has been successfully applied to continuum states. The complex rotated Schrödinger equation is then solved numerically by expanding the wavefunctions over a complete basis set. Photoionization cross-sections are obtained from the complex eigenvalues and 'rotated' eigenfunctions of the Hamiltonian.

We calculated photoionization spectra at optical wavelengths in a strong magnetic field with parallel electric fields of various strengths. The electric field modulates the ionization spectra strongly through an onsetting resonance structure (Fig. 2), which behaves smoothly by changing the electric field strength. The electric fields in the atmosphere of the white dwarf are generated by free electrons and ions in the stellar plasma and hence distributed statistically. Taking this statistical origin into account the strong resonance features are smeared out by the electric field distribution and the opacities can be approximated by straight lines over the relatively small wavelength range in question.

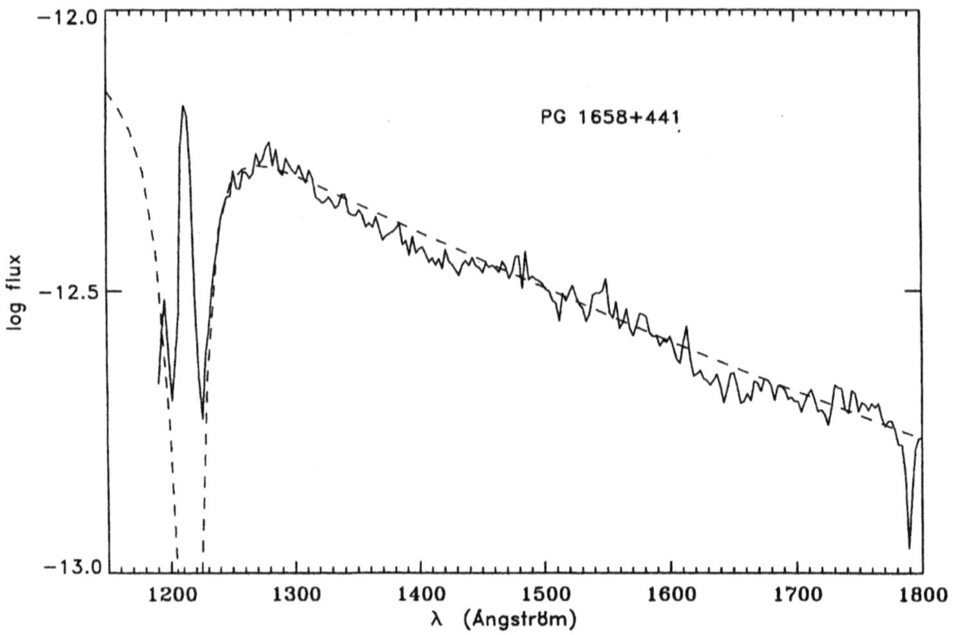

Figure 3. UV-spectrum of the magnetic white dwarf PG 1658+441. Superimposed on the absorption line of the white dwarf is the geocoronal Lyman-α line. The dashed line is the model spectrum assuming a dipole geometry for the magnetic field with a field strength of 280 T and an inclination angle of 50°.

CALCULATION OF MODEL SPECTRA

The temperature and pressure structure for DA white dwarf atmospheres is taken from zero magnetic field LTE models (Koester, private communication). This is a valid approximation for magnetic field strengths below 10000 T (e.g. Wickramasinghe and Martin, 1986). The line absorption coefficients are calculated using the new wavelengths and oscillator strengths for the hydrogen atom in parallel magnetic and electric fields. Wavelengths and oscillator strengths for perpendicular electric and magnetic fields are approximated by their values at zero electric field strength (Faßbinder and Schweizer, 1996). For the distribution of the electric field we use the Holtsmark probability function (Mozer and Baranger, 1960). The wavelengths and oscillator strengths for bound-free transitions are calculated in the approximation of Lamb and Sutherland (1974) for low magnetic field strengths. Finally the radiative transfer equations for polarized light including magneto-optical effects are analytically solved using an algorithm as described by Martin and Wickramasinghe (1979).

COMPARISON TO OBSERVATIONS

In Fig. 3 we show the UV-spectrum of the magnetic white dwarf PG 1658+441. For the model spectrum we assumed a dipole geometry for the magnetic field with a field strength of 280 T and an inclination angle of 50° as derived from model fits to circular polarization and flux spectra in the visual spectral range (Friedrich et al., 1996). The broadening of the Lyα line is due to electric fields in the atmosphere of the white dwarf and is well reproduced by the model spectrum.

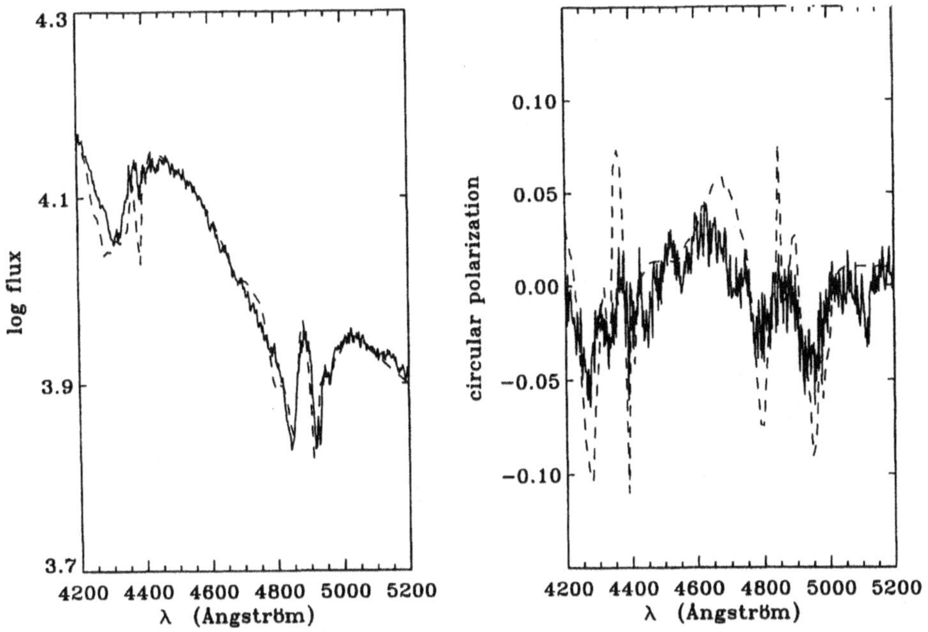

Figure 4. Flux (left) and circular polarization (right) spectrum of the magnetic white dwarf KPD 0253+5052. The dashed line is the model spectrum calculated assuming an offset dipole geometry for the magnetic field with a dipole field strength of 800 T and an inclination angle of 46°.

In Fig. 4 we show the best fit model determined by means of a χ^2 fit for one of our phaseresolved flux and circular polarization spectra of the rotating magnetic white dwarf KPD 0253+5052. It is an offset dipole geometry for the magnetic field with offsets of 0.2 and 0.12 white dwarf radii along and perpendicular to the magnetic axis, respectively, and a dipole field strength of 800 T. The inclination of the rotation axis is 20° and the colatitude 50°. It is obvious especially from the polarization spectrum that there are clear deviations between the radiative transfer model and observation. They might be caused in the first place by the Holtsmark distribution, not valid for magnetic fields, and secondly by the approximation for the bound-free opacities.

In Fig. 5 we show two prominent absorption lines at 4123Å and 5830Å in the spectrum of the highly magnetic white dwarf Grw+70°8247. They are caused by stationary Hβ and Hα components, respectively. The lack of model atmospheres appropriate for a magnetic field strength of about 35000 T prevents the calculation of model spectra, but qualitatively the onset of the absorption edge about 10Å beyond the theoretical line position calculated for the pure magnetic field (solid line) can be explained by the influence of strong electric fields (dashed line) of the order of $5 \cdot 10^7$ V/m. The broad red wings are due to the variation of the magnetic field strength over the surface of the white dwarf.

CONCLUSIONS

We could show that the electric field has strong influence on both bound-bound and bound-free transitions of the hydrogen atom. Generally the observed spectra can be well reproduced by model spectra calculated with a radiative transfer code. But deviations between theory and observation are obvious especially in circular polarization spectra,

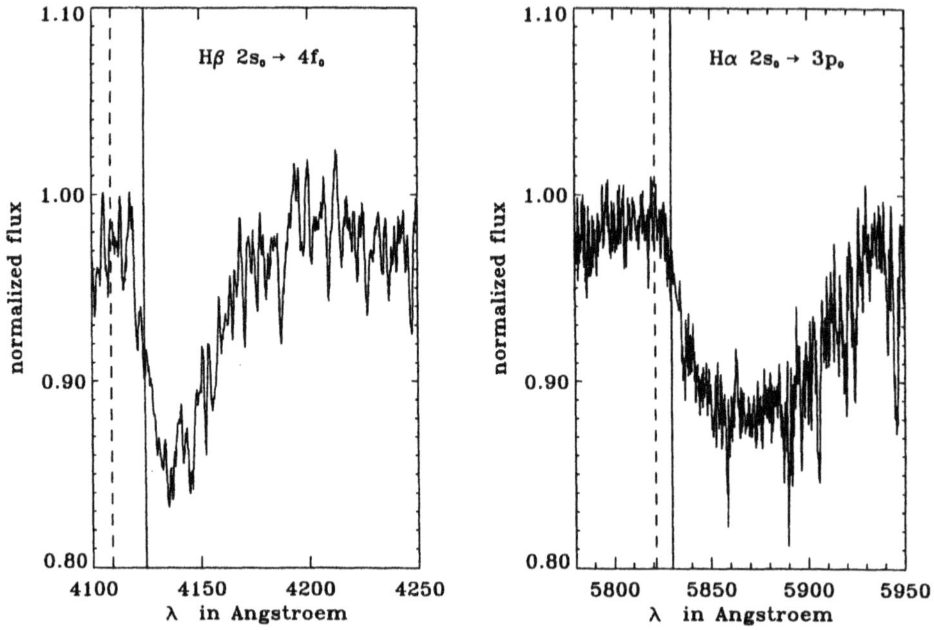

Figure 5. Stationary Hβ (4123Å) and Hα (5830Å) lines in the spectrum of the highly magnetic white dwarf Grw+70°8247. Solid lines mark the theoretical line position calculated for pure magnetic fields, dashed lines the line position with an additional electric field of the order of $5 \cdot 10^7$ V/m.

which are very sensitive to errors in the modelling. These deviations might be mainly due to the lack of an appropriate electric field distribution. This problem will be tackled as well as the calculation of the bound-free transitions for relevant electric and magnetic field strengths of magnetic white dwarfs.

Acknowledgements

This work was supported by the Deutsche Forschungsgemeinschaft.

REFERENCES

Faßbinder, P., Schweizer, W., 1996, Hydrogen atom in very strong magnetic and electric fields, *Phys. Rev. A* 53:2135

Friedrich, S., Östreicher, R., Schweizer, W., 1996, Observation of flux and circular polarization spectra of white dwarfs with low magnetic fields, *A&A* 309:227

Lamb, F.K. and Sutherland, P.G., 1974, Continuum polarization in magnetic white dwarfs, *in*: "Physics of Dense Matter", Hansen C.J., ed., D. Reidel, Dordrecht, Netherlands

Martin, B., Wickramasinghe, D.T., 1979, Solutions for radiative transfer in magnetic atmospheres, *MNRAS* 189:883

Merani, N., Main, J., Wunner, G., 1995, Balmer and Paschen bound-free opacities for hydrogen in strong white dwarf magnetic fields, *A&A* 298:193

Mozer, B., Baranger, M., 1960, Electric field distributions in an ionized Gas, *Phys. Rev.* 118:626

Ruder, H., Wunner, G., Herold, H., Geyer, F., 1994, "Atoms in Strong Magnetic Fields", Springer, Berlin

Seipp, I., Schweizer, W., 1996, Electric fields for hydrogen bound-free transitions in magnetic white dwarfs, *A&A* 318:990

Wickramasinghe, D.T., Martin, B., 1986, Magnetic blanketing in white dwarfs, *MNRAS* 223:323

HELIUM DATA FOR STRONG MAGNETIC FIELDS OBTAINED BY FINITE ELEMENT CALCULATIONS

Moritz Braun[1], Wolfgang Schweizer[2,3], and Heiko Elster[3]

[1]Physics Department
University of South Africa
P O Box 392
Pretoria 0003, South Africa
[2]Fakultät für Physik und Astronomie
Theoretische Physik I
Ruhr-Universität Bochum
44780 Bochum, Germany
[3]Institut für Astronomie und Astrophysik
Abteilung Theoretische Astrophysik
Universität Tübingen
72076 Tübingen, Germany

INTRODUCTION

Since the discovery of huge magnetic fields in the vicinity of white dwarfs ($B \approx 10^2 - 10^5 T$) and neutron stars ($B \approx 10^7 - 10^9 T$) much work has been done in calculating atomic energy values and transition probabilities in the atmosphere of these compact objects. At these field strengths the magnetic forces outweigh the Coulomb binding forces, even for low-lying energies.
Whereas the extensive calculations for the hydrogen atom in strong magnetic fields (Ruder et al., 1994) have resulted in a much better understanding of the spectra of hydrogen-dominated white dwarfs, there are still magnetic white dwarfs like the GD229 with unexplained absorption spectra, for which transitions of neutral He are considered to be important (Schmidt and Latter, 1990). Therefore the properties of the helium atom in strong magnetic fields are of great relevance.
Only recently we succeeded for the first time to calculate bound/bound transitions in strong magnetic fields relevant for white dwarf stars by a combination of the hyperspherical close coupling (Zhou and Li, 1994) and finite element methods (Braun et al., 1993). In this contribution we present results obtained for wavelengths of selected dipole transitions in neutral helium at field strengths of up to $1.88\,10^5$ Tesla.

THEORY AND METHOD

We consider a system consisting of two electrons and a nucleus of charge Ze in a homogeneous magnetic field B along the z-axis. If we use Z-scaled atomic units, i.e. as energy unit Z^2 Rydberg and as length unit $a_{\rm Bohr}/Z$, and if we neglect the finite mass of the nucleus, the Hamiltonian reads

$$H = \sum_{i=1}^{2}\left[-\nabla_i^2 - \frac{2}{|\boldsymbol{r}_i|} + \beta_Z^2(x_i^2 + y_i^2)\right] + \frac{2}{Z|\boldsymbol{r}_1 - \boldsymbol{r}_2|} + 2\beta_Z\left[\boldsymbol{L}_z + g_e\boldsymbol{S}_z\right], \tag{1}$$

where $\beta_Z = B/(Z^2 \times 4.7010\ 10^5\ \text{Tesla})$ and g_e is the g-factor of the electron. However, the finite mass $m_{\rm nuc}$ of the nucleus can be taken into account if the units are appropriately rescaled.

The Hamiltonian (1) is invariant under rotation around to the z-axis and inversion with respect to the origin. Therefore the conserved quantum numbers of the Hamiltonian are

- the z-component M of the total angular momentum,
- The parity P,
- the total spin S,
- its z-component M_S.

We describe the system by 3 internal coordinates and the 3 Eulerian angles α, β and γ, that describe the orientation of the body frame with respect to the laboratory frame. As internal coordinates, we choose the hyperradius

$$R = \sqrt{\xi_1^2 + \xi_2^2}\,, \tag{2}$$

the hyperangle

$$\phi = \arctan\left(\frac{\xi_2}{\xi_1}\right), \tag{3}$$

and the angle u between $\boldsymbol{\xi}_1$ and $\boldsymbol{\xi}_2$ with respect to the Jacobi vectors

$$\boldsymbol{\xi}_1 = \frac{1}{\sqrt{2}}(\boldsymbol{r}_1 - \boldsymbol{r}_2) \tag{4}$$

and

$$\boldsymbol{\xi}_2 = \frac{1}{\sqrt{2}}(\boldsymbol{r}_1 + \boldsymbol{r}_2). \tag{5}$$

The z'-axis is chosen to coincide with $\boldsymbol{\xi}_1$ while $\boldsymbol{\xi}_2$ is taken to lie in the $x' - z'$ plane. The choice of Jacobi vectors instead of the radius vectors has the advantage of making the symmetry requirements for the wave function more tractable. Since they are connected to the radius vectors by an orthogonal transformation, the form of the diamagnetic Hamiltonian is left unchanged.

Introducing the reduced wave function

$$\Phi(R, \phi, u, \Omega) = R^{\frac{5}{2}}\ \sin\phi\cos\phi\ \Psi\,, \tag{6}$$

where Ω stands for the Eulerian angles, expanding in terms of the eigenfunctions of the symmetric top of definite parity (Braun, 1993) up to a maximum J-value $J_{\max}$

$$\Phi = \sum_{J=|M|}^{J_{\max}} \sum_{Q} \Phi(R, \phi, u)_{J,Q} |PJMQ\rangle_{\Omega} |SM_S\rangle \,, \tag{7}$$

and projecting onto the $|PJMQ\rangle_{\Omega}$ we obtain the following system of coupled partial differential equations for the functions of the internal coordinates.

$$\begin{aligned}
&\Bigg[-\frac{\partial^2}{\partial R^2} + \frac{1}{R^2}\Bigg(-\frac{1}{4} - \frac{\partial^2}{\partial \phi^2} - \frac{1}{\sin^2\phi \cos^2\phi \sin u}\frac{\partial}{\partial u}\sin u \frac{\partial}{\partial u} \\
&+\frac{Q^2}{\sin^2\phi\cos^2\phi\sin^2 u} - \frac{2Q^2 - J(J+1)}{\cos^2\phi}\Bigg) \\
&+\frac{C(\phi,u)}{R}\Bigg]\Phi_{JQ} + \\
&\frac{1}{R^2}\sum_{Q'} \frac{\langle PJMQ|2iJ_2|PJMQ'\rangle \frac{\partial}{\partial u} + \langle PJMQ|2J_1J_3|PJMQ'\rangle \cot u}{\cos^2\phi}\Phi_{JQ'} \\
&+\beta_{\mathrm{Z}}^2 R^2 \sum_{J'Q'}\Big(\frac{2}{3}\delta_{JJ'}\delta_{QQ'} + \sqrt{\frac{8\pi^2}{5}}\sum_{k=0}^{2}\langle PJMQ| + 20k|PJ'MQ'\rangle p_k(\phi,u)\Big)\Phi_{J',Q'} \\
&+2\beta_{\mathrm{Z}}\left[M + g_{\mathrm{e}}M_S\right]\Phi_{JQ} = E\Phi_{JQ}
\end{aligned} \tag{8}$$

Those functions were determined by expanding them in terms of the adiabatic eigenfunctions, leading to a system of unidimensional differential equations which was finally solved by using the finite element method.

RESULTS WITHOUT MAGNETIC FIELD

To test the reliability our method we compared our energies against the numerically exact results for field-free non-relativistic helium by Pekeris et al.(1971) in tables 1 and 2. The agreement is good to very good for the S-states while fair for the P-states. Thus one can expect our wavelengths to be accurate to about 1%. Considering the strong variation of the magnetic field within a neutron star due to the dipole field assumed, this accuracy should be sufficient for a reliable calculation of model-spectra. As a further, more sensitive, test of our numerical method we also checked our results

Table 1. Comparison of the energies obtained for singlet-states of helium with the values of Pekeris et al. (1971)

	E	ΔE
1^1S	-1.4518580	$4.2 \cdot 10^{-6}$
2^1S	-1.072969	$1.8 \cdot 10^{-5}$
3^1S	-1.030601	$3.5 \cdot 10^{-5}$
4^1S	-1.0153	$1.2 \cdot 10^{-3}$
2^1P	-1.0610	$9.2 \cdot 10^{-4}$

Table 2. Comparison of the energies obtained for triplet-states of helium with the values of Pekeris et al. (1971)

	E	ΔE
2^3S	-1.0876149	$2.1 \cdot 10^{-7}$
3^3S	-1.0343412	$3.3 \cdot 10^{-6}$
4^3S	-1.01735	$9.1 \cdot 10^{-4}$
5^3S	-1.0026	$8.7 \cdot 10^{-3}$
2^3P	-1.0656	$9.8 \cdot 10^{-4}$

for oscillator strengths at zero field. In table 3 we show our results for the oscillator strengths at zero field. We also give those obtained by watanabe et al. (1992). It is seen that the agreement is fair to good except for $2^1P_0 \rightarrow 2^1S_0$ which is due to the small energy difference involved. Howeve. the corresponding wavelength is very large and thus unimportant for astrophysical applications.

Table 3. Comparison of the oscillator strengths obtained at zero field with those of Watanabe et al.(1992)

$\Delta M = 0,\ \beta_Z = 0$		
	This work	Tang et al.
$2^1P_0 \rightarrow 1^1S_0$	0.267	0.276
$3^1P_0 \rightarrow 2^1S_0$	0.144	0.149
$2^1P_0 \rightarrow 2^1S_0$	0.423	0.377
$3^1P_0 \rightarrow 1^1S_0$	0.070	0.074

WAVELENGTHS OF SELECTED DIPOLE TRANSITIONS

We have calculated the wavelengths of dipole transitions between a number of ^1P and ^1S as well as ^3P and ^3S- states for $\beta_Z \leq 0.1.$, i.e. for field strengths up to $1.88\,10^5$ Tesla.

In figures 1 to 3 we show the wavelengths obtained for a few selected transitions between singlet states with $\Delta M = -1$. It is seen that as function of the magnetic field strength, the wavelengths vary by approximately a factor of 2. We also see that those transitions exhibit wavelength maxima translating into stationary lines. The maxima are at $\beta_Z =$ 0.008 ,0.018 and 0.016 respectively corresponding to field strengths of $1.5\,10^4$, $3.4\,10^4$ and $3.0\,10^4$ Tesla. The corresponding wavelengths are 4820, 7730 and 2250 Å.

ACKNOWLEDGEMENTS

This work was supported by the Deutsche Forschungsgemeinschaft (DFG) and the Foundation for Research Development(FRD) of South Africa.

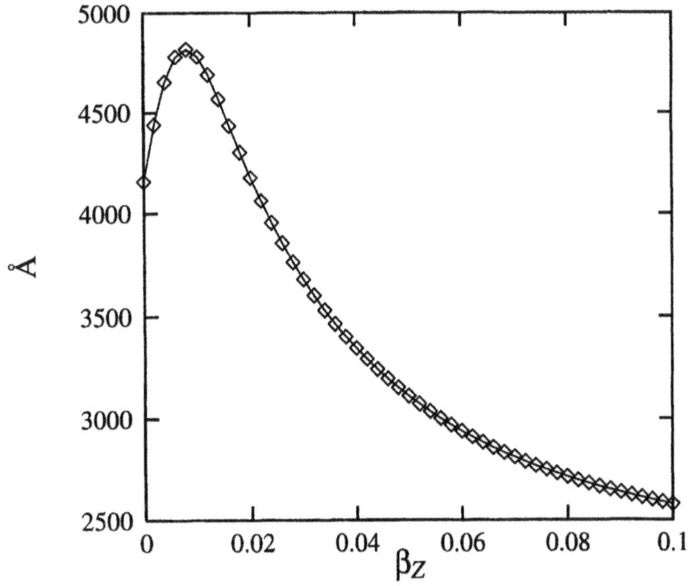

Figure 1. Wavelength of the $\Delta M = -1$ transition $3^1P_1 \rightarrow 2^1S$ for $\beta_Z \leq 0.1$.

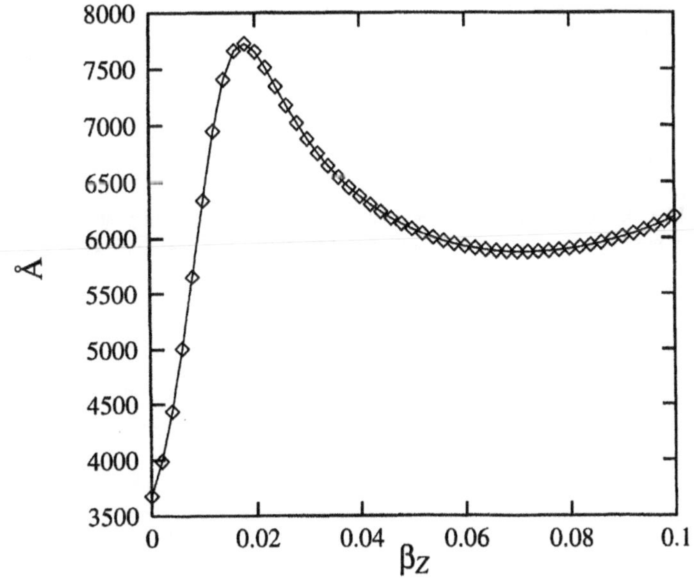

Figure 2. Wavelength of the $\Delta M = -1$ transition $4^1P_1 \rightarrow 3^1S$ for $\beta_Z \leq 0.1$.

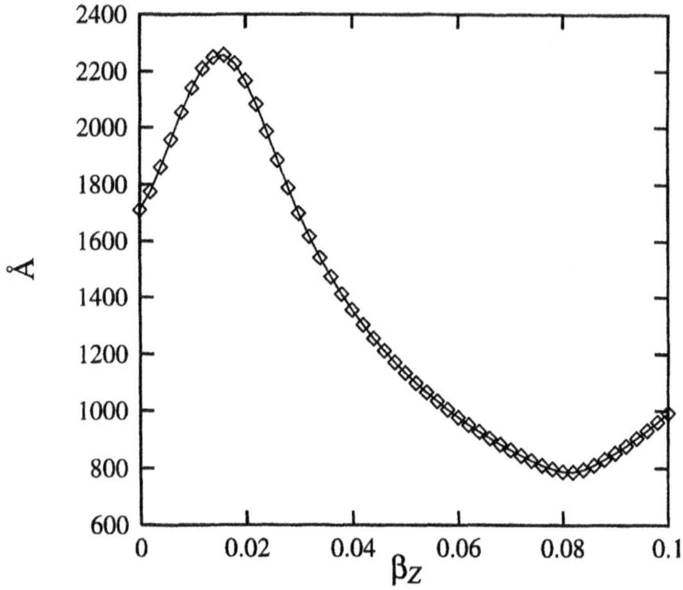

Figure 3. Wavelength of the $\Delta M = -1$ transition $5^1P_1 \to 3^1S$ for $\beta_Z \leq 0.1$.

REFERENCES

Accad, Y., Pekeris, C. L. and Schiff, B. 1971 *Phys. Rev. A* 4:516

Braun, M., 1993, "Application of the Method of Finite Elements to the Two-Electron-Problem in a Strong Magnetic Field", PhD-Thesis, University of Tübingen

Braun, M., Schweizer, W, and Herold, H., 1993, Finite element calculations for S states of helium, *Phys. Rev. A* 48:1916

Ruder, H., Wunner, G. Herold, H., and Geyer. F., 1994, "Atoms in Strong Magnetic Fields" Springer-Verlag Berlin,

Schmidt, G.D., and Latter, W.B., 1990, On the spectrum and field strength of the magnetic white dwarf GD229, *Astrophys. J.* 350:758

Tang. J. , Wanatabe S., and Matsuzawa M., 1992, Dipole density function and evaluation of the oscillator strengths of He, *Phys. Rev. A* 46:3758

Zhou, Y., and Li, C.D., 1994, Hyperspherical close-coupling calculation of positronium formation cross sections in positron-hydrogen scattering at low energies, *J. Phys. B* 27:5065

THE SPECTRUM OF ATOMIC HYDROGEN IN MAGNETIC AND ELECTRIC FIELDS OF WHITE DWARF STARS

Peter Faßbinder,[1] and Wolfgang Schweizer[1,2]

[1]Institut für Astronomie und Astrophysik
Abteilung Theoretische Astrophysik
Universität Tübingen
72076 Tübingen, Germany
[2]Fakultät für Physik und Astronomie
Theoretische Physik I
Ruhr-Universität Bochum
44780 Bochum, Germany

INTRODUCTION

Despite its long history, the problem of a hydrogen atom in external magnetic and electric fields is still of significant interest. Besides the fundamental questions associated with this nonintegrable system, decisive stimulus came from the discovery of huge magnetic and electric fields in compact astrophysical objects, such as white dwarf stars (Kemp et al., 1970) with field strengths of the order of $B \approx 10^2 - 10^5$ T and $F \approx 10^6 - 10^9$ V/m (Friedrich et al., 1996). At these extremely high field strengths the Lorentz force acting on an atomic electron equals or exceeds the Coulomb binding force even for low-lying states, and a recalculation of the atomic structure becomes necessary. Whereas the diamagnetic hydrogen atom has been treated comprehensively for all relevant field strengths (Wunner et al., 1989), there are still many open questions associated with the general case of magnetic *and* electric fields.

It is the purpose of this work, to examine the influence of the additional electric field on the hydrogen atom in the atmosphere of magnetic white dwarf stars. To this end, we solved the Schrödinger equation for a hydrogen atom in strong magnetic and electric fields and obtained accurate values for the wavelengths and oscillator strengths of low-lying bound-bound transitions. We discuss various aspects of the influence of the electric field on the calculated spectra, especially with respect to *stationary lines*, which provide a direct connection to astrophysical observations. Our calculations cover the whole range of magnetic and electric field strengths relevant to white dwarf stars.

Atoms and Molecules in Strong External Fields
Edited by Schmelcher and Schweizer, Plenum Press, New York, 1998

METHOD

The non-relativistic single particle Hamiltonian of a hydrogen atom in an external magnetic field $\mathbf{B}$ and electric field $\mathbf{F}$ reads:

$$H = \underbrace{\frac{|\mathbf{p}|^2}{2m_e}}_{\text{kin.}} - \underbrace{\frac{e^2}{|\mathbf{r}|}}_{\text{Coul.}} + \underbrace{\frac{e}{2m_e}\mathbf{B}\cdot\mathbf{L}}_{\text{paramagn.}} + \underbrace{\frac{e^2}{8m_e}(\mathbf{B}\times\mathbf{r})^2}_{\text{diamagn.}} + \underbrace{e\,\mathbf{F}\cdot\mathbf{r}}_{\text{electr.}} \,. \tag{1}$$

It has been shown (Faßbinder and Schweizer, 1996a), that for most field strengths relevant to white dwarf stars the influence of the electric field component perpendicular to the magnetic field is negligible compared to the influence of its parallel component. Therefore, we can restrict ourselves to parallel external fields. Then the Hamiltonian (1) becomes rotational invariant and the three-dimensional Schrödinger equation reduces to an effective two-dimensional problem.

Using a discretization method first applied to the hydrogen atom in external fields by Melezhik (1993, 1997), the problem is transformed into a system of unidimensional differential equations, which we solved with the finite-element method. This approach proved to be very suitable for our purposes (Faßbinder and Schweizer, 1996a), and we are able to calculate accurate atomic data for all magnetic field strengths relevant to white dwarf stars. Especially in the intermediate magnetic field range, in which the transition from a spherical problem (dominated by the Coulomb potential) to a cylindrical problem (dominated by the magnetic field) takes place and in which most stationary components of the hydrogen atom have their extrema, it yields much better results than expanding the wave functions into a set of global basis functions. The finite proton mass is taken into account by using appropriate scaling laws (Pavlov-Verevkin & Zhilinskii 1980).

STATIONARY HYDROGEN LINES OF WHITE DWARF STARS

In figure 1 the wavelengths of all Lyman α, Balmer α and Balmer β transitions of the hydrogen atom in parallel magnetic and electric fields are shown as functions of the magnetic field strength from $B \approx 4.7 \cdot 10^2$ T to $B \approx 9.4 \cdot 10^4$ T. Solid lines correspond to transitions already present in the spectrum of the diamagnetic hydrogen atom, dashed transitions are due to the additional electric field. Most of the transitions are strongly dependent on the magnetic field strength and change their positions by up to several thousand Å within one order of magnitude in B. Taking into account that the magnetic field strength varies at least by a factor of two across the surface of a magnetic white dwarf star, it is obvious that most lines are broadened to invisibility in the spectra. However, for magnetic fields stronger than $B \approx 10^3$ T some of the transitions run through maxima or minima of the wavelength and thus can produce sharp absorption features – the stationary components of the hydrogen atom.

On a dense mesh of electric field strengths we calculated the magnetic field strengths, the wavelengths and the oscillator strengths of the extrema of all stationary lines of Lyman α, Balmer α and Balmer β – the most important features in the optical and UV region of the spectra. In order to investigate the influence of the additional electric field on the wavelengths and the oscillator strengths, the results are compared with the well known case of the diamagnetic hydrogen atom. Detailed tables with accurate values for the atomic data are published in Faßbinder and Schweizer (1996b).

Out of the 15 Balmer α transitions allowed in the diamagnetic case, five transitions exhibit stationary behaviour. Since two of them run through two extrema in

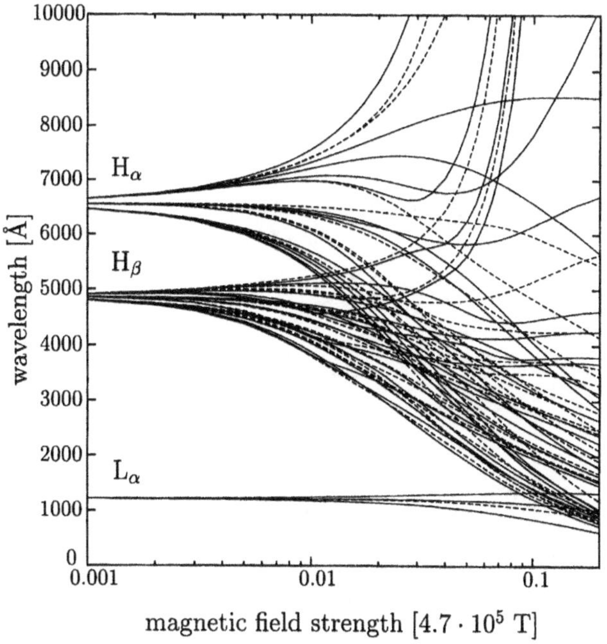

Figure 1. Wavelengths of all Lyman α, Balmer α and Balmer β transitions of the hydrogen atom in parallel magnetic and electric fields.

the magnetic field range relevant to white dwarf stars, there are altogether seven stationary components. In an additional electric field one further Balmer α transition ($2p'_0 \rightarrow 3p'_{-1}$) becomes dipol-allowed and stationary. In figure 2 the wavelengths of all stationary lines of Balmer α are shown for different electric field strength ranging from $F = 0$ to $F = 10^8$ V/m. All states are labeled according to the quantum numbers of the field free limit. The wavelengths of the stationary components of Balmer α are already slightly shifted for an electric field strength of $F = 10^7$ V/m, and the differences compared to the diamagnetic hydrogen atom can get as large as 30 Å for an electric field strength of $F = 10^8$ V/m. There is no definite direction of the shifts detectable; we get blue as well as red shifts. But there are clear differences between the transitions. Whereas the influence of the electric field on the $2p'_{-1} \leftrightarrow 3d'_{-2}$- and on the $2p'_0 \leftrightarrow 3d'_{-1}$-transition is almost negligible, the $2p'_{+1} \leftrightarrow 3s'_0$- and the $2s'_0 \leftrightarrow 3p'_0$-transition show a stronger shift of the wavelength of their stationary components. Note, that this differences can be qualitatively understood. For the Balmer transition $2p'_{-1} \leftrightarrow 3d'_{-2}$, e.g., the magnetic quantum number m has for both states its minimum value. As m is conserved, for states with angluar momentum $l = n - 1$ and $m = -l$ inter-l-mixing is forbidden and only inter-n-mixing is possible.

The influence of the electric field on the transition probabilities of the stationary components of Balmer α is rather small for most field strengths and most transitions. Merely the transition between the states with field free quantum numbers $2p_{+1}$ and $3s_0$ exhibits a significant decrease in the oscillator strength for very strong electric fields. Since the $2p'_0 \leftrightarrow 3p'_{-1}$ transition is forbidden in the diamagnetic case, its oscillator strengths vanishes for $F = 0$. Under the influence of an additional electric field it becomes an allowed transition and the oscillator strength of the stationary component increases with increasing field strength. But even for an electric field strength of $F = 10^8$ V/m, it amounts to only approximately ten percent of the oscillator strength of the

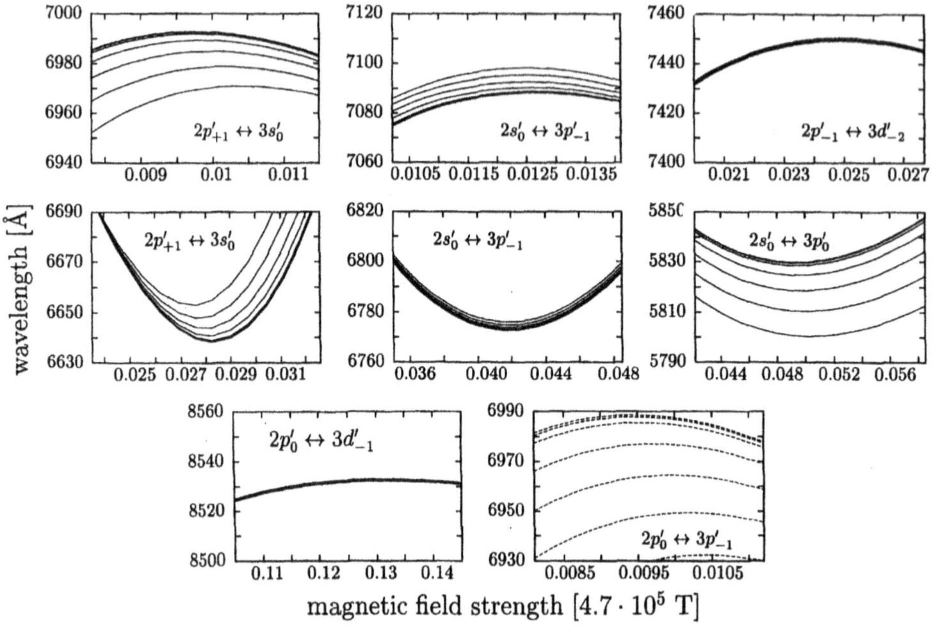

Figure 2. Stationary Lines of Balmer α for electric field strengths from $F = 0$ to $F = 10^8$ V/m.

weakest Balmer α transition of the diamagnetic hydrogen atom.

In figure 3 the absorption-feature of the stationary line $2s'_0 \leftrightarrow 3p'_0$ is shown. As a first approximation we used the Holtsmark-distribution (Lang, 1980) for the electric field strength. Compared to the diamagnetic case ($F = 0$), the absorption feature is broadened and blue-shifted if the additional electric field is considered. A mean electric field strength of $F_0 = 10^7$ V/m causes a shift of approximately 5 Å, a value comparable to observational estimates (Friedrich et al., 1994)

The atomic data obtained for the hydrogen atom in magnetic and electric fields are a further step towards a complete understanding of the spectra of DA magnetic white dwarf stars. In order to get a clearer evidence about the strength of the electric fields in the atmosphere of magnetic white dwarf stars, detailed calculations of synthetic spectra using the correct electric field distribution are necessary.

ELECTRONIC WAVEPACKET EVOLUTION

As a further application of the numerical method described in Faßbinder and Schweizer (1996a), we investigated the dynamics of electronic wave packets in Rydberg atoms. (For an overview of experimental and numerical methods describing wavepackets see Alber and Zoller, 1991). In a recent paper, Naudeau et al. (1997) presented experimental results of wavepacket dynamics of electrons bound in the non-separable potential of cesium in a static external electric field. By selecting a particular value of the electric field strength, they could control the structure of the recurrence spectra and enhance the interaction of the wavepacket with the cesium core. The laser spectrum was windowed so that only very few states around $n_{\rm eff} = 24$ are excited, thereby minimizing dispersion of the wavepacket. In this section, we present a realistic numerical treatment of the problem by simulating the time evolution of Rydberg wavepackets

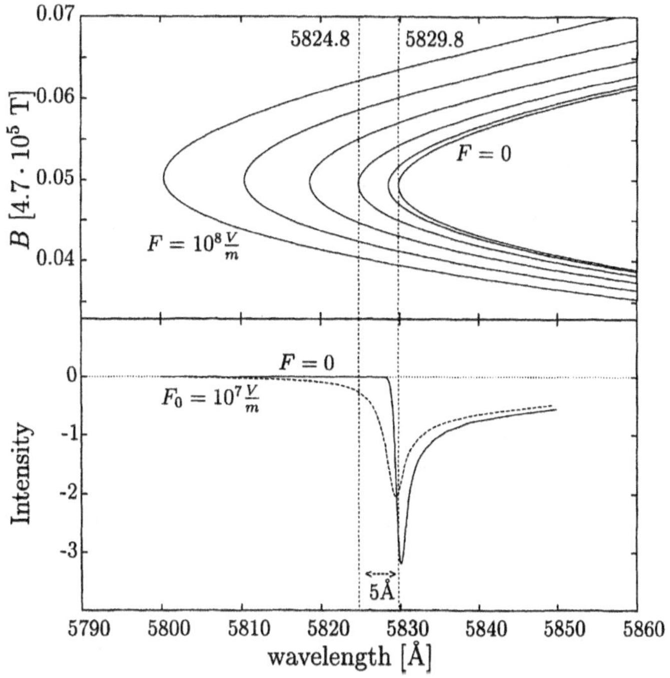

Figure 3. Absorption feature of the stationary line $2s'_0 \leftrightarrow 3p'_0$ using a Holtsmark-distribution for the electric field strength.

through direct solution of the time dependent Schrödinger equation on a spatial grid.

The influence of the non-hydrogenic core on the outer electron is represented by an exponential extension to the Coulomb term (Hanssen et al., 1979):

$$V(r) = -\frac{e^2}{r}\{1 + (Z-1)\exp(-a_1 r) + a_2 r \exp(-a_3 r)\} \quad . \tag{2}$$

Z is the nuclear charge and the coefficients a_i are optimized numerically so as to reproduce the field-free energy levels and hence quantum defects of the alkali atom. For cesium ($Z = 55$) we obtain $a_1 = 3.3$, $a_2 = 11.0$ and $a_3 = 1.51$ (Schweizer and Faßbinder, 1997). In order to describe the time evolution of a wavepacket, we need to solve the time dependent Schrödinger equation. For the spatial integration we used the same discretization technique as for the stationary problem described in Faßbinder and Schweizer (1996a). The wave function is then represented by a set of N functions

$$\psi_k(t) = \psi(\mathbf{r}_k, t) \quad , \quad k = 1, \ldots, N \quad , \tag{3}$$

and the Schrödinger equation reduces to a system of coupled ordinary differential equations. For the approximation of the time-development operator we used the Cayley form

$$e^{-\frac{i}{\hbar}\delta t H} \approx \left(1 + \frac{i}{2\hbar}H\delta t\right)^{-1}\left(1 - \frac{i}{2\hbar}H\delta t\right) \quad . \tag{4}$$

In figure 4 we show the square of the autocorrelation function $| < \psi(t)|\psi(0) > |^2$ in dependence of t for two electric field strengths. The structure of the recurrence spectrum depends strongly on the electric field strengths. For $F = 774$ V/cm, all odd angular return peaks are suppressed. For this particular F the odd angular returns

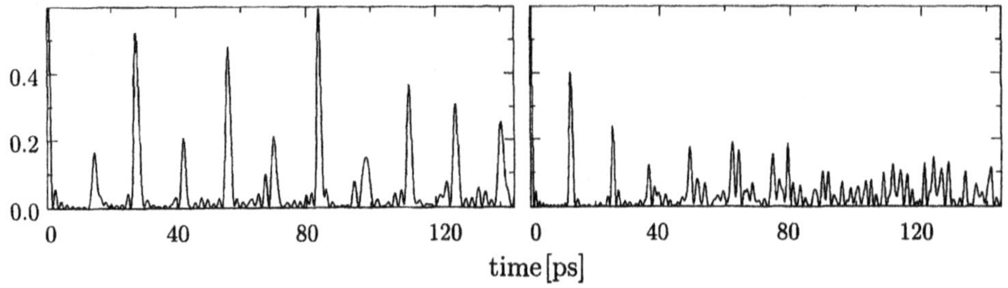

Figure 4. Calculated recurrence spectra of an electronic wavepacket in cesium subjected to a static electric field. We show the square of the autocorrelation function $| < \psi(t)|\psi(0) > |^2$ as a function of time for $F = 774$ V/cm (left) and $F = 830$ V/cm (right).

occur while the radial wavepacket is at its outer turning point and hence the overlap with the initial wavepacket is minimized. The case of $F = 830$ V/cm is just the opposite: Each angular return coincides with a radial return, which maximizes the interaction of the wavepacket with the cesium core. As a result, the return peaks diminish much faster in intensity (Faßbinder and Schweizer, 1997).

ACKNOWLEDGEMENTS

This work was supported by the Deutsche Forschungsgemeinschaft (DFG) and the Deutscher Akademischer Austauschdienst (DAAD).

REFERENCES

Alber, G., and Zoller, P., 1991, Laser excitation of electronic wave packets in rydberg atoms, *Phys. Rep.* 199:231.

Faßbinder, P., and Schweizer, W., 1996a, Hydrogen atom in very strong magnetic and electric fields, *Phys. Rev. A* 53:2135.

Faßbinder, P., and Schweizer, W., 1996b, Stationary hydrogen lines in magnetic and electric fields of white dwarf stars, *Astron. Astrophys.* 314:700.

Faßbinder, P., and Schweizer, W., 1997, Numerical simulation of electronic wavepacket evolution, *SFB 382* Report Nr. 78, *Phys. Rev. A* submitted.

Friedrich, S., Östreicher, R., Ruder, H., and Zeller, G., 1994, Line broadening in the presence of strong magnetic fields, *Astron. Astrophys.* 282:179.

Friedrich, S., Östreicher, R., and Schweizer, W., 1996, Observation of flux and polarization spectra of white dwarfs with low magnetic fields, *Astron. Astrophys.* 309:227.

Hanssen, J., McCarrol, R., and Valiron, P., 1979, Model potential calculations of the Na-He system, *J. Phys. B* 12:899.

Kemp, J. C., Swedlund, J. B., Landstreet, J. D., and Angel, J. R. P., 1970, *Astrophys. J.* 161:L77.

Lang, K. R., 1980, "Astrophysical Formulae," Springer, Berlin.

Melezhik, V. S., 1993, Three-dimensional hydrogen atom in crossed magnetic and electric fields, *Phys. Rev. A* 48:4528.

Melezhik, V. S., 1997, A computational method for quantum dynamics of a three-dimensional atom in strong fields, these proceedings.

Naudeau, M. L., Sukenik, C. I., and Bucksbaum, P. H., 1997, Core-scattering of Stark wavepackets, *Phys. Rev. A* submitted.

Pavlov-Verevkin, V. B., and Zhilinskii, B. I., 1980, Neutral hydrogen-like system in a magnetic field, *Phys. Lett. A* 78:244.

Schweizer, W., and Faßbinder, P., 1997, Numerical studies of model potentials for alkali atoms, preprint.

Wunner, G., Schweizer, W., and Ruder, H., 1989, Hydrogen atoms in strong magnetic fields – in the laboratory and in the cosmos, *in:* "The Hydrogen Atom," F. Bassini et al., ed., Springer, Berlin.

NEUTRON STAR ATMOSPHERES

George Pavlov[1,2]

[1]Pennsylvania State University,
525 Davey Lab, University Park, PA 16802, U.S.A.
[2]Ioffe Physical-Technical Institute, 194021, St.-Petersburg, Russia

OBSERVATIONAL MANIFESTATIONS OF NEUTRON STARS

When the thermonuclear fuel of a massive star is exhausted at the end of its evolution, the star explodes as a supernova releasing enormous amount of energy, $\sim 10^{52}$ erg. This explosion is accompanied by the collapse of the presupernova core which results in formation of either a black hole or a neutron star (NS), depending on core's mass. NSs are formed if the mass does not exceed a limit, $\sim 2-3M_\odot$, above which the pressure of degenerate nuclear matter (presumably neutrons) is not able to resist gravitational forces and maintain a stable configuration. Typical masses of NSs are expected to be in the range $0.5-2.5\ M_\odot$, quite common for ordinary stars. Their radii, however, should be incredibly small, $7-20$ km, depending on mass and (poorly known) equation of state of the superdense ($\sim 10^{15}$ g cm^{-3}) nuclear matter.

According to various estimates, the total number of NS in our Galaxy is $\sim 10^8-10^9$. Observational manifestations of NSs are quite different for *isolated neutron stars* (INSs) and *accreting neutron stars* (ANSs) in close binaries. A substantial fraction of observed INSs is comprised of ~ 750 rotation-powered *radio pulsars* (Taylor et al., 1993). Periods of the radio pulses, ~ 1 ms – 1 s, coincide with the NS rotation periods, and the radio waves is generated by relativistic electrons and/or positrons accelerated in the NS magnetospheres, with typical fields $B \sim 10^8 - 10^{13}$ G. The relativistic particles are responsible for γ radiation observed from at least 5 radio pulsars (Thompson et al., 1994). About 30 radio pulsars have been detected in X-rays (e. g., Becker & Trümper, 1997); the radiation from at least 3 of them is dominated by a thermal component emitted from the whole NS surface, with temperatures $\sim (2-10)\times 10^5$ K (Ögelman 1995). A few of the X-ray radiating radio pulsars show thermal-like radiation of higher temperatures, $\sim (1-5)\times 10^6$ K, from an area much smaller than that of the NS surface; this radiation is attributed to polar caps heated by the backward accretion of relativistic particles from the pulsar magnetospheres. Finally, at least 7 radio pulsars have been detected in the optical-UV range (Caraveo, 1997), a thermal component (i. e., the NS surface) is seen in 3 of them.

X-ray detections of several *radio-silent* INSs have been reported recently (see Caraveo et al., 1996). Since these objects do not show any activity and are not visible

Atoms and Molecules in Strong External Fields
Edited by Schmelcher and Schweizer, Plenum Press, New York, 1998

outside the X-ray range, their radiation almost certainly originates from the NS surface. Although interpretation of these observations is sometimes ambiguous, at least 4 radio-silent INSs have been found (3 of them in supernova remnants).

The most prominent types of ANSs are (accreting) X-ray pulsars and X-ray bursters (about 30 and 40 such objects have been detected, respectively — see, e. g., Lewin et al., 1995). X-ray pulsars are NSs with high magnetic fields, $\sim 10^{12}$ G, which force the matter from the secondary companion of the binary to accrete onto the NS polar regions. The balance between heating (due to release of the gravitational energy of the accreting matter) and radiative cooling occurs at a temperature of $\sim 10^8$ K, much higher than that at the polar caps of radio pulsars. X-ray bursters are believed to have much lower magnetic fields, $\sim 10^9 - 10^{10}$ G, which are not able to canalize the accretion. As a result, the whole NS surface is heated, but its temperature is lower, $\sim 10^7$ K. When a substantial amount of the accreting matter is accumulated, high pressure in bottom layers of the accreted envelopes makes thermonuclear burning unstable, so that a thermonuclear explosion occurs (X-ray burst), often accompanied by ejection of the accreted envelope. Although the X-ray radiation observed from most ANSs is generated close to the NS surfaces, not the surface per se is observed, but rather the hot accreting plasma.

As mentioned above, it is commonly accepted that NSs possess very high magnetic fields. The field magnitudes were first estimated assuming that the magnetic flux is conserved during the evolution from ordinary stars to NSs, including the collapse. This assumption results in typical fields $B \sim 10^{11} - 10^{13}$ G, depending on the field in the (ordinary) progenitor star, $B \sim 1 - 10^3$ G. According to radio pulsar models, the field of a pulsar, with the period P and its time derivative $\dot{P}$, can be estimated as $B = (3Ic^3 P\dot{P}/8\pi^2 R^6)^{1/2}$, where R and I are the NS radius and moment of inertia, and for most pulsars this estimate coincides with that predicted from the flux conservation. An exception is so-called millisecond (recycled) pulsars — very old objects for which the fields are much lower, $\sim 10^8 - 10^9$ G. Direct measurements of the magnetic fields are possible via observations of the electron cyclotron lines in spectra of X-ray pulsars; these lines were indeed detected in 9 X-ray pulsars, and the corresponding field values were found in the range $(1 - 4) \times 10^{12}$ G (Lewin et al., 1995). Much greater fields, up to 10^{15} G, were predicted by some models of soft gamma repeaters (e. g., Thompson and Duncan, 1995); these models, however, are still speculative and require more observational confirmations. Thus, the presence of very high magnetic fields in radiating layers of many NSs has been firmly established. Since such fields cannot be achieved in terrestrial laboratories, NSs *provide the unique opportunity to test atomic physics in extremely high magnetic fields.*

This brief sketch of observational manifestations of NSs shows that surface layers have been directly observed for about a dozen of INSs, and the number will certainly grow in near future. Main information about the NS surface is obtained from observations in soft and medium X-rays ($\sim 0.1 - 10$ keV), although UV and optical observations are also useful in some cases. Observations of NSs are potentially very important for elucidating poorly known properties of these objects. For instance, studying *thermal history* of INSs (dependence of the NS temperature on star's age) enables one to infer equation of state of the superdense matter and actual composition of the NS interiors — soft equations of state and "exotic" composition (e. g., quark-gluon plasma or π-meson condensate) result in accelerated cooling of NSs (e. g., Umeda et al., 1994). The thermal history depends also on properties of nucleon superfluidity which, for instance, may decelerate fast cooling and even lead to additional heating due to dissipation of the energy of differential rotation between the NS crust and interior superfluid (e. g.,

Shibazaki and Lamb, 1989). These properties are closely connected with the NS mass and radius which affect the spectra and light curves of the NS radiation (see below) — for instance, smaller radii are associated with softer equations of state and exotic interior compositions. Finally, direct investigation of physical properties and chemical composition of the NS surface layers is itself of great interest for understanding the nature of NSs and their environment. To properly interpret current and future observations of the NS radiation, adequate models of this radiation should be constructed.

ATMOSPHERES OF ISOLATED NEUTRON STARS: AN OVERVIEW

As any real objects, NSs are not black bodies, and their *thermal radiation is not blackbody radiation.* Properties of thermal radiation from a star are determined by thin plasma layers (atmospheres) at its surface. Since the atmospheric temperature varies with depth, and atmospheric opacity depends on frequency of radiation, we see layers of different temperatures at different frequencies, so that not only spectral features (e. g., absorption spectral lines) are displayed in stellar spectra, but also the shape of spectral continuum differs from the blackbody spectrum. NS atmospheres, however, are very different from those of usual stars. The immense gravity makes the atmospheres very thin, $\sim 0.1 - 10$ cm, and dense, up to $\sim 1 - 100$ g cm^{-3}, so that *nonideality* of the atmospheric matter becomes important and leads to pressure ionization of atomic species and smoothing spectral features. The gravity also leads to stratification of the atmospheres — heavy elements sink down to bottom layers so that upper layers, which determine properties of the outgoing radiation, are comprised of the lightest element present. The huge magnetic fields make NS atmospheres essentially *anisotropic*, so that absorption and emission of photons depend on the direction of the photon wave vector. Radiation in the strongly magnetized atmospheres propagates in two polarization (normal) modes, with quite different opacities. The electron cyclotron energy, $\hbar\omega_{Be} = \hbar eB/m_e c = 11.6\,(B/10^{12}$ G) keV, greatly exceeds the thermal energy kT, so that the transverse motion of electrons is quantized. Structure of atoms and ions is so much distorted by the magnetic fields that spectral features associated with the atoms have quite different energies and strengths, and ionization equilibrium of the atmospheric plasma is shifted from that at $B = 0$. Thus, the standard techniques widely used for modeling stellar atmospheres (e. g., Mihalas 1978) should be considerably modified to aplly them to the NS atmospheres.

An approach for modeling NS atmospheres with strong magnetic fields was described in detail by Pavlov et al. (1995). It involves, as for usual stellar atmospheres, simultaneous solving of equations of hydrostatic equilibrium, energy balance, ionization equilibrium and radiative transfer, complemented by calculations of spectral opacities. However, instead of one equation of radiative transfer for the intensity of radiation, two (generally coupled) equations for intensities of two polarization modes are to be solved, and reliable opacities for strongly magnetized, partially ionized, nonideal plasma are to be used. These opacities should be obtained from quantum-mechanical calculations of bound-bound, bound-free and free-free radiative transitions (spectral lines, photoionization and bremsstrahlung), as well as photon scattering. A first step for such calculations is investigation of atomic structure (energy levels and wave functions) in strong magnetic fields, the main subject of this workshop. Once the opacities are known, one can use the atmosphere modeling codes to calculate the atmospheric structure (dependences of temperature, density, etc, on depth) and spectral intensities of the outgoing polarization modes for given input parameters: chemical composition, effective temperature (defined as $\sigma T_{\rm eff}^4 = \int F(\nu)\, d\nu$; σ is the Stefan-Boltzmann constant, and $F(\nu)$ the local spectral flux), local magnetic field $\boldsymbol{B}$, and gravitational acceleration g. To obtain the total flux of radiation directly comparable with observations, the local intensities should be further integrated over the visible NS surface, with allowance for the gravitational redshift and bending of photon trajectories in the strong gravitational field (the latter effect makes partly visible the back NS hemisphere).

LOW-FIELD ATMOSPHERES

As mentioned above, magnetic fields of recycled pulsars and old INSs are relatively low, $B \sim 10^8 - 10^9$ G. Although such fields are much higher than those achieved in laboratories or observed in magnetic white dwarfs, they cannot change considerably properties of X-ray radiation emergent from the NS surface — the corresponding electron cyclotron energies, $\sim 1 - 10$ eV, are much lower than X-ray energies or energies of atomic levels involved in interaction with X-ray radiation. First models of the low-field atmospheres were calculated by Romani (1987). Further works (Rajagopal and Romani, 1996; Zavlin et al., 1996) used improved opacities (Iglesias and Rogers, 1996) for pure hydrogen, helium and iron compositions. These works showed that spectra of radiation emerging from light-element (hydrogen or helium) atmosphere are much harder (less steep) than the blackbody spectra at $h\nu \gtrsim kT_{\rm eff}$. This means that fitting observed spectra with the blackbody models (a widely accepted approach in X-ray astronomy) yields "blackbody temperatures" exceeding the true effective temperatures by a factor of $1.5 - 3$, which makes great difference for the comparison with the models of NS cooling. The low-field models have been successfully applied to interpretation of soft X-ray observations of the nearest ($d = 180$ pc), low-field ($B \sim 3 \times 10^8$ G) millisecond pulsar J0437–4715 (Pavlov et al., 1996b; Zavlin and Pavlov, 1997a). Fitting its X-ray spectra and light curves with the atmosphere models showed that its radiation can be interpreted as emitted from two polar caps, with temperatures $\simeq 1 \times 10^6$ K and radii $\simeq 1$ km, covered with *pure hydrogen*. The analysis of the light curves with allowance for bending of photon trajectories enabled Pavlov & Zavlin (1997) to constrain the NS mass-to-radius ratio: $1.4 < (M/M_\odot)(10 \text{ km}/R) < 1.6$. These results demonstrate importance of NS atmosphere models for elucidating NS properties from observational data.

FIRST GENERATION OF MAGNETIC HYDROGEN ATMOSPHERES

First reliable models of magnetic hydrogen atmospheres have been constructed recently (see Pavlov et al., 1995, for a review). These models are based upon simplified opacities of strongly magnetized, partially ionized hydrogen plasma. These opacities include free-free and bound-free transitions and Thomson scattering in strong magnetic fields, but they do not include the bound-bound transitions, neglect effects of atomic motion (motional Stark effect), and use a simplified model for ionization equilibrium. Nevertheles, the models have been useful to obtain a qualitative picture for magnetic effects on the emergent radiation, and they are accurate enough in the case of high effective temperatures, $\gtrsim 10^6$ K, when the hydrogen is almost completely ionized even in the very strong magnetic fields. The spectra of the local fluxes at the NS surface deviate substantially from the blackbody spectrum (Shibanov et al., 1992): they are harder (although not as much as the nonmagnetic hydrogen spectra) at high frequencies, show photoionization edges and proton cyclotron absorption lines. The spectra depend on B and the angle Θ_B between the magnetic field and the normal to the NS surface, which means that the radiative flux emitted by a rotating NS is pulsed if even the effective temperature is uniform along the NS surface. Angular distribution of the local intensity shows a sharp peak along the magnetic field and a broader peak at intermediate angles, the widths of the peaks depend on photon energy (Pavlov et al., 1994). Integration of the local intensities over the NS surface (Zavlin et al., 1995a) should take into account not only the non-uniformity of the magnetic field (e. g., dipole geometry), but also the temperature non-uniformity — the magnetic poles are hotter than the equator because the transverse thermal conductivity is suppressed by strong magnetic fields (Shibanov and Yakovlev 1996). The integration should be performed with allowance for the gravitational bending of photon trajectories (e. g., Zavlin et al., 1995b) which, as a rule, smoothes the light curves (Shibanov et al., 1995).

The model X-ray spectra have been compared with those observed from some radio pulsars and radio-silent INSs (Anderson et al., 1993; Meyer et al., 1994; Page et al., 1995; Zavlin and Pavlov, 1997b). An important result of the atmosphere fits is that they yield effective temperatures considerably lower than those obtained from the blackbody fits, usually below the temperatures predicted by the standard NS cooling models (see Introduction). However, in many cases both the blackbody and atmosphere models give statistically acceptable fits because of the narrowness of the X-ray energy range and poor spectral resolution of current X-ray detectors. Whether the hydrogen atmosphere models indeed describe adequately the observed radiation can be understood from the analysis of multiwavelength (optical through X-ray) observations (Pavlov et al., 1996c) and from taking into account additional astrophysical arguments (e. g., fitting with different models yields different values of the interstellar hydrogen column density, n_H, often known from independent observations). The most convincing case, where the hydrogen atmosphere model not only gives an excellent fit, but also is compatible with all other data for the object, is the INS in the supernova remnant PKS 1209–52 (Zavlin and Pavlov, 1997b). The fits of the spectra of active radio pulsars with the hydrogen atmosphere models, although statistically acceptable, are less convincing because they yield the R/d ratio systematically greater than expected from independent estimates. This could be partly explained by the low accuracy of the first generation models at relatively low effective temperatures, but the discrepancy is too high in some cases. In addition, one can naturally assume that active, young radio pulsars are not able to accrete the interstellar matter, so that their surfaces are depleted of hydrogen. Although final conclusions cannot be made until the next generation of the hydrogen models is developed, and observations with higher spectral resolution are carried out, investigations of NS atmospheres consisting of heavier elements are certainly warranted.

HEAVY-ELEMENT ATMOSPHERE MODELS

A first attempt to construct magnetic heavy-element atmosphere model was pursued by Miller (1992) who computed not only hydrogen, but also helium, carbon and nitrogen models. He used approximate energy levels and cross sections calculated by Miller and Neuhauser (1991) who employed a multiconfigurational Hartree-Fock code developed by Neuhauser et al. (1986). This code computes the wave functions and state energies in the "adiabatic approximation", which assumes that the Landau number N is a "good quantum number" and considers only states with $N = 0$, neglecting any admixture of states with $N > 0$. Although this approximation is good enough for hydrogen, at $B \gtrsim 3 \times 10^{11}$ G and $\omega \ll \omega_{Be}$, it becomes progressively worse with increasing the effective ion charge, especially for calculating the cross sections of radiative transitions. The models of Miller (1992) have some other flaws — e. g., they neglect all radiative processes except for the bound-free transitions and use photoionization cross sections averaged over polarization, which may cause substantial errors because the (extraordinary) mode with small absorption coefficient gives the main contribution to the outgoing flux. Indeed, comparison of Miller's spectra for hydrogen atmospheres with more accurate results described above shows considerable discrepancy. Nevertheless, these models have demonstrated that one should expect quite different spectra from light-element (H, He) and heavy-element atmospheres — in the latter case the spectral continuum is much closer to the blackbody spectrum, and numerous spectral features are seen at X-ray energies.

Since the supernova explosion has a mass cut in the iron layer, it is quite probable that surfaces of many NSs, particularly young, high-field pulsars, are comprised of pure iron. First magnetic iron atmosphere models were presented by Rajagopal et al. (1997). They used the same code of Neuhauser et al. (1986) to calculate the energies and wave functions of iron ions for subsequent calculations of the ionization equilibrium and radiative transitions for two polarization modes. The models are inevitably very approximate: for instance, applicability of the adiabatic approximation is even more questionable for iron than for carbon and nitrogen, the nonideality effects on the equation of state and radiative transitions (virtually neglected in this paper) are particularly severe in iron atmospheres, etc. However, the models can be considered as a starting pont for future work on atmosphere modeling and a baseline for comparison with the magnetic hydrogen atmosphere models. As expected, the emergent spectra folded with the responses of low-resolution X-ray detectors are fairly close to the blackbody spectra, and blackbody fits of such spectra may yield "blackbody temperatures" even lower than the effective temperatures, in sharp contrast with the hydrogen atmospheres. The spectra have a wealth of spectral features observable with X-ray detectors to be launched in near future. Although developing new, more accurate iron atmosphere models is an extremely challenging problem, its solution would be of great importance for adequate interpretation of the NS thermal radiation.

CURRENT WORK ON OPACITIES

One of the important issues neglected in the current atmosphere models is the motional Stark effect: the electric field induced by the center-of-mass (CM) motion breaks the cylindrical symmetry and distorts the atomic structure (Gor'kov and Dzialoshinskii, 1968; Herold et al., 1981; Johnson et al., 1983). For small values of the generalized transverse momentum $P_\perp$ of the atom, the dependence of the energy levels on $P_\perp$ can be interpreted in terms of a mass anisotropy — the atom becomes "heavier" when it moves across the magnetic field, the transverse mass grows with B and is higher for more excited states (Vincke and Baye, 1988; Pavlov and Mészáros, 1993). For very large values of $P_\perp$, the so-called decentering occurs — atomic electron is localized in a shallow potential well far from the nucleus (Burkova et al., 1976; Baye et al., 1992). At intermediate values of $P_\perp$, the dependence on $P_\perp$ is complicated; in particular, neighboring atomic levels approach each other showing anticrossings (Vincke et al., 1992; Potekhin, 1994). As shown by Pavlov and Mészáros (1993), such behavior leads to a number of spectroscopic and thermodynamic effects: additional shifts and broadening of spectral lines and photoionization edges, violation of the usual selection rules for radiative transitions, modification of the ionization balance of strongly magnetized plasmas. In particular, the magnetic broadening, caused by different dependences of the atomic levels on $P_\perp$, was shown to exceed the collisional and Doppler broadenings by orders of magnitude for conditions expected in atmospheres of strongly magnetized

NSs.

The effect of the CM motion is especially complicated for charged ions (Baye and Vincke, 1986; Schmelcher and Cederbaum, 1991, 1997; Bezchastnov, 1995). When the net charge differs from zero, one cannot separate the CM and internal motion (Avron et al., 1978), so that even in the simplest case of hydrogen-like ions an accurate solution for the ion structure should be based on a two-particle Schrödinger equation. First numerical solutions of such kind were obtained by Bezchastnov et al. (1997) who used a two-particle basis set of Bezchastnov (1995) to calculate the energies and wave functions for once-ionized helium. Since the motion of an ion as a whole is quantized, the level energies depend on a new discrete quantum number, instead of $P_\perp$. Although radiative transitions have not yet been calculated, the inferred behavior of the ion structure allows one to predict new spectroscopic effects (e. g., a fine structure of spectral lines and photoionization edges) peculiar to charged ions.

Bound-bound transitions in strong magnetic fields have been thoroughly investigated by the Tübingen group (see Ruder et al., 1994) for hydrogen and helium atoms at rest. The oscillator strengths for moving hydrogen atoms were calculated by Potekhin (1994) and Pavlov and Potekhin (1995) for a broad range of the transverse generalized momenta. In the latter paper the bound-bound absorption coefficients were calculated for a thermal, partially ionized plasma. In particular, the magnetic broadening was investigated with account for decentered states of moving atoms. Due to this broadening, spectral lines are smeared over wide, overlapping frequency bands.

The photoionization cross sections for hydrogen atoms at rest in strong magnetic fields were studied in the adiabatic approximation by Potekhin and Pavlov (1993). The cross sections are highly anisotropic and polarization dependent, being strongly reduced when the radiation is polarized perpendicular to the magnetic field. On the other hand, the conclusion of some earlier works that the cross sections for photons polarized perpendicular to the magnetic field vanish identically in the most important frequency range, $\omega < \omega_{Be}$, was show to be erroneous. Potekhin et al. (1997) extended calculations beyond the adiabatic approximation and showed that nonadiabatic corrections are substantial for $\omega \gtrsim 0.3\omega_{Be}$. The non-adiabatic treatment of the continuum includes coupling between closed and open channels, which leads to the autoionization of quasi-bound energy levels associated with the electron cyclotron (Landau) excitations and gives rise to Beutler-Fano resonances of the photoionization cross sections. Effects of atomic motion on the bound-free transitions were studied by Potekhin and Pavlov (1997). They employed exact numerical treatment of both initial and final states of moving hydrogen atoms, taking into account the quasi-bound states as well as coupling of different ionization channels. They folded the cross sections with the thermal distribution of atoms and found that the averaged cross sections differ substantially from those of atoms at rest, showing strong magnetic broadening of the photoionization edges. The decentered states of the atoms give rise to a low-energy component of the averaged cross sections which grows with increasing temperature and decreasing density.

In the aforementioned papers the bound-bound and bound-free absorption coefficients were calculated for three "basic" polarizations: linear polarization along the magnetic field and two circular (right and left) polarizations across the magnetic field. To apply the results for solving the radiative transfer in a magnetized atmosphere, these coefficients are to be combined into two absorption coefficients of the (normal) polarization modes, which requires the polarization vectors of the modes to be known for different frequencies and directions of propagation. Although this problem has been solved a long ago for a completely ionized plasma (Gnedin and Pavlov 1973), it is more complicated when bound atoms and ions are involved. Bulik and Pavlov (1996) solved this problem by constructing the polarizability tensor of a magnetized hydrogen gas from the basic absorption coefficients with the aid of the Kramers-Kronig relations. They showed that the spectral features and anisotropy associated with the bound-bound and bound-free transitions are manifested in the polarization properties of the normal modes, which affects substantially the transfer of radiation in NS atmospheres.

An important ingredient of atmosphere models is ionization equilibrium which is strongly affected by both the strong magnetic field (Gnedin et al., 1974) and the high density of the plasma (e. g., Hummer and Mihalas, 1988). In the current atmosphere models the ionization equilibrium is considered with account for magnetic effects, but the effects of the atomic motion are neglected. Including these effects for relatively low temperatures, when the decentered states are not populated, is simple once the energy levels of moving atoms are calculated. At higher temperatures the magnetic and density effects are closely connected because the partition functions diverge unless both the summation over discrete levels and integration over the generalized momentum are truncated to take into account that highly excited atoms are destroyed by neighboring particles. An approximate solution of this problem was presented by Lai and Salpeter (1995).

FUTURE INVESTIGATIONS

Observational Prospects

Most spectral X-ray observations of NS have been done with the use of proportional counters, with quite modest energy resolution (e. g., the Positional Sensitive Proportional Counter onboard the *ROSAT* X-ray observatory has only 5 independent energy channels in the range 0.1 – 2.4 keV). This hampers the comparison of the NS atmosphere models with observations. Fortunately, a new type of detectors, Charge Coupling Devices (CCD), with much better energy resolution has recently become available for X-ray spectroscopy. First X-ray observatory equipped with CCD detectors, *ASCA*, was launched in 1993, and first X-ray spectra with moderate resolution (mainly, of hot plasmas in supernova remnants and interstellar medium) have been obtained. *ASCA* has detected about a dozen of INSs, but no conclusive evidence about spectral features in their spectra has been obtained because the effective area of the detectors is too small to collect enough quanta, and the angular resolution of *ASCA* telecopes is low, ~ 3 arcmin, so that the images of point sources are contaminated by the background. Several new X-ray observatories, with CCD detectors of greater effective areas and higher spectral resolution and X-ray telescopes of a much better spatial resolution, will be launched in a few years: *AXAF* (1998), *XMM* (2000), *ASTRO-E* (2001). For instance, *AXAF* telescopes will provide subarcsecond angular resolution, and its CCD camera, in combination with gratings, will be able to resolve spectral features of brighter INSs with an accuracy of a few eV in the range of 0.2 – 10 keV. Such capabilities will yield new observational data of unprecedented quality, so that the direct comparison of theoretical results for atoms and molecules in strong magnetic fields will become possible, for the first time.

A promising new approach is investigation of INSs in the optical-UV range (e. g., Pavlov et al., 1996a). Thermally radiating INSs are extremely faint in this range, and observing their spectra and light curves is a challenge, even with large telescopes. However, first positive results have been obtained (e. g., Caraveo, 1997). In particular, unusual spectral features have been reported in the spectra of Geminga (tentatively interpreted as an ion cyclotron line — see Bignami et al., 1996) and, perhaps, PSR

0656+14 (Pavlov et al., 1997). New instruments installed at the focal plane of the *Hubble Space Telescope* in 1997 will enable us to reliably separate thermal and non-thermal components of the INS optical-UV radiation, to study periodicities, and to obtain more accurate photometry with narrower filters. The optical-UV observations are particularly useful when combined with X-ray observations (Pavlov et al., 1996c).

Theoretical Problems to Be Solved

The improved capabilities of the new X-ray space observatories and optical telescopes warrant more extensive and intensive theoretical investigations of the NS atmospheres. As discussed above, such investigations are impossible without reliable opacities of strongly magnetized, dense plasmas of various chemical compositions, and these opacities cannot be calculated without the corresponding atomic data. Most urgent problems to be solved for NS atmosphere modeling are the following.

For *hydrogen atmospheres*, the problem which almost has not been explored is the structure, number density, and radiative transitions of hydrogen *molecules, atomic and molecular ions and chains.* Although a number of calculations of the binding energies has been performed for these species, and crude estimates of the dissociation equilibrium have been obtained (Lai and Salpeter, 1997), a lot of work remains to be done to obtain detailed and reliable results. The problem of radiative transitions of these species has not been even touched, to the best of my knowledge.

Another challenging problem, actually relevant to any NS atmospheres, is calculation of the *ionization equilibrium and equation of state in the dense, magnetized plasmas.* The most direct approach would be to use the so-called "physical picture", in which all electrons, free and bound, and nuclei are considered as individual objects, an activity expansion of the grand canonical ensemble is used, and complicated numerical methods are applied to calculate the composition and the state of the plasma (e. g., Iglesias and Rogers, 1996; Pierleoni et al., 1994). This approach is computationally expensive, and it requires generalization for the case of strong magnetic fields. An alternative, simpler approach employs the "chemical picture", in which only few types of species (atoms, ions, molecules and free electrons) are considered as individual objects, and the state of the plasma is obtained from minimization of the free energy with respect to the densities of different species. This approach is limited by relatively low densities, and attempts to obtain self-consistent results for arbitrary plasma density failed even for the field-free hydrogen plasmas.

Since it is very probable that surfaces of many observable INSs are comprised of iron, comprehensive investigations of *iron ions* are greatly desirable. In particular, it would be useful to extend the multi-configurational Hartree-Fock calculations beyond the adiabatic approximation which fails for strongly charged iron ions even at relatively high magnetic fields. Such calculations would be very useful for further applications to studying opacities of the iron plasma and developing iron atmosphere models.

The chemical composition of NS atmospheres is actually unknown (in fact, it is one of the problems to be solved). Since some of the (hydrogen-depleted) supernova ejecta may eventually accrete onto the compact core remnant, the surfaces of some NSs may be covered by abundant elements different from hydrogen or iron. The most abundant elements which may be expected at the NS surface are helium, carbon, nitrogen, oxygen, etc, and it would be interesting to calculate the structure and radiative transitions of the corresponding ions in strong magnetic fields. Perhaps, self-consistent models of helium atmospheres should be developed first, with allowance for the effects of motion of atoms and ions. These results could be further used as a basis for understanding more complicated atmospheres comprised of heavier elements.

To conclude, many challenging problems of atomic and statistical physics are associated with the NS atmospheres, and solution of these problems is urgently needed for further progress of our understanding of physics and astrophysics of NSs.

ACKNOWLEDGEMENTS

I am greatly indebted to many colleagues who participated in joint exploration of the NS atmospheres and related problems. I am particularly grateful to Slava Zavlin, Yura Shibanov, Alexander Potekhin and Victor Bezchastnov for their cooperation. I acknowledge the kind support from WE-Heraeus-Stiftung and DFG. This work was partially supported by NASA Grant NAG5-2807.

REFERENCES

Anderson, S. B., Córdova, F. A., Pavlov, G. G., Robinson, C. R., and Thompson, R. J., 1993, ROSAT High Resolution Imager observations of PSR 0656+14, *Astrophys. J.* 414:867.

Avron, J. E., Herbst, I. W., and Simon B., 1978, Separation of center of mass in homogeneous magnetic fields, *Ann. Phys. (NY)* 114:431.

Baye, D., Clerbaux, N., and Vincke, M., 1992, Delocalized states of atomic hydrogen in crossed electric and magnetic fields, Phys. Lett. A 166:135.

Baye, D., and Vincke, M., 1986, Centre-of-mass energy of hydrogenic ions in a magnetic field, *J. Phys. B* 19:4051.

Becker, W., and Trümper, J., 1997, The X-ray luminosity of rotation-powered neutron stars, *Astron. Astrophys.*, in press.

Bezchastnov, V. G., 1995, A new basis of the Landau states for a hydrogen-like ion moving in a strong magnetic field, *J. Phys. B* 28:167.

Bezchastnov, V. G., Pavlov, G. G., and Ventura, J., 1997, Discrete energy spectrum of He^+ moving in a strong magnetic field, *J. Phys. B*, submitted.

Bignami, G. F., Caraveo, P. A., Mignani, R., Edelstein, J., and Bowyer, S., 1996, Multiwavelength data suggest a cyclotron feature on the hot thermal continuum of Geminga, *Astrophys. J.* 456:L111.

Bulik, T., and Pavlov, G. G., 1996, Polarization modes in a strongly magnetized hydrogen gas, *Astrophys. J.* 469:373.

Burkova, L. A., Dzialoshinskii, I. E., Drukarev, G. P., and Monozon, B. S., 1976, Hydrogen-like systems in crossed electric and magnetic fields, *Zh. Exp. Teor. Fiz. (Sov. Phys. JETP)* 71:526.

Caraveo, P. A., 1997, Neutron star astronomy, *Adv. Space Res.*, in press.

Caraveo, P. A., Bignami, G. F., and Trümper, J., 1996, Radio-silent isolated neutron stars as a new astronomical reality, *Ann. Rev. Astron. Astrophys.* 7:209.

Gnedin, Yu. A., and Pavlov, G. G., 1973, Transfer equations for normal waves and polarization of radiation in an anisotropic medium, *Zh. Exp. Teor. Fiz. (Sov. Phys. JETP)* 65:1806.

Gnedin, Yu. A., Pavlov, G. G., and Tsygan, A. I., 1974, Photoeffect in strong magnetic fields and X-ray radiation of neutron stars, *Zh. Exp. Teor. Fiz. (Sov. Phys. JETP)* 66:421.

Gor'kov, L. P., and Dzialoshinskii, I. E., 1967, Contribution to the theory of the Mott exiton in a strong magnetic field, *Zh. Exp. Teor. Fiz. (Sov. Phys. JETP)* 53:717.

Herold, H., Ruder, H., and Wunner, G., 1981, The two-body problem in the presence of a homogeneous magnetic field, *J. Phys. B* 14:751.

Hummer, D. G., and Mihalas, D., 1988, The equation of state for stellar envelopes. I – An occupation probability formalism for the truncation of internal partition functions, *Astrophys. J.* 331:794.

Iglesias, C. A., and Rogers, F. J., 1996, Updated OPAL opacities, *Astrophys. J.* 464:943.

Johnson, B. R., Hirschfelder, J. O., and Yang, K.-H., 1983, Interaction of atoms, molecules and ions with constant electric and magnetic fields, *Rev. Mod. Phys.* 55:109.

Lai, D., and Salpeter, E. E., 1995, Motion and ionization equilibrium of hydrogen atoms in a superstrong magnetic field, *Phys. Rev. A* 52:2611.

Lai, D., and Salpeter, E. E., 1997, Hydrogen phases on the surface of a strongly magnetized neutron star, *Astrophys. J.*, accepted

Lewin, W. H. G., van Paradijs, J., and van den Heuvel, E. P. J., 1995, "X-ray Binaries", Cambridge Univ. Press, Cambridge.

Meyer, R. D., Pavlov, G. G., and Mészáros, P., 1994, Soft X-ray spectral fits of Geminga with model neutron star atmospheres, *Astrophys. J.* 433:265.

Mihalas, D., 1978, "Stellar Atmospheres", W. H. Freeman, San Francisco.

Miller, M. C., 1992, Model atmospheres for neutron stars, *Mon. Not. Roy. Astron. Soc.* 255:129.

Miller, M. C., and Neuhauser, D., 1991, Atoms in very strong magnetic fields, *Mon. Not. Roy. Astron. Soc.* 253:107.

Ögelman, H., 1995, X-ray observations of cooling neutron stars, *in* "The Lives of the Neutron Stars," M. A. Alpar, Ü. Kiziloğlu and J. van Paradijs, eds., Kluwer, Dordrecht.

Page, D., Shibanov, Yu. A., and Zavlin, V. E., 1996, Temperature, distance and cooling of the Vela pulsar, *in* "Röntgenstrahlung from the Universe," Zimmermann, H. U., Trümper, J., and Yorke, H., eds., MPE Report 263, MPE, Garching.

Pavlov, G. G., and Mészáros, P., 1993, Finite-velocity effects on atoms in strong magnetic field and implications for neutron star atmospheres, *Astrophys. J.* 416:752.

Pavlov, G. G., and Potekhin, Y. A., 1995, Bound-bound transitions in strongly magnetized hydrogen plasma, *Astrophys. J.* 450:883.

Pavlov, G. G., and Zavlin, V. E., 1997, Mass-to radius ratio for the millisecond pulsar J0437–4715, *Astrophys. J.*, submitted.

Pavlov, G. G., Shibanov, Yu. A., Ventura, J., and Zavlin, V. E., 1994, Model atmospheres and radiation of magnetic neutron stars: Anisotropic thermal emission, *Astron. Astrophys.* 289:837.

Pavlov, G. G., Shibanov, Yu. A., Zavlin, V. E., and Meyer, R. D., 1995, Neutron star atmospheres, *in* "The Lives of the Neutron Stars," M. A. Alpar, Ü. Kiziloğlu and J. van Paradijs, eds., Kluwer, Dordrecht.

Pavlov, G. G., Stringfellow, G. S., and Córdova, F. A., 1996a, Hubble Space Telescope observations of isolated pulsars, *Astrophys. J.* 467:370.

Pavlov, G. G., Zavlin, V. E., Becker, W., and Trümper, J., 1996b, Soft X-ray and EUV radiation from polar caps of the millisecond pulsar J0437-4715, *Bull. AAS* 28:947.

Pavlov, G. G., Zavlin, V. E., Trümper, J., and Neuhäuser, R., 1996c, Multiwavelength observations of isolated neutron stars as a tool to probe the properties of their surfaces, *Astrophys. J.* 472:L33.

Pavlov, G. G., Welty, A. D., and Córdova, F. A., 1997, Hubble Space Telescope observations of the middle-aged pulsar 0656+14, *Astrophys. J.*, submitted.

Pierleoni, C., Bernu, B., Ceperley, D. M., and Magro, W. R., 1994, Equation of state of the hydrogen plasma by path integral Monte Carlo simulation, *Phys. Rev. Lett.* 73:2145.

Potekhin, A. Y., 1994, Structure and radiative transitions of the hydrogen atom moving in a strong magnetic field, *J. Phys. B* 27:1073.

Potekhin, A. Y., and Pavlov, G. G., 1993, Photoionization of the hydrogen atom in a strong magnetic field, *Astrophys. J.* 407:330.

Potekhin, A. Y., and Pavlov, G. G., 1997, Photoionization of hydrogen in atmospheres of magnetic neutron stars, *Astrophys. J.* 483:414.

Potekhin, A. Y., Pavlov, G. G., and Ventura, J., 1997, Ionization of the hydrogen atoms in strong magnetic fields: Beyond the adiabatic approximation, *Astron. Astrophys.* 317:618.

Rajagopal, M., and Romani, R., 1996, Model atmospheres for low-field neutron stars, *Astrophys. J* 461:327.

Rajagopal, M., Romani, R., and Miller, M. C., 1997, Magnetized iron atmospheres for neutron stars, *Astrophys. J.* 479:347.

Romani, R., 1987, Model atmospheres for cooling neutron stars, *Astrophys. J.* 313:718.

Ruder, H., Wunner., G., Herold, H., and Geyer, F., 1994, "Atoms in Strong Magnetic Fields," Springer, Berlin–Heidelberg.

Schmelcher, P., and Cederbaum, L. S., 1991, Interaction of the Landau orbitals of atomic ions in a magnetic field with electronic motion, *Phys. Rev. A* 43:287.

Schmelcher, P., and Cederbaum, L. S., 1997, Two interacting charged particles in strong static fields: A variety of two-body phenomena, *Structure and Bonding* 86:27.

Shibanov, Yu. A., and Yakovlev, D. G., 1996, On cooling of magnetized neutron stars, *Astron. Astrophys.* 309:171.

Shibanov, Yu. A., Zavlin, V. E., Pavlov, G. G., and Ventura, J., 1992, Model atmospheres of magnetic neutron stars: I. The fully ionized case, *Astron. Astrophys.* 266:313.

Shibanov, Yu. A., Pavlov, G. G., Zavlin, V. E., Qin, L, and Tsuruta, S., 1995, Anisotropic cooling and atmospheric radiation of neutron stars with strong magnetic field, *Ann. NY Acad. Sci.*, 759:291.

Shibazaki, N., and Lamb, F. K., 1989, Neutron star evolution with internal heating, *Astrophys. J.* 346:808.

Taylor, J. H., Manchester, R. N., and Lyne, A. G., 1993, Catalog of 558 pulsars, *Astrophys. J. Suppl.* 88:529.

Thompson, C., and Duncan, R. C., 1995, The soft gamma repeaters as very strongly magnetized neutron stars. I – Radiative mechanism for outbursts, *Mon. Not. Royal Astron. Soc.*, 275:255.

Thompson, D. J., et al. 1994, EGRET high-energy gamma-ray pulsars. I – Young spin-powered pulsars, *Astrophys. J.* 436: 229.

Umeda, H.,Tsuruta, S., and Nomoto, K., 1994, Nonstandard thermal evolution of neutron stars, *Astrophys. J.* 433:256.

Vincke, M., Le Dourneuf, M., and Baye, D., 1992, Hydrogen atom in crossed electric and magnetic fields: transition from weak to strong electron-proton decentering, *J. Phys. B* 25:2787.

Zavlin, V. E., and Pavlov, G. G., 1997a, Soft X-rays from polar caps of the millisecond pulsar J0437–4715, *Astron. Astrophys*, submitted.

Zavlin, V. E., and Pavlov, G. G., 1997b, Neutron star in the supernova remnant PKS 1209–52, *Astrophys. J.*, submitted

Zavlin, V. E., Pavlov, G. G., Shibanov, Yu. A., and Ventura, J., 1995a, Thermal radiation from rotating single neutron star: Effect of the magnetic field and surface temperature distribution, *Astron. Astrophys.* 297:441.

Zavlin, V. E., Shibanov, Yu. A., and Pavlov, G. G., 1995b, Effect of the neutron star gravitational field on radiation from the hot polar spots of radio pulsars, *Astron. Lett.* 21:168.

Zavlin, V. E., Pavlov, G. G., and Shibanov, Yu. A., 1996, Model neutron star atmospheres with low magnetic fields: I. Atmospheres in radiative equilibrium, *Astron. Astrophys.*, 315:141.

HYDROGEN ATOMS IN NEUTRON STAR ATMOSPHERES: ANALYTICAL APPROXIMATIONS FOR BINDING ENERGIES

Alexander Y. Potekhin*

Ioffe Physico-Technical Institute, St.-Petersburg 194021, Russia

INTRODUCTION

Since the first observations of neutron stars thirty years ago, they have affected many branches of physics. These extremely compact stars serve as natural physical laboratories for probing the properties of matter under extreme physical conditions. In particular, more than half of them possess magnetic fields $B > 10^{12}$ G.

Despite their name, neutron stars consist not only of neutrons. They have a crust containing ionized iron, heavier elements, and exotic neutron-rich nuclei,[1] above which lie liquid and gaseous outer envelopes, which are thought to be composed of iron or lighter elements.[2] The atmosphere, that affects the spectrum of outgoing thermal radiation, likely consists of hydrogen, the most abundant element in the Universe, which might be brought to the star surface by fall-out of circumstellar medium. Neutral atoms can provide an appreciable contribution to the atmospheric opacity.

Apart from the physics of neutron stars, quantum mechanical calculations of strongly magnetized hydrogen atoms find application also in the physics of white dwarf stars[3,4] and in the solid state physics.[5] Because of this practical demand, hydrogen in strong magnetic fields has been well studied in the past two decades.[6] The peculiarity of the problem for neutron stars is that an atom cannot be considered abstractedly from its thermal motion. Indeed, neutron star atmospheres are hot ($T \sim 10^5 - 10^6$ K), so that typical kinetic energies of the atoms are non-negligible in comparison with typical binding energies. Taking the thermal motion into account is highly non-trivial, because an atom moving across magnetic field is equivalent to an atom placed in orthogonal electric and magnetic fields, so that the cylindrical symmetry is broken.

At $\gamma \gg 1$, where $\gamma \equiv \hbar\omega_c/2$ Ryd $= B/2.35 \times 10^9$ G $\gg 1$ and ω_c is the electron cyclotron frequency, the collective motion effects[7,8] become especially pronounced. In particular, so-called decentered states (with the electron localized mostly in the "magnetic well" aside from the Coulomb center) are likely to be populated even at the relatively high densities $\rho > 10^{-2}$ g cm^{-3} typical of neutron star atmospheres. These exotic states have been predicted two decades ago by Burkova et al.[9] and studied recently by other authors.[10–12]

*e-mail: palex@astro.ioffe.rssi.ru

Collective-motion effects on the usual "centered" states have been first considered in frames of the theory of perturbation.[8,13] Non-perturbative results covering both centered and decentered states were subsequently presented for binding energies and wavefunctions,[14,15] oscillator strengths,[15] spectral line shapes,[16] and photoionization cross sections.[17] None of these data, however, has been published in an easy-to-use form of tables or analytical expressions.

In this contribution I propose approximate analytical expressions for the binding energies of the hydrogen atom arbitrarily moving in a magnetic field typical of neutron stars, $300 \leq \gamma \leq 10^4$. This range is physically distinguished, since at weaker fields the spectrum is strongly complicated by multiple narrow anticrossings,[14] while the upper bound, $\gamma \sim 10^4$, corresponds to the onset of non-negligible relativistic effects.[18]

THEORETICAL FRAMEWORK

Motion of the hydrogen atom in a magnetic field can be conveniently described by the pseudomomentum $\mathbf{K} = m_p\dot{\mathbf{r}}_p + m_e\dot{\mathbf{r}}_e - (e/c)\mathbf{B}\times(\mathbf{r}_e - \mathbf{r}_p)$, where the subscript $i = e$ or $i = p$ indicates electron or proton, respectively, $\dot{\mathbf{r}}_i = -(i\hbar/m_i)\nabla_i - (q_i/m_i c)\mathbf{A}(\mathbf{r}_i)$ is the velocity operator, m_i the mass, $q_p = -q_e = e$ the charge, and $\mathbf{A}(\mathbf{r})$ the vector potential of the field. Gorkov and Dzyaloshinskii[19] have shown that in the representation in which all components of $\mathbf{K}$ have definite values, the relative motion can be described in terms of a one-particle Hamiltonian which depends on $\mathbf{K}$.

It is convenient to describe the centered states of the atom using the relative coordinate $\mathbf{r}^{(0)} = \mathbf{r}_e - \mathbf{r}_p$ as independent variable and the axial gauge of the vector potential, $\mathbf{A}(\mathbf{r}) = \frac{1}{2}\mathbf{B}\times\mathbf{r}$. For the decentered states, the "shifted" representation[19] is more convenient. In the latter representation, the independent variable is $\mathbf{r}^{(1)} = \mathbf{r}_e - \mathbf{r}_p - \mathbf{r}_c$ and the gauge is $\mathbf{A}(\mathbf{r}) = \frac{1}{2}\mathbf{B}\times(\mathbf{r} - [(m_p - m_e)/m_\mathrm{H}]\,\mathbf{r}_c)$. Here, $\mathbf{r}_c = \frac{c}{eB^2}\mathbf{B}\times\mathbf{K}$ is the relative guiding center, and $m_\mathrm{H} = m_p + m_e$.

Let us assume that $\mathbf{B}$ is directed along the z-axis. The z-component of the pseudomomentum corresponding to the motion along the field yields the familiar term $K_z^2/2m_\mathrm{H}$ in the energy, while the transverse components $\mathbf{K}_\perp$ produce non-trivial effects. Therefore we assume $K_z = 0$ and $\mathbf{K}_\perp = \mathbf{K}$ hereafter.

If there were no Coulomb attraction, then the transverse part of the wavefunction could be described by a Landau function $\Phi_{ns}(\mathbf{r}_\perp^{(1)})$, where $\mathbf{r}_\perp^{(1)}$ is the projection of $\mathbf{r}^{(1)}$ in the (xy)-plane. The energy of the transverse excitation is

$$E_{ns}^\perp = [n + (m_e/m_p)(n+s)]\hbar\omega_c, \tag{1}$$

where the zero-point and spin terms are disregarded.

A wavefunction ψ_κ of an atomic state $|\kappa\rangle$ can be expanded over the complete set of the Landau functions

$$\psi_\kappa^{(\eta)}(\mathbf{r}^{(\eta)}) = \sum_{ns} \Phi_{ns}(\mathbf{r}_\perp^{(\eta)})\, g_{n,s;\kappa}^{(\eta)}(z), \tag{2}$$

where $\eta = 0$ or 1 indicates the conventional or shifted representation, respectively (a generalization to arbitrary η proved to be less useful[15]). The one-dimensional functions $g_{ns;\kappa}^{(\eta)}$ are to be found numerically. The adiabatic approximation used in early works[9,19] corresponds to retaining only one term in this expansion.

A bound state can be numbered[15] as $|\kappa\rangle = |n_\kappa, s_\kappa, \nu, \mathbf{K}\rangle$, where n_κ and s_κ relate to the leading term of the expansion (2), and ν enumerates longitudinal energy levels

$$E_{n_\kappa,s_\kappa,\nu}^{\|}(K) = E_\kappa - E_{n_\kappa s_\kappa}^\perp \tag{3}$$

Table 1. Parameters of the approximation (4) at $10^{-1} \le \gamma \le 10^4$.

s	0	1	2	3	4	5	6	7
p_1	15.55	0.5332	0.1707	0.07924	0.04696	0.03075	0.02142	0.01589
p_2	0.3780	2.100	4.150	6.110	7.640	8.642	9.286	9.376
p_3	2.727	3.277	3.838	4.906	5.787	6.669	7.421	8.087
p_4	0.3034	0.3092	0.2945	0.2748	0.2579	0.2431	0.2312	0.2209
p_5	0.4380	0.3784	0.3472	0.3157	0.2977	0.2843	0.2750	0.2682

and controls the z-parity: $g^{(\eta)}_{n,s;\kappa}(-z) = (-1)^\nu g^{(\eta)}_{n,s;\kappa}(z)$. For the non-moving atom at $\gamma > 1$, the states $\nu = 0$ are tightly bound in the Coulomb well, while the states $\nu \ge 1$ are hydrogen-like, with binding energies below 1 Ryd. The states with $n \neq 0$ belong to continuum at $\gamma > 0.2$ and will not be considered here.

At small pseudomomenta K, the states $\nu = 0$ remain tightly bound and centered, the mean electron-proton separation $\bar{x}$ being considerably smaller than r_c (for the hydrogen-like states $\nu \ge 1$, however, $\bar{x}$ is close to r_c at any K). The larger K, the greater is the distortion of the wavefunction towards $\mathbf{r}_c$, caused by the motion-induced electric field in the co-moving reference frame, until near some K_c transition to the decentered state occurs, and the character of the motion totally changes. With further increasing K, the transverse velocity decreases and tends to zero, whereas the electron-proton separation increases and tends to r_c. Thus, for the decentered states, the pseudomomentum characterizes electron-proton separation rather than velocity.

At very large K the longitudinal functions become oscillator-like, corresponding to a wide, shallow parabolic potential well.[9] For a fixed ν, this limit is reached at $K \gg (\nu+\frac{1}{2})^2\hbar/a_B$, where a_B is the Bohr radius. Still at arbitrarily large K, there remain infinite number of bound states with high values of ν whose longitudinal wavefunctions are governed by the Coulomb tail of the effective one-dimensional potential.[15]

The decentered states of the atom at $K > K_c \sim 10^2$ au have relatively low binding energies and large quantum-mechanical sizes, $l \sim K/\gamma$ au; therefore they are expected to be destroyed by collisions with surrounding particles in the laboratory and in the white-dwarf atmospheres. In neutron-star atmospheres at $\gamma \sim 10^3$, however, the decentered states may be significantly populated. This necessitates inclusion of the entire range of K below and above K_c in the consideration.

ANALYTICAL APPROXIMATIONS

Binding Energies of the Non-Moving Hydrogen Atom

Extensive tables of binding energies of the hydrogen atom at rest with respect to the magnetic field have been presented by Rösner et al.[20] and supplemented by other authors.[21–23] Recently, the accuracy $\sim 10^{-12}$ Ryd has been achieved.[24] In the astrophysics, a lower accuracy is usually sufficient, and simple analytical estimates are often desirable.

For this reason, we have constructed a fit to $E^{(0)}$, where $E^{(0)}_{ns\nu} \equiv -E^{\parallel}_{ns\nu}(0)$, in a possibly widest range of γ. For the tightly-bound states, we have

$$E^{(0)}_{0s0}(\gamma) = \ln\left(\exp\left[(1+s)^{-2}\right] + p_1\left[\ln(1+p_2\sqrt{\gamma})\right]^2\right) + p_3\left[\ln(1+p_4\gamma^{p_5})\right]^2 \text{ Ryd.} \tag{4}$$

The parameters $p_1 - p_5$ depend on s; they are listed in table 1. This fit is accurate to within 0.1–1% at $\gamma = 10^{-1} - 10^4$, and it also provides the correct limits at $\gamma \to 0$.

Table 2. Parameters of (5) at $1 \le \gamma \le 10^4$.

ν	1	2	3	4	5	6
a_ν	0.785	0.578	0.901	0.631	0.970	0.660
b_ν	1.724	0.765	1.847	0.717	1.866	0.693

For the hydrogen-like states, we use the asymptotic result[25]

$$E_{ns\nu}^{(0)} = \frac{1 \text{ Ryd}}{(N+\delta)^2}, \text{ where } \begin{cases} N = (\nu+1)/2, & \delta \sim \gamma^{-1} \text{ for odd } \nu, \\ N = \nu/2, & \delta \sim (\ln\gamma)^{-1} \text{ for even } \nu. \end{cases} \tag{5}$$

We have obtained the following fits to the quantum defect δ: for odd ν, $\delta = (a_\nu + b_\nu\sqrt{\gamma} + 0.077\gamma)^{-1}$, where $a_\nu \approx 1$ and $b_\nu \approx 2$; and for even ν, $\delta = [a_\nu + 1.28 \ln(1 + b_\nu \gamma^{1/3})]^{-1}$, where $a_\nu \approx \frac{2}{3}$ and $b_\nu \approx \frac{2}{3}$. More accurate values of a_ν and b_ν are given in table 2. At $1 \le \gamma \le 10^4$, errors of these approximations lie within $\sim 10^{-3}$.

Binding Energies of the Moving Hydrogen Atom

For the moving hydrogen atom in a strong magnetic field, the first analytical fit to $E(K)$ has been published by Lai and Salpeter.[26] It is rather accurate for the ground state at $K < K_c$ but cannot be applied to excited or decentered states.

We describe the longitudinal energy (3) by the formula

$$|E_{ns\nu}^{\parallel}(K)| = \frac{E_{ns\nu}^{(1)}(K)}{1 + (K/K_c)^{1/\alpha}} + \frac{E_{ns\nu}^{(2)}(K)}{1 + (K_c/K)^{1/\alpha}}. \tag{6}$$

The two-term structure of (6) is dictated by the necessity to describe the two physically distinct regions of K below and above K_c. The parameter α has the meaning of the width of the transition region near K_c in logarithmic scale of pseudomomenta.

For the tightly-bound states, we parameterize the dependencies $E^{(j)}(K)$ as follows:

$$E_{0s0}^{(1)}(K) = E_{0s0}^{(0)} - \frac{K^2}{2m_{\text{eff}} + q_1 K^2 / E_{0s0}^{(0)}}, \qquad E_{0s0}^{(2)}(K) = \frac{2 \text{ Ryd}}{\sqrt{r_*^2 + r_*^{3/2}} + q_2 r_*}, \tag{7}$$

where $r_* = r_c/a_B = K/(\gamma \text{ au})$, q_1 and q_2 are dimensionless fitting parameters, and m_{eff} is the effective mass which is close to (but not necessarily coincident with) the transverse effective mass $M_{ns\nu}^{\perp}$ obtained by the perturbation technique. At $\gamma \ge 300$, we put $q_1 = \log_{10}(\gamma/300)$ if $s = 0$ and $q_1 = 0.5$ otherwise, $q_2 = 0.158\,[\ln((1{+}0.1s)\gamma/215)]^{2/5}$, and $\alpha = 0.053 \ln(\gamma/150)$. For the effective mass, we have $m_{\text{eff}} = m_{\text{H}}\,[1 + (\gamma/\gamma_0)^{c_0}]$, where $c_0 = 0.937{+}0.038 s^{1.58}$ and $\gamma_0 = 6150(1{+}0.0389 s^{3/2})/[1{+}7.87 s^{3/2}]$. For the critical pseudomomentum, we have $K_c = [c_1 + \ln(1 + \gamma/\gamma_1)]\sqrt{2m_{\text{H}} E^{(0)}}$. The parameters c_1 and γ_1 take on the values $c_1 = 0.81, 1.09, 1.18, 1.24$ and $\gamma_1 = (8.0, 3.25, 2.22, 1.25) \times 10^4$ for $s = 0, 1, 2, 3$, respectively. For $s \ge 4$, we put $c_1 = 0.93 + 0.08s$ and $\gamma_1 = 6500$.

In figure 1 the above fitting formulae are compared with our numerical results[15] and with the previous approximations.[26] The figure demonstrates that the present approximations are valid at any K from 0 to infinity. Appreciable discrepancies occur only in narrow ranges of K near anticrossings.

Now let us turn to the hydrogen-like states. Their binding energies are approximated by the same formula (6) but with slightly different expressions for $E^{(1)}$ and $E^{(2)}$. For these states, $M_{ns\nu}^{\perp}$ exceeds m_{H} by orders of magnitude, and the perturbation method fails already at small K,[13] rendering the notion of the effective mass practically useless

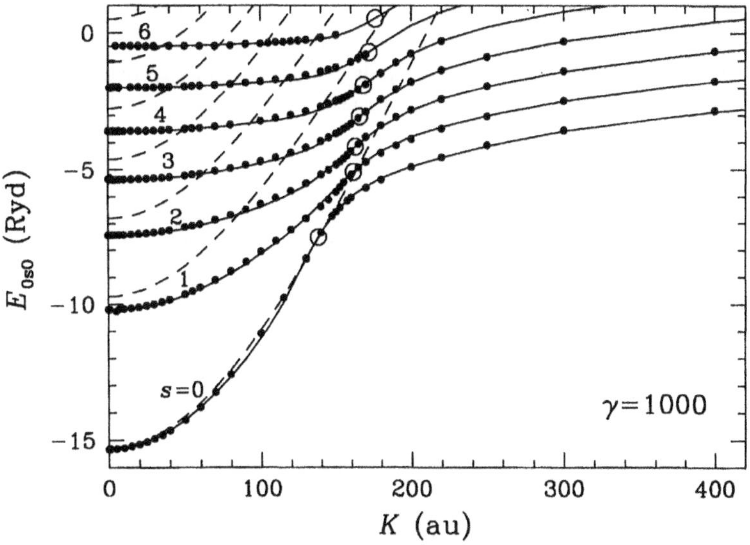

Figure 1. Energy spectrum of the hydrogen atom moving across strong magnetic fields. Numerical values (dots) are compared with the present analytical approximations (full lines) and with previously published[26] ones (dashed lines).

for the fitting. Thus we consider m_{eff} as effectively infinite and put $E^{(1)}_{0s\nu}(K) = E^{(0)}_{0s\nu}$ ($\nu \geq 1$). Furthermore, the transition region is not well defined, and therefore K_c and α lose their clear meaning and become mere fitting parameters. For odd states, we have, approximately, $K_c = (\nu^{5/4}\gamma/170)^{0.9}\sqrt{2m_{\text{H}}E^{(0)}}$ and $\alpha = 0.66 + \nu/20$. For even states, $K_c = \nu\sqrt{(\gamma/600)m_{\text{H}}E^{(0)}}$ and $\alpha = 0.66$.

The function $E^{(2)}(K)$ that describes the longitudinal energy at large K is now

$$E^{(2)}_{0s\nu}(K) = \left\{(2\ \text{Ryd})^{-1}\left[r_*^2 + (2\nu+1)r_*^{3/2} + q_2 r_*\right]^{1/2} + 1/E^{(0)}_{0s\nu}\right\}^{-1}, \tag{8}$$

with $q_2 = \nu^2 - 1$ for odd ν, and $q_2 = \nu^2 + 2^{\nu/2}\log_{10}(\gamma/300)$ for even ν (at $\gamma \geq 300$). The first and second terms in the square brackets ensure the correct asymptotic behavior.[15]

CONCLUDING REMARKS

The analytical approximations for binding energies presented in this contribution depend continuously on two arguments — magnetic field strength and transverse pseudomomentum. They are accurate, typically, within a few parts in 100–1000. The accuracy can be improved by almost an order of magnitude by optimizing the parameters m_{eff}, K_c, α, q_1, q_2 in equations (6)–(8) separately at each discrete value of γ. Tables of such optimized parameters have been obtained and will be published elsewhere, together with analytical approximations of geometrical sizes of various quantum-mechanical states of the moving atom and oscillator strengths of radiative transitions among them. The atomic sizes play important role in distribution of atoms over quantum states in a plasma and in their contribution to the plasma absorption coefficients. For example, a size of an atom may be used to evaluate effects of "unbounding" of electrons caused by random charge distribution in the plasma. For non-magnetized hydrogen plasma, an approximate treatment of these effects was revised recently;[27] for the strong magnetic fields analogous work is under way. Eventually, the analytical estimates of K-dependencies of the binding energies, atomic sizes, and transition rates help to generalize previously developed models of fully ionized atmospheres of magnetic neutron stars[28] to the more realistic case of partially ionized atmospheres.

REFERENCES

1. C.J. Pethick and D.G. Ravenhall, Matter at large neutron excess and the physics of neutron-star crusts, *Ann. Rev. Nucl. Part. Sci.* 45:429 (1995)
2. See for example G. Chabrier, A.Y. Potekhin, and D.G. Yakovlev, Cooling neutron stars with accreted envelopes, *Astrophys. J.* 477:L99 (1997), and references therein
3. G. Wunner and H. Ruder, Atoms in strong magnetic fields, *Phys. Scr.* 36:291 (1987)
4. P. Fassbinder and W. Schweizer, Stationary hydrogen lines in magnetic and electric fields of white dwarf stars, *Astron. Astrophys.* 314:700 (1996)
5. T.O. Klaassen, J.L. Dunn and C.A. Bates, Shallow donor states in a magnetic field, *this volume*
6. For review, see H. Ruder, G. Wunner, H. Herold, and F. Geyer. "Atoms in Strong Magnetic Fields," Springer-Verlag, Berlin (1994)
7. J.E. Avron, I.W. Herbst, and B. Simon, Separation of center of mass in homogeneous magnetic fields, *Ann. Phys. (N.Y.)* 114:431 (1978)
8. M. Vincke and D. Baye, Centre-of-mass effects on the hydrogen atom in a magnetic field, *J. Phys. B: At. Mol. Opt. Phys.* 21:2407 (1988)
9. L.A. Burkova, I.E. Dzyaloshinskii, S.F. Drukarev, and B.S. Monozon, Hydrogen-like system in crossed electric and magnetic fields, *Sov. Phys.-JETP* 44:276 (1976)
10. I. Dzyaloshinskii, Effects of the finite proton mass of a hydrogen atom in crossed magnetic and electric fields: a state with giant electric dipole moment, *Phys. Lett.* A165:69 (1992)
11. D. Baye, N. Clerbaux, and M. Vincke, Delocalized states of atomic hydrogen in crossed electric and magnetic fields, *Phys. Lett.* A166:135 (1992)
12. P. Schmelcher, Delocalization of excitons in a magnetic field, *Phys. Rev.* B48:14642 (1993)
13. G.G. Pavlov and P. Mészáros, Finite-velocity effects on atoms in strong magnetic fields and implications for neutron star atmospheres, *Astrophys. J.* 416:752 (1993)
14. M. Vincke, M. Le Dourneuf, and D. Baye, Hydrogen atom in crossed electric and magnetic fields: transition from weak to strong electron-proton decentring, *J. Phys. B: At. Mol. Opt. Phys.* 25:2787 (1992)
15. A.Y. Potekhin, Structure and radiative transitions of the hydrogen atom moving in a strong magnetic field, *J. Phys. B: At. Mol. Opt. Phys.* 27:1073 (1994)
16. G.G. Pavlov and A.Y. Potekhin, Bound-bound transitions in strongly magnetized hydrogen plasma, *Astrophys. J.* 450:883 (1995)
17. A.Y. Potekhin and G.G. Pavlov, Photoionization of hydrogen in atmospheres of magnetic neutron stars, *Astrophys. J.* 483:414 press (1997)
18. Z. Chen and S.P. Goldman, Relativistic and nonrelativistic finite-basis-set calculations of low-lying levels of hydrogenic atoms in intense magnetic fields, *Phys. Rev.* A45:1722 (1992)
19. L.P. Gorkov and I.E. Dzyaloshinskii, On the theory of the Mott exciton in a strong magnetic field, *Sov. Phys.-JETP* 26:449 (1968)
20. W. Rösner, G. Wunner, H. Herold, and H. Ruder, Hydrogen atoms in arbitrary magnetic fields. I. Energy levels and wave functions, *J. Phys. B: At. Mol. Phys.* 17:29 (1984)
21. D. Wintgen and H. Friedrich, Matching the low-field region and the high-field region for the hydrogen atom in a uniform magnetic field, *J. Phys. B: At. Mol. Phys.* 19:991 (1986)
22. M.V. Ivanov, The hydrogen atom in a magnetic field of intermediate strength, *J. Phys. B: At. Mol. Opt. Phys.* 21:447 (1988)
23. J. Xi, L. Wu, X. He, and B. Li, Energy levels of the hydrogen atom in arbitrary magnetic fields, *Phys. Rev.* A46:5806 (1992)
24. Yu.P. Kravchenko, M.A. Liberman and B. Johansson, Exact solution for a hydrogen atom in a magnetic field of arbitrary strength, *Phys. Rev.* A54:287 (1996)
25. L.K. Haines, D.H. Roberts, One-dimensional hydrogen atom, *Am. J. Phys.* 37:1145 (1969)
26. D. Lai and E.E. Salpeter, Motion and ionization equilibrium of hydrogen atoms in a superstrong magnetic field, *Phys. Rev.* A52:2611 (1995)
27. A.Y. Potekhin, Ionization equilibrium of hot hydrogen plasma, *Phys. Plasmas* 3:4156 (1996)
28. Yu.A. Shibanov, V.E. Zavlin, G.G. Pavlov, and J. Ventura, Model atmospheres and radiation of magnetic neutron stars. I – The fully ionized case, *Astron. Astrophys.* 266:313 (1992)

ABSORPTION OF NORMAL MODES IN A STRONGLY MAGNETIZED HYDROGEN GAS

Tomasz Bulik[1], George Pavlov[2]

[1]Nicolaus Copernicus Astronomical Center
Bartycka 18, 00716 Warsaw, Poland
[2]Department of Astronomy and Astrophysics
The Pennsylvania State University
525 Davey Lab., University Park, PA 16802, USA

INTRODUCTION

Propagation of radiation in a magnetized medium can be described in terms of two normal modes (NM) with different polarizations and absorption coefficients which can be found from the polarizability tensor. The antihermitian part of this tensor is simply connected, in the dipole approximation, with the absorption coefficients μ_α for three "basic" polarizations of photons: linear polarization along the magnetic field ($\alpha = 0$), and right and left circular polarizations transverse to the field ($\alpha = \pm 1$). However, to calculate the NM polarizations and absorption coefficients for an arbitrary direction of propagation, both hermitian and antihermitian parts of the polarizability tensor are needed. A convenient approach for calculating the hermitian part and the NM absorption coefficients μ_j ($j = 1, 2$) is described in detail by Bulik and Pavlov[1]. Briefly, we start from the absorption coefficients μ_α and use the Kramers-Kronig transformations to obtain the hermitian part of the polarizability tensor. Given the full polarizability tensor we then proceed to calculate the NM polarizations for all frequencies and directions. Finally, we use the NM polarizations and the full polarizability tensor to calculate the absorption coefficients for arbitrary energy and direction of propagation.

A novel feature of the calculations presented here is that we take into account bound-bound and bound-free transitions for the left circular polarization which are strictly forbidden by the dipole selection rules for non-excited atoms at rest and become allowed only due to the violation of the cylindrical symmetry of the atomic structure caused by the atomic motion. Account for these transitions is important, in spite of their small oscillator strengths, when radiation propagates in a direction close to that of the magnetic field. In this case, the (extraordinary) NM is left-circularly polarized, so that the gas would be completely transparent for this mode unless the motion-induced violation of the selection rules is taken into account. Here we consider the atomic motion in the perturbation approach (cf. Pavlov and Mészáros[2]). Throughout this paper we will follow the notations introduced by Bulik and Pavlov[1].

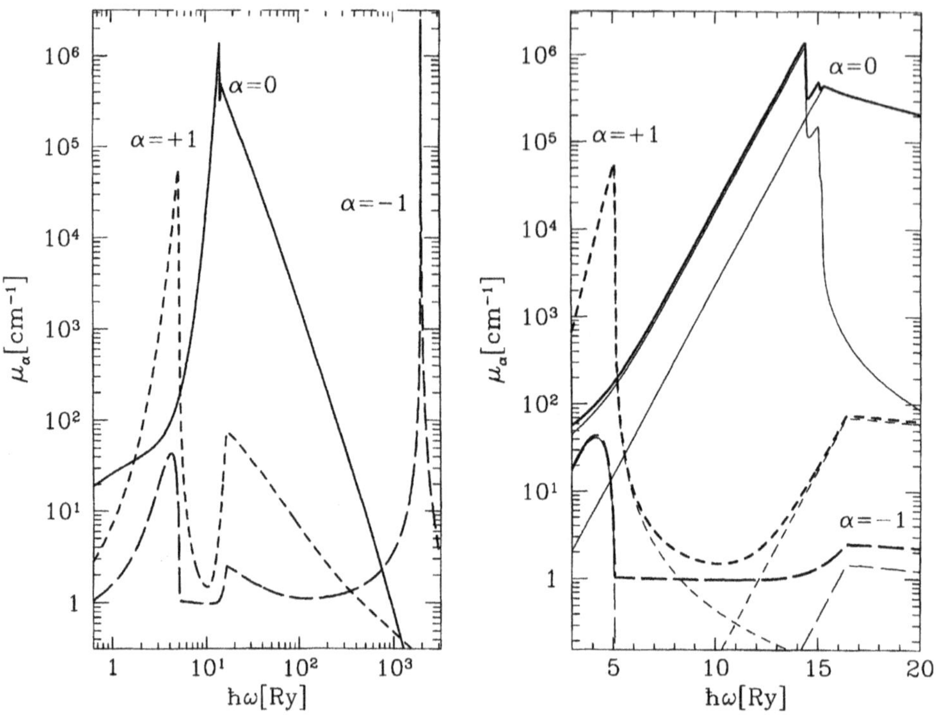

Figure 1. Frequency dependences of the absorption coefficients μ_α for three polarizations: $\alpha = 0$ (linear polarization along the magnetic field B) and $\alpha = \pm 1$ (circular polarizations in the plane perpendicular to B). The magnetic field is $B = 2.35 \times 10^{12}$ G, the temperature is $kT = 1$ Ry, and the density is $\rho = 1\,\mathrm{g\,cm}^{-3}$. The right panel shows the detailed structure of μ_0 and $\mu_{\pm 1}$ in the range of bound-bound resonances and bound-free edges. The contributions of the bound-bound and bound-free transitions are shown separately (thin-lines).

ABSORPTION COEFFICIENT FOR THE LEFT POLARIZATION

We consider the case of low temperatures, when all atoms are in the lowest discrete state, $|Ns\nu\rangle = |000\rangle$, and the absorption of the photon occurs via bound-free (photoionization) or bound-bound transitions, $\mu_\alpha = \mu_\alpha^{(\mathrm{bf})} + \mu_\alpha^{(\mathrm{bb})}$. Calculation of the coefficients for permitted transition, μ_0 and μ_{+1}, was described in detail by Bulik and Pavlov[1]. The dependence of the *bound-free* cross section for the left polarization, σ_{-1}, on the transverse component of the generalized momentum, $P_\perp$, was derived by Pavlov and Mészáros[2] with the perturbation approach ($\sigma_{-1} \propto P_\perp^2$; see their eq. [3.18]) and confirmed by numerical calculations of Potekhin and Pavlov[3]. The non-zero cross section σ_{-1} arises due to the motion-induced admixture of the state $|010\rangle$ to the ground state $|000\rangle$:

$$\psi_{000}^{(1)} = \frac{i\omega_{B0} a_L (P_x + iP_y)}{\sqrt{2}\epsilon_{000,010}} \psi_{010}^{(0)}\,, \tag{1}$$

where P_x and P_y are the components of the generalized momentum of the atom, $\omega_{B0} = eB/Mc$, $M = m_p + m_e$ is the atomic mass, $a_L = (c\hbar/eB)^{1/2}$ is the magnetic length, and $\epsilon_{000,010}$ is the energy of transition between the ground and first excited states at $P_\perp = 0$ (for instance, $\epsilon_{000,010} = 5.14$ Ry $= 70$ eV for $B = 2.35 \times 10^{12}$ G). We obtain the

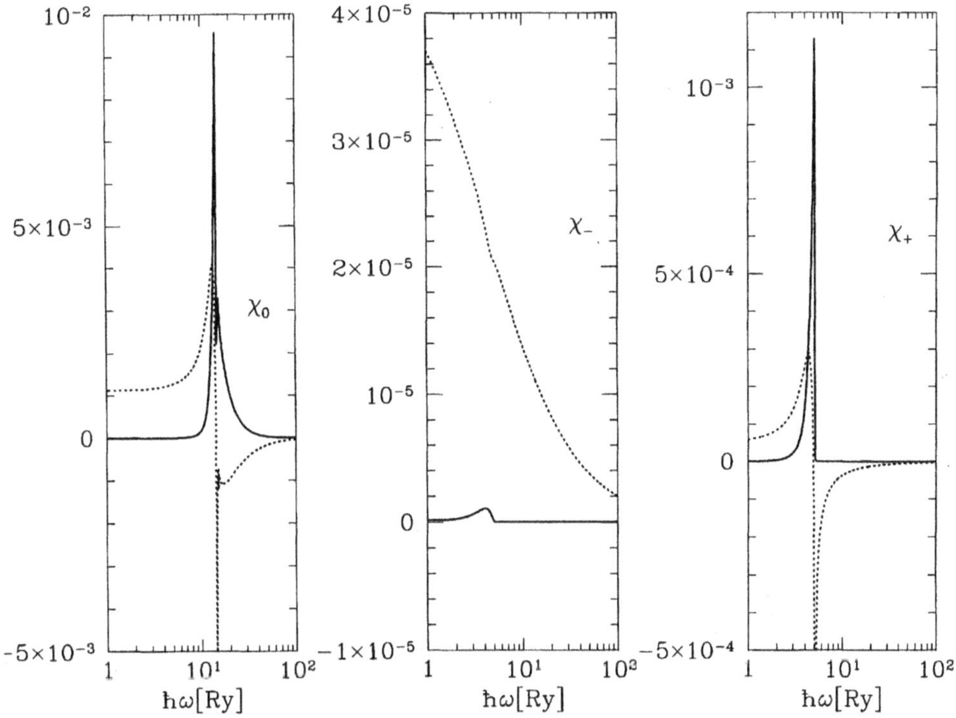

Figure 2. Hermitian (dotted lines) and antihermitian (solid lines) parts of the three components of the polarizability tensor: χ_0 and $\chi_\pm$.

absorption coefficient by averaging the cross section over the thermal distribution[3]:

$$\mu_{-1}^{(\mathrm{bf})} = \frac{\hbar\omega_{B0}kT}{2\epsilon^2_{000,010}}\mu_{+1}^{(\mathrm{bf})} \,. \tag{2}$$

The main *bound-bound* transition from the ground level for the left polarization is that to the state $|010\rangle$, and it is important to take into account, together with the correction (1), the admixture of the ground state to the excited state:

$$\psi_{010}^{(1)} = \frac{i\omega_{B0}a_L(P_x - iP_y)}{\sqrt{2}\epsilon_{000,010}}\psi_{000}^{(0)} \,. \tag{3}$$

Only these admixed states, $\psi_{000}^{(1)}$ and $\psi_{010}^{(1)}$, contribute to the matrix element for absorption of the left-polarized photons, so that the matrix element is proportional to $P_\perp^2$ (unlike the bound-free transition, for which it is proportional to $P_\perp$), and the oscillator strength is

$$f_{-1} = \left(\frac{P_\perp^2}{2M}\frac{\hbar\omega_{B0}}{\epsilon^2_{000,010}}\right)^2 f_{+1} \,. \tag{4}$$

In the perturbation approach, the dependence of atomic energies on $P_\perp$ can be described in terms of the transverse mass $M_{Ns\nu}$: $\epsilon_{Ns\nu}(P_\perp) = \epsilon_{Ns\nu}^{(0)} + P_\perp^2/(2M_{Ns\nu})$. The transverse mass is higher for more excited states; for instance[2], $M_{000} = 1.27M$ and $M_{010} = 2.37M$ for $B = 2.35 \times 10^{12}$ G (i. e., $\gamma \equiv \hbar\omega_{Be}/(2\,\mathrm{Ry}) = 1000$). This means that the energy of the transition between two discrete states decreases with increasing $P_\perp$, which leads to the magnetic broadening of spectral lines[2]. For instance, the transition $|000\rangle \to |010\rangle$

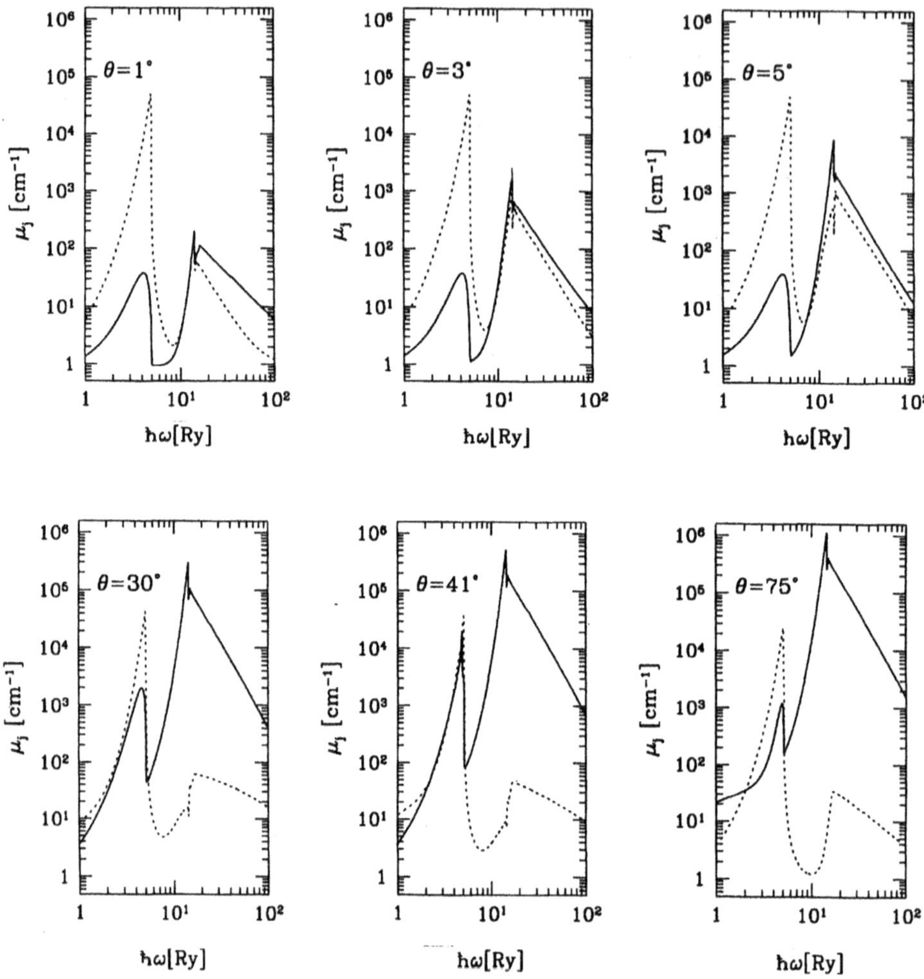

Figure 3. Frequency dependence of the absorption coefficients for six angles of propagation with respect to the magnetic field B.

at a frequency $\omega < \epsilon_{000,010}$ is provided by atoms whose transverse momentum satisfies the equation

$$P_\perp^2 = \frac{2(\epsilon_{000,010} - \hbar\omega) M_{000} M_{010}}{M_{010} - M_{000}} . \tag{5}$$

Substituting equations (5) and (4) into the general equation for the absorption coefficient (Pavlov and Potekhin[4]), which includes averaging over the generalized momentum, we obtain

$$\mu_{-1}^{(\mathrm{bb})} = \left[\frac{(\epsilon_{000,010} - \hbar\omega)\, \hbar\omega_{B0}\, m_{000}\, m_{010}}{\epsilon_{000,010}^2 (m_{010} - m_{000})} \right]^2 \mu_{+1}^{(\mathrm{bb})} , \tag{6}$$

where $m_{Ns\nu} = M_{Ns\nu}/M$. Thus, equations (2) and (6) express the absorption coefficient μ_{-1} in terms of the known μ_+.

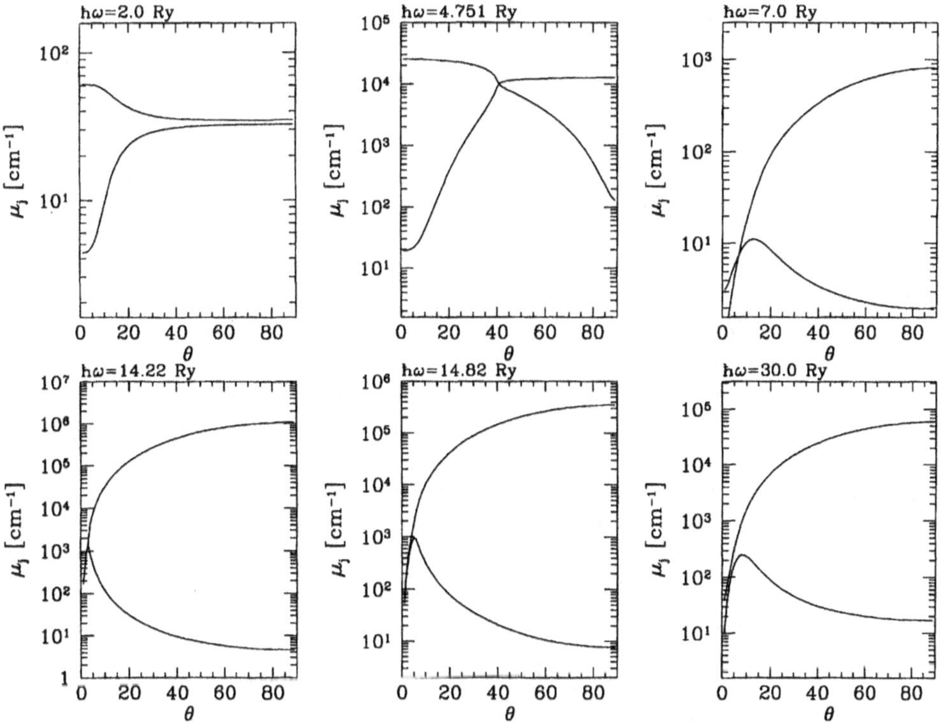

Figure 4. Angular dependences of the absorption coefficients for six characteristic photon energies. Note the differences in the vertical scale of the plots.

RESULTS AND DISCUSSION

As an illustrative example, we consider the case when $B = 2.35 \times 10^{12}$ G, $kT = 1$ Ry, and $\rho = 1\ \mathrm{g\,cm^{-3}}$ (the density determines an additional collisional broadening of spectral lines, see Bulik and Pavlov[1]). We present the basic absorption coefficients for the three polarization modes in Figure 1. The contributions of the bound-bound and bound-free transitions are shown separately by thin lines. The absorption coefficient μ_{-1} includes the bound-bound and bound-free transitions to states with $N = 0$, described by equations (2) and (6), and also the very strong transition $|000\rangle \rightarrow |1\,-10\rangle$ ($\epsilon_{000,1-10} = 2004$ Ry) to the excited Landau level $N = 1$. In Figure 2 we show the hermitian and antihermitian components of the polarizability tensor obtained with the Kramers-Kronig transformation. The motion-induced bound-bound and bound-free transitions for $\alpha = -1$ alter the antihermitian part of the polarizability tensor χ^{A}_{-1} significantly, but they only slightly modify the hermitian part χ^{H}_{-1}, which is dominated by the effects of permitted transitions for $\alpha = +1$ at negative frequencies. Polarizations of the NMs are determined mainly by the dominant χ_0 and χ_{+1} (note the vertical scales in Figure 2) and are almost not affected by the changed μ_{-1}. For instance, the "critical" angles, $\theta_{c1} = 40.28°$ and $\theta_{c2} = 3.42°$, and frequencies, $\hbar\omega_{c1} = 4.753$ Ry and $\hbar\omega_{c2} = 14.23$ Ry, at which two NMs coincide with each other, differ from those obtained without the motion-induced transitions[1] by less than 0.2%.

We present plots of absorption coefficients as a function of frequency for a few characteristic angles in Figure 3 (cf. Figures 6 and 7 of Bulik and Pavlov[1]). The NM polarizations are almost circular for small angles; however, they become non-orthogonal in the vicinity of the critical frequency ω_{c2}. Therefore the absorption coefficients follow

μ_{+1} and μ_{-1} at energies below ≈ 10 Ry, and they coincide around 14 Ry due to the strong non-orthogonality of NMs in this region. The frequency dependence of the absorption coefficients at intermediate and large angles is shown in the bottom panels of Figure 3. A typical behavior for propagation angles $\theta_{c2} < \theta < \theta_{c1}$ is presented in the lower left panel for $\theta = 30°$. Here the modes are linearly polarized, except for the region around ω_{c1}. For propagation near the angle θ_{c1}, the modes almost coincide in the vicinity of ω_{c1}, and they are almost linear everywhere else. Finally, the case of propagation at a large angle with respect to the magnetic field is presented for $\theta = 75°$. Here the modes are almost linear at all frequencies, except for a narrow region of increased ellipiticity around $\hbar\omega \approx 5$ Ry.

The angular dependence of the absorption coefficients for a few characteristic energies is presented in Figure 4 (cf. Figure 8 of Bulik and Pavlov[1]). The effects of the bound-bound and bound-free transitions for μ_{-1} are seen in the plots for $\hbar\omega = 2$ Ry and $\hbar\omega = 4.751$ Ry, where the low-absorption mode flattens for small angles of propagation, contrary to the cse when the motion-induced transitions for μ_{-1} were neglected. The behavior of the absorption coefficients around the critical points is clearly seen in the plots for $\hbar\omega = 4.751$ Ry and $\hbar\omega = 14.22$ Ry. The mode polarizations are linear and coinciding at the the critical points, and the absorption coefficients for the two modes also coincide there. At frequencies above ω_{c2} the modes are linearly polarized for all the angles except for a narrow cone of small θ. At these frequencies the absorption coefficient for the ordinary (high-absorption) mode grows with the angle of propagation, while that for the extraordinary mode first increases, when the polarization is circular, then reaches a maximum, and finally decreases when the polarization becomes linear with increasing θ.

The inclusion of the motion-induced bound-bound and bound-free transitions for $\alpha = -1$ "patches a hole" in modeling atmospheres of neutron stars. With these transitions neglected, the model low-temperature atmospheres are nearly transparent for small angles of propagation: the left-polarized radiation leaks out from very deep layers without interaction with outer layers, which limits the modeling to relatively high temperatures. A second effect is that spectral lines should be produced for propagation along the field in both modes, whereas only the ordinary mode would exhibit such lines with these transitions neglected.

Acknowledgements

This work was partially supported by NASA grant NAG5-2807 and PSCSR grant 2P03D00911.

REFERENCES

1. T. Bulik and G. G. Pavlov, Polarization modes in a strongly magnetized hydrogen gas, *Astrophys. J.* 469:373 (1996).
2. G. G. Pavlov and P. Mészáros, Finite-velocity effects on atoms in strong magnetic fields and implications for neutron star atmospheres, *Astrophys. J.* 416:752 (1993).
3. A. Y. Potekhin and G. G. Pavlov, Photoionization of hydrogen in atmospheres of magnetic neutron stars, *Astrophys. J.* 483:414 (1997).
4. G. G. Pavlov and A. Y. Potekhin, Bound-bound transitions in strongly magnetized hydrogen plasma, *Astrophys. J.* 450:883 (1995).
5. A. Y. Potekhin, Structure and radiative transitions of the hydrogen atom moving in a strong magnetic field, *J. Phys. B: At. Mol. Opt. Phys.* 27:1073 (1994).

ELECTRONIC STRUCTURE OF LIGHT ELEMENTS IN STRONG MAGNETIC FIELDS

Patrice Pourre, Philippe Arnault, François Perrot

CEA Limeil-Valenton, 94195 Villeneuve-St-Georges Cedex, France

The properties of atoms in strong magnetic fields are relevant to several domains such as astrophysics, condensed matter physics, or plasma physics.[1, 2]

In the past, only hydrogen and helium have been studied for magnetic fields of arbitrary strength.[3] Indeed, the calculation of the electronic structure of atoms in a regime where Coulomb and magnetic effects are of equal importance is a challenge in many respects, including the breaking of the spherical symmetry of the atomic wavefunctions and the influence of the magnetic field on the exchange and correlation energy.[4, 5] A non-perturbative approach is needed, in which case the problem may be solved in two dimensions. The lack of data for light elements other than helium has motivated recent Hartree-Fock (HF) computations for lithium and carbon.[12]

In order to tackle this problem, we tried different approaches. In one of these, the quantum mechanical equations of density functional theory are solved using the finite element method. In another approach, within the framework of the HF method, the wavefunctions are expanded in an original basis set of numerical wavefunctions including variational parameters.

We present these methods and preliminary results for total energies of various configurations of atomic helium in strong magnetic fields.

KOHN-SHAM EQUATIONS AND THE FINITE ELEMENT METHOD

In current-density functional theory (CDFT),[4] the ground-state energy E of a many-electron system, subject to external scalar and vector potentials $V(\mathbf{r})$ and $\mathbf{A}(\mathbf{r})$, is uniquely determined as a functional of the ground-state density $n(\mathbf{r})$ and of the canonical paramagnetic current $\mathbf{j}_p(\mathbf{r})$. Writing the kinetic contribution to the functional E as the Hartree-Fock form of the kinetic energy, T_s^{HF}, enables one to reduce the unknown part of this functional to a term representing the exchange-correlation energy, E_{xc}. As T_s^{HF} is the expectation value of an operator between one-electron orbitals, one must introduce such orbitals and the eigenvalue equations with which to calculate them. These equations are the Kohn-Sham (KS) one-electron equations.[7] For CDFT, the KS equations are[4]

$$\left[\frac{1}{2m}\left\{-i\hbar\,\nabla + e\,\mathbf{A}_{eff}(\mathbf{r})\right\}^2 + V_{eff}(\mathbf{r})\right]\,\psi_n(\mathbf{r}) = e_n\,\psi_n(\mathbf{r}) \tag{1}$$

where the effective scalar potential is given by

$$V_{eff}(\mathbf{r}) = V(\mathbf{r}) + e^2 \int d^3\mathbf{r}' \frac{n(\mathbf{r}')}{|\mathbf{r}-\mathbf{r}'|} + \frac{\delta E_{xc}}{\delta n(\mathbf{r})}[n, \mathbf{j}_p] + \frac{e^2}{2m}\left[\mathbf{A}^2(\mathbf{r}) - \mathbf{A}^2{}_{eff}(\mathbf{r})\right] \quad (2)$$

and where an effective vector potential appears

$$e\,\mathbf{A}_{eff}(\mathbf{r}) = e\,\mathbf{A}(\mathbf{r}) + \frac{\delta E_{xc}}{\delta \mathbf{j}_p(\mathbf{r})}[n, \mathbf{j}_p] \quad (3)$$

For the sake of simplicity, we confined ourselves to spin-compensated systems. Otherwise, the spin densities and their coupling with the magnetic field must be introduced. The density which minimizes the total energy is generally equal to the sum of the squares of the eigenfunctions ψ_n of the N lowest states. The paramagnetic current is obtained from the same orbitals and is related to the physical current density by $\mathbf{j}(\mathbf{r}) = \mathbf{j}_p(\mathbf{r}) + \frac{e}{m}\, n(\mathbf{r})\, \mathbf{A}(\mathbf{r})$.

To solve the KS equations, one must resort to an approximation for the unknown exchange-correlation energy functional $E_{xc}[n, \mathbf{j}_p]$. In CDFT, a local approximation (LA) may be defined. Introducing the vorticity $\mathbf{v}(\mathbf{r}) = \nabla \times \frac{\mathbf{j}_p(\mathbf{r})}{n(\mathbf{r})}$ and provided that $n(\mathbf{r})$ and $\mathbf{v}(\mathbf{r})$ are sufficiently slowly varying, one has

$$E_{xc}^{\mathrm{LA}}[n, \mathbf{v}] \approx \int d^3\mathbf{r}\; n(\mathbf{r})\; \epsilon_{xc}[n, \mathbf{v}] \quad (4)$$

Indeed, $E_{xc}[n, \mathbf{j}_p]$ depends on $\mathbf{j}_p$ only through the combination defining the vorticity $\mathbf{v}$.[4] $\epsilon_{xc}[n, \mathbf{v}]$ is the exchange-correlation energy per electron of a uniform electron gas in a uniform magnetic field $\mathbf{B}_0 = -\frac{m}{e}\,\mathbf{v}$.

The KS equations within the local approximation become usable as soon as $\epsilon_{xc}[n, \mathbf{B}_0]$ is given. Recently, this functional has been studied by Skudlarksi and Vignale[5] within the random phase approximation (RPA). They expressed the variation of the xc energy due to the magnetic field $\Delta\epsilon_{xc}[n, B_0] = \epsilon_{xc}[n, B_0] - \epsilon_{xc}[n, 0]$ by introducing another variable which is related to B_0 and n. This variable is the occupation factor of the lowest Landau subband, λ, whose square is proportional to the Fermi energy μ_0 of the uniform electron gas, $\mu_0 = \lambda^2\omega_c$, where $\omega_c = eB/m$ is the cyclotron frequency. We used the analytical approximation derived by Perrot and Grimaldi[8] to calculate λ as a function of B_0 and n.

Skudlarksi and Vignale[5] stressed that $\Delta\epsilon_{xc}$ exhibits an approximate scaling law for $\lambda \geq 0.1$ and in the range of densities characterized by $r_s = 0.5-6$, where $r_s = \left(\frac{3}{4\pi n}\right)^{1/3}$ in units of a Bohr radius. This approximate scaling law mimics the rigorous scaling property of the local exchange energy $\epsilon_x[r_s, 0]$

$$\Delta\epsilon_{xc}[r_s, \lambda] \approx \frac{\Delta\epsilon_{xc}[1, \lambda]}{r_s} \quad (5)$$

Indeed, for the spin-compensated case,[9] the exchange energy per electron is $\epsilon_x[r_s, 0] = -\frac{0.916}{r_s}\; Ry$.

Skudlarksi and Vignale[5] gave an analytical fit of $\Delta\epsilon_{xc}[1, \lambda]$ (Figure 1). A noticeable fact is that, for a large range of densities and magnetic field intensities, $\Delta\epsilon_{xc}[1, \lambda]$ is small in comparison with $\epsilon_x[1, 0]$. Therefore, it is worthwhile to treat the contribution from $\Delta\epsilon_{xc}[n, B_0]$ as a *perturbation*, at least as a first step toward a more self-consistent solution. The domain of validity of this approximation can be obtained from figure 2 representing isovalues of the occupation factor, λ, vs. the radius r_s and the magnetic field, B_0.

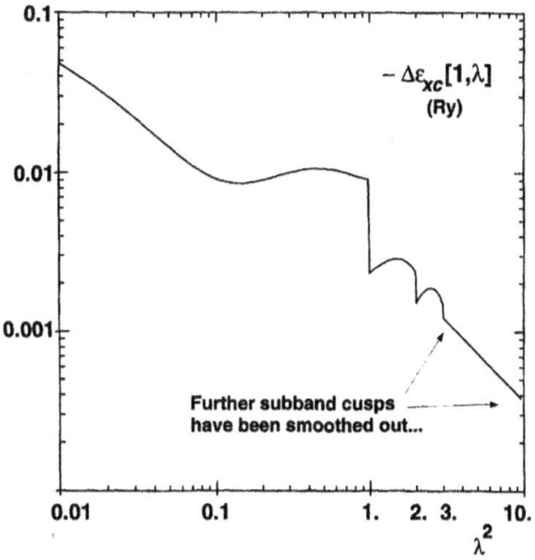

Figure 1. Numerical fit of Skudlarski and Vignale.

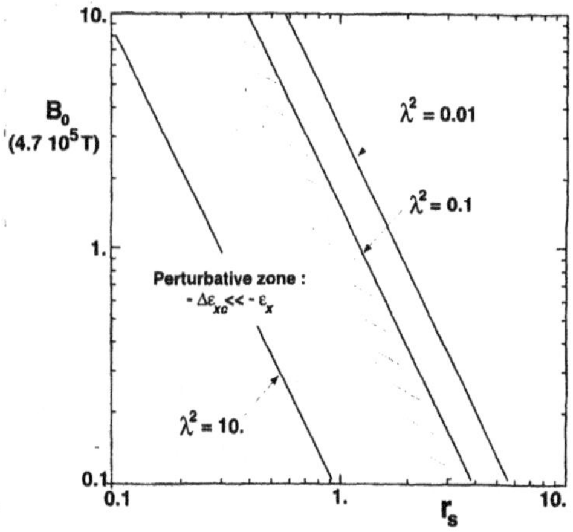

Figure 2. Domain of validity of the fit : $\lambda^2 \geq 0.01$ and of a perturbative treatment of $\Delta\epsilon_{xc}$.

Considering $\Delta\epsilon_{xc}[n, B_0]$ as a *perturbation* leads to a drastic simplification of the KS equations of the CDFT. Indeed, neglecting the current dependent part of the xc energy, the KS equations contain only a scalar exchange-correlation potential, which only depends on the electronic density. We used the Vosko-Wilk-Nusair formula for the correlation energy.[10, 11, 12] Lastly, using the symmetric gauge where $\mathbf{A} = \frac{1}{2}\,\mathbf{B} \times \mathbf{r}$ enables one to search for *real* KS orbitals.

Finally, we used the finite element method to solve the resulting KS equations. When expressed in two dimensions (using the axial symmetry around the magnetic field direction) this allows to handle directly the breaking of spherical symmetry of the atomic wavefunctions as the magnetic field increases.[13]

Numerical results for the $(1s)^2$ and $(1s)(2p_{-1})$ configurations of Helium at various strengths of the magnetic field are compared with other calculations in the last section. For these first results, we made an assumption which will be discussed elsewhere. It consists in extending the local approximation to vanishing physical current $\mathbf{j}$, allowing one to identify B_0 with B.

HARTREE-FOCK EQUATIONS WITH AN ORIGINAL BASIS SET

Hartree-Fock equations originate from the assumption of a single Slater determinant form of the N-electron wavefunction which is the solution of the Schrödinger equation (see reference 3 for instance)

$$\sum_{i=1}^{N} \mathcal{H}_{0,i} \mid \psi > + \sum_{i<j}^{N} \left(\frac{2}{r_{ij}}\right) \mid \psi > = E_{tot} \mid \psi > \tag{6}$$

with

$$\mathcal{H}_{0,i} = -\Delta_i - \frac{2Z}{r_i} + 2\gamma\,(L_{zi} + 2S_{zi}) + \gamma^2 r_i^2 \sin^2\theta_i \tag{7}$$

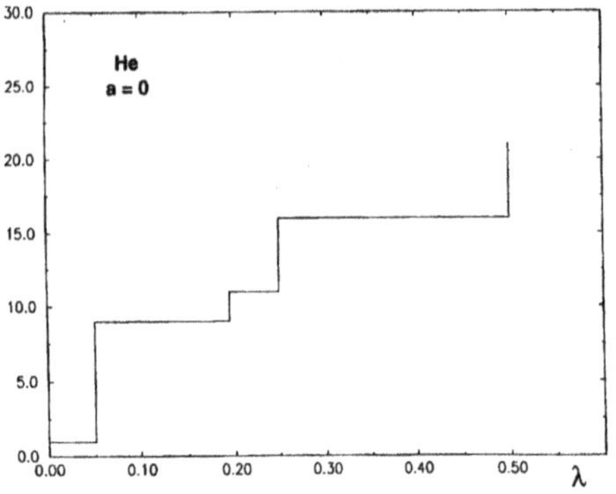

Figure 3. Number of basis functions obtained for a given value of λ.

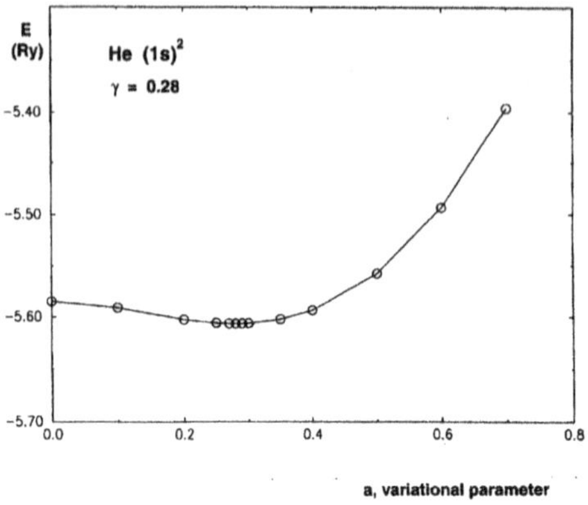

Figure 4. Variation of the total energy with respect to a.

where $\gamma = B/4.701\ 10^5\ T$ and atomic Rydberg units are used.

Usually, the one-electron components $\mid \psi_i >$ of the Slater determinant $\mid \psi >$ are projected on a basis of spin orbitals $\{\mid \phi_j >\}$

$$\mid \psi_i > = \sum_j \alpha_{ij} \mid \phi_j > \tag{8}$$

with $\vec{\alpha_i}$ the eigenvectors of the resulting Hartree-Fock matrix.

Whereas, in conventional HF methods, the basis functions are Slater or gaussian-type orbitals, we used an original basis set composed of *radial Kohn-Sham wavefunctions*, $R_k^{\rm KS}(r)$, computed without magnetic field. In this way we introduced the main part of the radial wavefunction behavior with few functions in the expansion. The angular part of the basis functions is represented by spherical harmonics with fixed quantum number m since the component of the orbital momentum parallel to the magnetic field is conserved. To reduce further the number of functions in the expansion, we simulated the effect of the magnetic field on the anisotropy of the wavefunction by multiplying the basis functions with an exponential factor $e^{-\frac{1}{2}a^2r^2\sin^2\theta}$, where a is a variational parameter. Finally, to preserve the orthonormality of the basis set, one must build $\phi_{jm}(\mathbf{r})$ as

$$\phi_{jm}(\mathbf{r}) = \sum_{k=(n,l)} \beta_{jk}\, R_k^{\rm KS}(r)\; Y_{lm}(\theta,\varphi)\; e^{-\frac{1}{2}a^2r^2\sin^2\theta} \tag{9}$$

where each $\vec{\beta_j}$ is an eigenvector of the overlap integrals matrix $< \phi_i \mid \phi_k >$.

However, this original basis expansion presents a shortcoming. Indeed, only a very limited number of KS orbitals $R_k^{\rm KS}$ can be generated since the KS potential has a finite range in the local density approximation (LDA). To get round this difficulty, we artificially substituted a Coulomb potential $V(r) = -\frac{2\lambda}{r}$ $(0 \le \lambda \le 1)$ of infinite range for the tail of the KS potential $v_{\rm KS}(r)$. Consequently, we got 2 parameters to minimize the total energy. The first a is used in the HF program via the exponential factor, the other λ is included in the KS program to generate the basis.

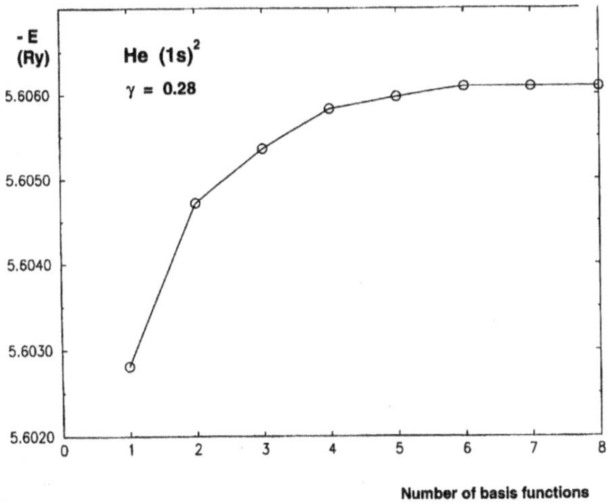

Figure 5. Convergence of the variational total energy as the number of basis functions increases.

Our procedure is as follows :

- First, considering the case without magnetic field we fix $a = 0$ and search for the best value of λ which produces the most important basis set and minimizes the HF total energy (Figure 3).
- Second, keeping this value of λ, we compute the optimal parameter a corresponding to the lowest HF total energy in the presence of the magnetic field (Figure 4).
- Finally, we check whether convergence has been reached with respect to the number of elements in the basis set (Figure 5).

Our results are presented in the next section.

HELIUM IN STRONG MAGNETIC FIELD

We present a comparison between our KS and HF energies together with HF results already published for the $(1s)^2$ (Table 1, Figure 6) and $(1s)$ $(2p_{-1}$ (Table 2, Figure 7) configurations of Helium at various strengths of the magnetic field. Whereas total energies differ between the KS and HF approaches, the contributions $E(B) - E(0)$ due to the magnetic field are much closer to each other (Figures 6 and 7). Indeed, it is well known that KS calculations in the LDA underestimate the self-exchange energy for atomic systems.[14] For instance, the exact value of the $(1s)^2$ ground state energy at $B = 0$ is 5.807 Ry, whereas the HF value is 5.723 Ry and the KS value is 5.669 Ry. Moreover, the discrepancy between the KS and HF values of the $(1s)$ $(2p_{-1})$ energy level at $B = 0$ is more pronounced because we did not take the spin polarization into account, leading to underestimating further the exchange energy. Work in progress aims at implementing a local spin density approximation with a self-interaction correction (SIC-LSD), and at studying ground states of heavier elements.

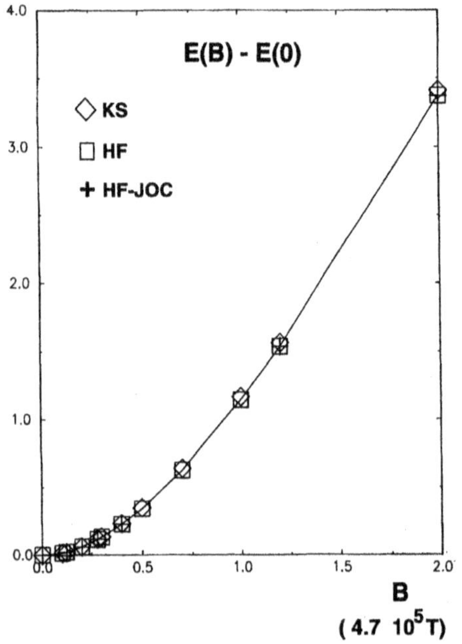

Figure 6. $(1s)^2$ configuration of He in strong magnetic field.

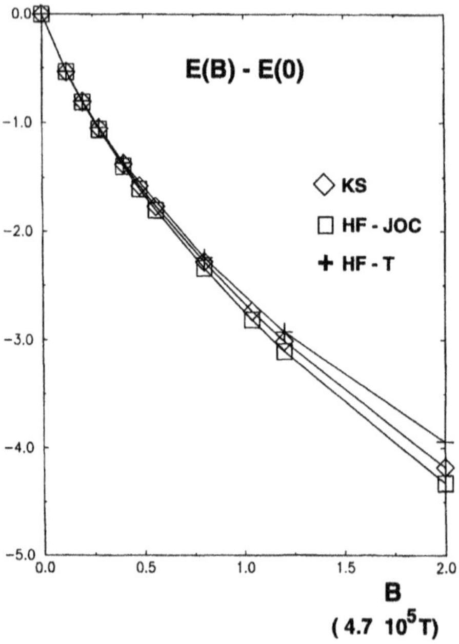

Figure 7. $(1s)(2p_{-1})$ configuration of He in strong magnetic field.

Table 1. $(1s)^2$ configuration of He in strong magnetic field.

B $(4.7\ 10^5\ T)$	$-E_{KS}$ (Ry)	$-E_{HF}$ (Ry)	$-E_{HF}^{JOC}$ (Ry) ref. 12
0.00	5.669	5.7231	5.7234
0.10	5.654	5.7075	
0.12	5.648	5.7007	5.7008
0.20	5.608	5.6620	5.6620
0.28	5.551	5.6061	5.6060
0.30	5.534	5.5896	
0.40	5.435	5.4939	5.4936
0.50	5.317	5.3780	
0.70	5.030	5.0964	
1.00	4.503	4.5780	
1.20	4.105	4.1841	4.1846
2.00	2.257	2.3499	2.3524

Table 2. $(1s)(2p_{-1})$ configuration of He in strong magnetic field.

B $(4.7\ 10^5\ T)$	$-E_{KS}$ (Ry)	$-E_{HF}^{JOC}$ (Ry) ref.12	$-E_{HF}^{T}$ (Ry) ref. 15
0.00	3.985	4.2628	
0.12	4.520	4.7984	4.7946
0.20	4.790	5.0732	5.0624
0.28	5.033	5.3212	5.3032
0.40	5.361	5.6598	5.6276
0.48	5.562	5.8686	
0.56	5.752	6.0660	
0.80	6.267	6.6034	6.5080
1.04	6.721	7.0768	
1.20	6.997	7.3678	7.1896
2.00	8.165	8.5900	8.2032

REFERENCES

1. F. S. Felber, M. M. Malley, F. J. Wessel, M. K. Matzen, M. A. Palmer, R. B. Spielman, M. A. Liberman and A. L. Velikovich, *Phys. Fluids* 31:2053 (1988).
2. F. S. Felber, M. A. Liberman and A. L. Velikovich, *Appl. Phys. Lett.* 46:1042 (1985).
3. H. Ruder, G. Wunner, H. Herold and F. Geyer. "Atoms in Strong Magnetic Fields," Springer, Berlin (1994).
4. G. Vignale and M. Rasolt, *Phys. Rev. B* 37:10685 (1988).
5. P. Skudlarski and G. Vignale, *Phys. Rev. B* 48:8547 (1993).
3. M. D. Jones, G. Ortiz and D. M. Ceperley, *Phys. Rev. A* 54:219 (1996).
7. W. Kohn and L. J. Sham, *Phys. Rev.* 140:A1133 (1965).
8. F. Perrot and A. Grimaldi, *J. Phys. : Condens. Matter* 7:6545 (1995).
9. O. Gunnarsson and B. I. Lundqvist, *Phys. Rev. B* 13:4274 (1976).
10. S. H. Vosko, L. Wilk and M. Nusair, *Can. J. Phys.* 58:1200 (1980).
11. J. M. MacLaren, D. P. Clougherty, M. E. McHenry and M. M. Donovan, *Comp. Phys. Comm.* 66:383 (1991).
12. D. M. Ceperley and B. J. Alder, *Phys. Rev. Lett.* 45:566 (1980).
13. J. Shertzer, *Phys. Rev. A* 39:3833 (1989).
14. J. P. Perdew and A.Zunger, *Phys. Rev. B* 23:5048 (1981).
15. G. Thurner, H. Korbel, M. Braun, H. Herold, H. Ruder and G. Wunner, *J. Phys. B* 26:4719 (1993).

FROM FIELD-FREE ATOMS TO FINITE MOLECULAR CHAINS IN VERY STRONG MAGNETIC FIELDS

Michel R. Godefroid

Laboratoire de Chimie Physique Moléculaire
Université Libre de Bruxelles, CP160/09
50 av. F. Roosevelt, B-1050, Brussels, Belgium

INTRODUCTION

We report on progress achieved in variational calculations of atomic structures of *field-free* atoms, from isotope shifts and hyperfine structures to radiative transition probabilities[1]. The role of electron correlation is investigated through the use of the multiconfiguration Hartree-Fock approximation. We then turn to the main topic of the meeting, ie. "Atoms and Molecules in Strong External Fields" by describing our Hartree-Fock study[2] on the stability of light homonuclear diatomic molecules and of finite hydrogen and helium molecular chains immersed in very intense magnetic fields relevant in the physics of neutron stars ($B = 2.35\ 10^{12}$ G $\equiv 2.35B_{12}$ or $\gamma = B\hbar^3/m^2ce^3 = 1000$).

FIELD-FREE ATOMS

The Multiconfiguration Hartree-Fock Approximation

In atomic structure calculations of field-free atoms, the multiconfiguration Hartree-Fock (MCHF) approximation is often used for calculating correlated wave functions and accurate spectroscopic properties[1]. In the non-relativistic MCHF approach[3] the wave function Ψ for a state labeled γLS, where γ represents the configuration and any other quantum numbers required to specify the state, is expanded in terms of configuration state functions (CSFs) with the same LS term.

$$\Psi(\gamma LS) = \sum_j c_j \Phi(\gamma_j LS). \tag{1}$$

The configuration state functions $\Phi(\gamma LS)$ are anti-symmetrized linear combinations of products of spin-orbitals

$$\phi_{nlm_lm_s}(\mathbf{q}) = \phi_{nlm_lm_s}(\mathbf{r},\sigma) = \frac{1}{r}P_{nl}(r)Y_{lm_l}(\theta,\varphi)\xi_{m_s}(\sigma) \tag{2}$$

Atoms and Molecules in Strong External Fields
Edited by Schmelcher and Schweizer, Plenum Press, New York, 1998

where the radial functions $P_{nl}(r)$ are represented by their numerical values at a number of gridpoints. In the multiconfiguration self-consistent field procedure *both* the orbitals and the expansion coefficients are optimized as to leave the total energy stationary. Once a set of radial orbitals has been obtained, a configuration interaction (CI) calculation can be performed. The wave function is also expanded in configuration state functions, but only the expansion coefficients $\{c_i\}$ are determined by diagonalizing the Hamiltonian matrix. For large expansions the iterative Davidson method can be used to determine a restricted number of the lowest eigenvalues and eigenvectors[4]. Using a sparse matrix representation, where only non-zero matrix elements are saved, large expansions can be used, the limit being set by the available disk space.

A new trend in atomic structure calculations is to assess the reliability of the theoretical calculations by "monitoring" the property as a function of the size of the orbital active set on which the CSF space is built within a given model. The reader is referred to previous work[1] and references therein for the isotope shift and transition probability calculations presented at the meeting. For the proceedings, we will focus on the evaluation of the hyperfine structure parameter of the ground state of nitrogen[5]. This is indeed an interesting difficult test case for which electron correlation is definitely the key of the problem.

A Nice Example: The Hyperfine Structure of the Nitrogen Ground State

The hyperfine structure of atomic energy levels is caused by the interaction between the electrons and the electromagnetic multipole moments of the nucleus. For the $1s^22s^22p^3\ ^4S^o_{3/2}$ nitrogen ground state, with a total orbital angular momentum equal to zero ($L = 0, J = S$), only the Fermi contact term contributes to the hyperfine interaction constant which is given (in MHz) by

$$A_J = 95.41067 \left(\frac{\mu_I}{I}\right) \left(\frac{g_e}{6}\right) \frac{1}{J} a_c \tag{3}$$

where

$$a_c = \langle \gamma LS(M_L = L)(M_S = S)| \sum_{i=1}^{N} 8\pi\delta^3(\mathbf{r}_i) s_{zi} |\gamma LS(M_L = L)(M_S = S)\rangle \tag{4}$$

and $g_e = 2.002\ 319\ 3$ is the electron spin g-factor. The isotope ^{14}N has a nuclear spin $I = 1$ with a magnetic dipole moment $\mu = 0.40376100(6)\ \mu_N$. In the one-configuration Hartree-Fock approximation for which only the dominant term (ie. $2p^3\ ^4S^o$) is kept in the expansion (1), the Fermi contact parameter a_c is strictly zero. In this case, the dominating contributions to the hyperfine interaction come from the spin-polarization of the closed $1s$ and $2s$ shells due to the Coulomb exchange interaction with the open shells. In the MCHF method, spin-polarization is described by including CSFs of the form $\Phi(1sns(^3S)2s^22p^3\ ^4S^o)$ and $\Phi(1s^22sns(^3S)2p^3\ ^4S^o)$ in the configuration expansion. The spin-up/spin-down asymmetry enters into this scheme through the expansion coefficients and the shapes of the polarization orbitals ns. The difficulties mainly arise from the large and cancelling spin-polarization contributions from the closed $1s$ and $2s$ shells.

As a starting point a number of variational MCHF calculations were performed. The configuration expansions for the MCHF calculations were obtained using the active space method, where CSFs of a specified parity and LS symmetry are generated by electron excitation from one or more reference configurations to an active set of orbitals. To describe the major correlation effects in the pair-correlation approximation,

Table 1. Total energies and hyperfine structure constants for $1s^22s^22p^3\ ^4S^o$ in ^{14}N.

active set	$A_{3/2}$ (MHz)	E (a.u.)	# CSFs
SD-MCHF			
HF	0.0	-54.400 934	1
3s2p1d/3s	8.632	-54.517 526	179
4s3p2d1f/3s	5.267	-54.558 194	438
5s4p3d2f1g/3s	4.924	-54.571 488	889
6s5p4d3f2g/3s	5.592	-54.576 998	1538
7s6p5d4f3g/3s	5.534	-54.579 219	2385
8s7p6d5f4g/3s	5.555	-54.580 233	3430
9s8p7d6f5g/3s	5.595	-54.580 674	4673
10s9p8d7f5g/3s	5.663	-54.580 892	5884
SDT- and SDTQ-CI			
SD{10s9p8d7f5g/3s} ∪ SDT			
3s2p1d/3s	7.488	-54.582 559	6 511
4s3p2d1f/3s	9.278	-	-
5s4p3d2f1g/3s	9.861	-54.584 457	20 792
6s5p4d3f1g/3s	10.169	-54.584 689	39 307
7s6p5d3f1g/3s	10.234	-54.584 771	58 878
SD{10s9p8d7f5g/3s} ∪ SDT{7s6p5d3f1g/3s} ∪ SDTQ			
4s3p2d1f/3s	10.395	-54.585 927	79 373
Experiment[6]	10.45		

all single (S) and double (D) excitations were allowed from the Hartree-Fock reference configuration to the active set of orbitals. The orbital set was then increased in a systematic way, allowing the convergence of the expectation values to be studied. Following the notation used in quantum chemistry, the active set is characterized by the number of orbitals of a certain symmetry. The set $3s2p1d$ for example, contains three s-orbitals, two p-orbitals and one d-orbital. In Table 1 the hyperfine coupling constant, the total energy and the total number of configurations are shown as a function of the increasing active set of orbitals. The three s-orbitals, preceeded by a slash in the table, have been added and optimized by allowing single excitations only for describing specifically the spin-polarization effects.

To investigate the influence of higher order correlation effects a number of configuration interaction calculations were performed in which CSFs generated by triple (T) and quadruple (Q) excitations from the reference configuration to the increasing active set were added to the largest expansion from the preceeding MCHF calculation. The merging of CSF lists is denoted by the union symbol (∪) in the table. As seen from the table the inclusion of CSFs generated by T excitations has a drastic effect on the hyperfine coupling constant, which is increased by as much as 80%. The effect of the included CSFs generated by Q excitations is in comparison very small and configurations obtained by higher excitations are believed to be small and have been neglected. The final result is in rather good agreement with experiment[6].

For treating atoms and molecules in a uniform magnetic field, we use the following Hamiltonian

$$H = \sum_i h_i^B + \sum_{i,\alpha} V_{i\alpha} + \sum_{i<j} V_{ij} + \sum_{\alpha<\beta} V_{\alpha\beta} \tag{5}$$

in which h_i^B is the hamiltonian of electron i in the homogenous and constant magnetic field chosen to be oriented along the z axis. The three potential terms represent the Coulomb interaction between the electron and nuclei labeled by latin and greek letters respectively. We are working in the "standard" Born-Oppenheimer approximation in which we solve the electronic Schrödinger equation for a given nuclear configuration, and add the corresponding nuclear repulsion energy to the total electronic energy to get the effective potential energy in which the nuclei move. This approximation is valid if the electronic energy-level spacings are large compared to the typical energy-level spacings associated with the nucleus motion[7]. We are interested in a Landau regime where the magnetic field is dominant and is only slightly perturbed by the Coulomb interactions. The total wave function Ψ can be approximated by a Slater determinant built on the N spatial one-electron functions ψ_i multiplied by the symmetric product of the anti-parallel spin functions. We assume that the transverse motion, perpendicular to the field, can be described by the transverse parameter $b_\perp$ fixed to the Larmor radius $b_\perp \equiv b_0 = \sqrt{\frac{2\hbar c}{eB}}$. The electrons will be confined in the ground Landau level correponding to the lowest radial quantum numbers $(n = 0, \Lambda \leq 0)$. In this adiabatic approximation, the motion perpendicular to the magnetic field is described by unperturbed magnetic Landau orbitals

$$\phi_{n=0,\Lambda}(\rho, \varphi; b_\perp) = \sqrt{\frac{1}{\pi \lambda! b_\perp^2}}\, e^{-i\lambda\varphi}\, e^{-\frac{\rho^2}{2b_\perp^2}} \left(\frac{\rho}{b_\perp}\right)^\lambda \tag{6}$$

where $\lambda \equiv |\Lambda|$. Each one-electron orbital can be factorized as

$$\psi_i(\mathbf{r}) = \phi_{0,\Lambda_i}(\rho, \varphi; b_0)\, \chi_{n_{\parallel_i}}^{\Lambda_i}(z) \tag{7}$$

and the Coulomb interaction only affects the longitudinal wave functions $\chi_{n_\parallel}^\Lambda(z)$.

The total energy can easily be evaluated from the total wave function by averaging the motion in the transverse plane. The Hartree-Fock equations can be derived[2] by applying the variational principle $\delta E = 0$ for any variation in the longitudinal one-electron wave functions $\delta\chi_{n_\parallel}^\Lambda(z)$. Though a numerical approach can be used for solving the one-dimensional HF equations in atoms[8] or molecules[9, 7] as in the field-free case[3], we used another computational strategy.

Hartree-Fock On A Mesh

In our approach, each individual wave function is discretized on a one-dimensional cartesian (or Fourier) mesh composed of N equidistant points

$$\chi_i(z) = \sum_{p=1}^{N} \chi_i(z_p) f_p(z) \tag{8}$$

where $\chi_i(z_p)$ are the values of the χ's at the mesh points z_p. The Lagrange functions $f_p(z)$ associated with the cartesian mesh[10],

$$f_p(z) = \frac{1}{N} \frac{\sin \pi(z-p)}{\sin[\pi(z-p)/N]} \tag{9}$$

Table 2. Atomic (E_A, E_B), molecular (E_{AB}) total energies and dissociation energies (E_D) in Rydberg and equilibrium internuclear distances (R in a_0) of ground states of homonuclear diatomic molecules. $B_{12} = 2.35$.

AB		R	E_{AB}	A	E_A	B	E_B	$E_A + E_B$	E_D
H_2	[01]	0.18	-35.8	[0]	-15.2	[1]	-11.2	-26.4	9.4
He_2	[0123]	0.17	-119.5	[01]	-55.4	[23]	-36.3	-91.7	27.8
Li_2	[012345]	0.19	-239.4	[012]	-118.4	[345]	-71.7	-190.1	49.3
Be_2	$[0_2 123456]$	0.26	-401.7	[0123]	-201.9	[0456]	-168.6	-370.5	31.2
B_2	$[0_2 1_2 2_2 3456]$	0.36	-605.2	[01234]	-304.2	[01256]	-292.2	-596.4	8.8
C_2	$[0_2 1_2 2_2 3_2 4_2 56]$	0.50	-845.5	[012345]	-424.3	[012346]	-420.5	-844.8	0.7

have very special properties:

$$f_p(z_q) = \delta_{pq} \quad ; \quad \int_{-\frac{N}{2}}^{\frac{N}{2}} f_p^*(z) f_q(z) dz = \delta_{pq} \tag{10}$$

The Hamiltonian matrix elements in the Lagrange basis can be evaluated using the Gaussian quadrature formula. The Gauss formula associated with the cartesian mesh is exact for the matrix elements of the kinetic energy operator and produces rather simple analytical expressions[10]. Though regularization techniques have been proposed[11], the one- and two-electron matrix elements of the potential terms are approximated with a cartesian mesh possibly denser than the discretization describing the individual wave functions.

The Homonuclear Diatomic Molecules

The lowest configurations of diatomic homonuclear molecules from H_2 to C_2 have been studied at $\gamma = 1000$ using the Hartree-Fock-on-a-mesh procedure described above. The main results are summarized in Table 2 in which the electronic configurations are characterized by the λ sequence $[(\lambda_1)_{\nu_1} \cdots (\lambda_n)_{\nu_n}]$ where ν gives the number of orbitals having a specific λ value. The evolution of the stability with increasing the nuclear charge can be understood from the analysis of the molecular configurations and their dissociation energy defined as $E_D = E_A + E_B - E_{AB}$. For the three first members of the series (H_2, He_2 and Li_2), the ground states correspond to configurations in which all electrons have different values of λ. The situation gradually changes for heavier molecules thanks to the growing nuclear attraction. From Be_2 the configuration with two $\lambda = 0$ electrons becomes definitely energetically more favorable than the configuration with all different λ's. The first excited orbitals $\lambda = 1, 2$ become competitive for higher Z and for C_2, the $\lambda = 0, 1, 2, 3$ and 4 first excited orbitals are all occupied in the ground state. The key of the molecular stability problem in the very intense magnetic field regime that we consider lies in the corresponding dissociation limit. The energetically most favorable partitions for the atomic dissociation products are reported in the same table. Multiple occurrence of the same λ values in the molecular configuration leads to less excited atoms as dissociation products and lowers the dissociation limit. The dissociation energy increases monotonically with the nuclear charge for molecular configurations containing only different values of λ and decreases dramatically when λ-excited orbitals become populated.

Our $B_{12} = 2.35$ results for the equilibrium interatomic distance, the ground state energy and the dissociation energy* are consistent with the values obtained by Lai and

*Note that the "dissociation energy" is defined by Lai and Salpeter[7] as the difference between the

Table 3. Molecular total energies ($E[01]$), dissociation energies (E_D), equilibrium internuclear distances (R in a_0) of the ground state [01] of H_2 and atomic energies ($E[\lambda]$). All energies in eV.

B_{12}	R	$E[01]$	$E[0]$	$E[1]$	$E[0]+E[1]$	E_D	ref.
1	0.24	-372.4	-162.2	-116.2	-278.5	93.9	Ortiz et al.[12]
2	0.20	-466	-198.5	-145.8	-344.3	121.7	Lai and Salpeter[7]
2.35	0.18	-487	-206.8	-152.4	-359.2	127.8	Demeur et al.[2]
5	0.15	-623	-257.1	-192.6	-449.7	173.3	Lai and Salpeter[7]

Table 4. Ground state configurations of H_n and He_n molecular chains at $\gamma = 1000$, equilibrium bond length (R in a_0), molecular total energy (E_{tot}), sum over the atomic total energies (Σ) according to the partition specified in the 5th column, dissociation E_D and binding energy per atom E_b. All energies are given in Rydberg.

	configuration	R	E_{tot}	partition	Σ	E_D	E_b
H_2	[01]	0.18	-35.8	0/1	-26.4	9.4	17.9
H_3	[012]	0.16	-56.3	0/1/2	-36.0	20.3	18.8
H_4	[0123]	0.15	-74.8	0/1/2/3	-44.5	30.3	18.7
H_5	[$0_2$123]	0.17	-93.8	0/0/1/2/3	-59.7	34.1	18.8
He_2	[0123]	0.17	-119.5	01/23	-91.7	27.8	59.8
He_3	[012345]	0.17	-175.6	05/14/23	-121.7	53.9	58.5
He_4	[$0_2$123456]	0.18	-236.8	05/14/23/06	-169.3	67.5	59.2

Salpeter[7] for H_2 at other magnetic field strengths (B_{12} = 2 and 5) and the B_{12} = 1 quantum Monte Carlo results of Ortiz et al.[12] (see Table 3). The overall increase of the atomic and molecular binding energies with increasing the magnetic field strength, the increase of the molecular stability and the contraction of the bond length appear very clearly from this table.

The H_n and He_n Finite Chains

We applied our "Hartree-Fock-on-a-mesh" approach to study the stability of hydrogen and helium linear chains[2] at $B_{12} = 2.35$. The ground state configurations are presented in Table 4 with their total energies (E_{tot}) and equilibrium nuclear interdistances. The dissociation (or atomization) energies E_D are calculated from the difference between the equilibrium total energy and the sum of the atomic total energies (Σ), according to the most stable λ-partition for the He_n chains. For the hydrogen chains, H_5 is the shortest chain in which a λ-duplication occurs. As found for the diatomic molecules, the nuclear charge has an efficient role in this process. For the He_n chains indeed, the occurrence of two $\lambda = 0$ already appears at $n = 4$.

As observed by Lai et al.[9], the binding energy per atom obtained by dividing the total energy of the chain by the number of atoms ($E_b \equiv E_{tot}/n$), stabilizes very quickly when increasing the number of atoms. This "saturation" phenomenon is observed for both kinds of chains. Condensation (or cohesion) energies (E_c) can be evaluated by substracting from the binding energy per atom the ground state energy of the isolated atom ($E_c = E_b - E_a$ with $E_a = E[0]$ for H_n and $E[01]$ for He_n). Using the multichannel density-functional method, Relovski and Ruder[13] obtained condensation

molecular ground state energy and the energy of the two isolated atoms, *both* in the ground state, while our definition corresponds to the depth of the energy curve.

Table 5. Bond lengths (R) (in a_0), binding (E_b) and condensation (E_c) energies (in Rydbergs) for hydrogenic chains

B_{12}	R	$E[0]$	E_b	E_c	ref.
2	0.17	-14.59	-18.15	3.56	Lai et al.[9]
2	0.177	-15.48	-21.0	5.52	Relovsky and Ruder[13]
2.35	0.17	-15.2	-18.8	3.6	Demeur et al.[2]
2.35	0.165	-16.21	-22.54	6.13	Relovsky and Ruder[13]

energies which are systematically much larger than the values obtained in the Hartree-Fock approximation[9, 2], as reported in Table 5.

For He_n chains, our $B_{12} = 2.35$ cohesion energy of 42-59 eV (depending on n) is consistent with the 20/146 eV and 25/149 eV found respectively by Lai et al.[9] and Neuhauser et al.[14] for $B_{12} = 1/5$.

CONCLUSION AND PERSPECTIVES

Our Hartree-Fock-on-a-(cartesian)-mesh approach produces results which are consistent with other recent work on diatomic molecules[7, 12] and molecular chains[9, 14]. The λ-duplication found in the ground state of homonuclear diatomic molecules beyond Li_2 and in molecular chains beyond H_4 or He_3 plays a crucial role in the stability (or unstability) of molecular species in the high-field regime.

As pointed out by Lai and collaborators[9, 7], configuration mixings can be important for getting the correct dissociation limit of the ground state of H_2. Though this intrinsic failure of the Hartree-Fock approximation does not affect Demeur et al.'s results[2], the energies of the dissociation atomic products being obtained independently of the molecular wave function, the one-configuration description should be improved for obtaining for instance the "good" potential governing the aligned vibration motion[7]. The large differences found between Hartree-Fock and multichannel density-functional condensation energies for hydrogenic and helium chains have been attributed by Relovsky and Ruder[13] not only to the use of different physical theories, but also to restrictions on the allowed single-particle wave functions in the Hartree-Fock scheme. The description of electron correlation being one of our major interests in the spectroscopy of field-free atoms as illustrated in the first part, we would like to go beyond the mean-field approximation by considering the multiconfiguration approach for very intense magnetic field situations. Interdisciplinary research is usually very rich and we will consider the new perspectives that the success of the B-spline approach in atomic structure calculations[15] might open.

Demeur et al.'s results[2] have been obtained by solving the fixed-nuclei electronic Hamiltonian, excluding mass effects. For H_2, the zero-point energies for the aligned and transverse vibrations have been estimated to be 9.5 eV and 21 eV respectively at $B_{12} = 5$, which represent together 18% of the corresponding dissociation energy $E_D =$ 173 eV (see previous footnote). In view of the relative magnitude of the electronic and nuclear vibration excitation energies, the adiabatic physical picture of the separation of the electronic and nuclear motion might become naïve and it would be rather interesting to investigate the use of the so-called "screened Born-Oppenheimer approximation" proposed by Schmelcher et al.[16].

REFERENCES

1. M. Godefroid, C. Froese Fischer and P. Jönsson, Multiconfiguration Hartree-Fock calculations of atomic properties in light atoms, *Physica Scripta* T65:70 (1996).
2. M. Demeur, P.-H. Heenen and M. Godefroid, Hartree-Fock study of molecules in very intense magnetic fields, *Phys. Rev. A* 49:176 (1994).
3. C. Froese Fischer, The MCHF atomic-structure package, *Comput. Phys. Commun.* 64:369 (1991).
4. A. Stathopoulos and C. Froese Fischer, A Davidson program for finding a few selected extreme eigenpairs of a large, sparse, real, symmetric matrix, *Comput. Phys. Commun.* 79:268 (1994).
5. M. Godefroid, G. Van Meulebeke, P. Jönsson and C. Froese Fischer, Large-scale MCHF calculations of hyperfine structures in nitrogen and oxygen, in preparation.
6. J.M. Hirsch, G.H. Zimmerman, III, D.J. Larson and N.F. Ramsey, Precision measurement of the hyperfine structure and g_J factor of atomic nitrogen 14, *Phys. Rev. A* 16:484 (1977).
7. D. Lai and E.E. Salpeter, Hydrogen molecules in a superstrong magnetic field: excitation levels, *Phys. Rev. A* 53:152 (1996).
8. P. Pröschel, W. Rösner, G. Wunner, H. Ruder, and H. Herold, Hartree-Fock calculations for atoms in strong magnetic fields. I: Energy levels of two-electron systems, *J. Phys. B : At. Mol. Phys.*, 15:1959 (1982).
9. D. Lai, E.E. Salpeter, and S.L. Shapiro, Hydrogen molecules and chains in a superstrong magnetic field, *Phys. Rev. A*, 45:4832 (1992).
10. D. Baye and P.-H. Heenen, Generalised meshes for quantum mechanical problems, *J. Phys. A : Math. Gen.*, 19:2041 (1986).
11. M. Vincke, L. Malegat and D. Baye, Regularization of singularities in Lagrange-mesh calculations, *J. Phys. B : At. Mol. Phys.*, 26:811 (1993).
12. G. Ortiz, M.D. Jones and D.M. Ceperley, Ground state of a hydrogen molecule in superstrong magnetic fields, *Phys. Rev. A*, 52:R3405 (1995).
13. B.M. Relovsky and H. Ruder, Multi-channel density-functional calculations for atoms and atomic chains in magnetic fields of compact stars, *Phys. Rev. A*, 53:4068 (1996).
14. D. Neuhauser, S.E. Koonin, and K. Langanke, Structure of matter in strong magnetic fields, *Phys. Rev. A*, 36:4163 (1987).
15. J. Sapirstein and W.R. Johnson, The use of basis splines in theoretical atomic physics, *J. Phys. B : At. Mol. Phys.*, 29:5213 (1996) and references therein.
16. P. Schmelcher, L.S. Cederbaum and U. Kappes, 1994, Molecules in magnetic fields: fundamental aspects, *in*: "Conceptual Trends in Quantum Chemistry" E.S. Kryachko, ed., Kluwer, Dortrecht (1994).

THE NATIONAL HIGH MAGNETIC FIELD LABORATORY – A PRECIS

Jack E. Crow,[1] John R. Sabin,[1,2] and Neil S. Sullivan[1,2]

[1]National High Magnetic Field Laboratory
Tallahassee, Florida 32310-4005
[2]Department of Physics
University of Florida
Gainesville, Florida 32611-8435

INTRODUCTION

The National High Magnetic Field Laboratory (NHMFL) is continuing the fine traditions established by the Francis Bitter Magnet Laboratory at MIT in Cambridge, Massachusetts, of providing access to high magnetic field facilities to all qualified users, national and international, developing magnet technology for the next generation of high field magnets, and developing an active in-house research program. In addition, the NHMFL is committed to fostering cooperation with the private sector, and with other national and international institutions in order to facilitate the development of new materials and technologies critical to the continued development in magnetic related research and technology. The NHMFL is jointly operated by the University of Florida (UF, Gainesville, Florida), Florida State University (FSU, Tallahassee, Florida), and Los Alamos National Laboratory (LANL, Los Alamos, New Mexico), with user activities at each of the three locations.

The main goal of the NHMFL is to provide facilities to investigate matter under extreme conditions of temperature, magnetic field and pressure, as well as to provide theoretical explanations and predictions of the phenomena encountered. To this end, the NHMFL operates continuous magnetic facilities including superconducting, hybrid and resistive magnets, and pulsed magnets driven by capacitive and motor-generator pulsed power supplies. Explosive flux compression magnets are also available on a limited access basis. In addition, the laboratory supports a variety of magnetic resonance instruments and systems. Many of these facilities are capable of operating over a large range of temperature, pressure, and sample size.

Atoms and Molecules in Strong External Fields
Edited by Schmelcher and Schweizer, Plenum Press, New York, 1998

Table 1. NHMFL Continuous Field Magnets

Superconducting Magnets

Field (T), Bore (mm)	Temperature	Research Projects
20, 52	20 mK	Magneto-optics, UV through far IR, magnetization, specific heat, transport, high pressure, NMR in highest fields – low resolution, mechanical stress testing
19.5, 52	0.4 – 300 K	
15, 45	10 mK	
14, 150	4.2 K	
19.5, 52	20 mK – 500 K	

Resistive and Hybrid Magnets

Field (T), Bore (mm)	Power (MW)	Research Projects
20, 50	8	Magneto-optics, UV through far IR, magnetization, specific heat, transport, high pressure, NMR in high fields – low to medium resolution
20, 200	20	
30, 32	15	
33, 32	20	
24.5, 32	15[a]	
24.5, 52	20[b]	
45/50, 32	20[c]	

[a]medium hmogenity, $\leq$ 20 ppm over 1 cm dsv.
[b]high homogenity, $\leq$ 1 ppm over 1 cm dsv, available September, 1999.
[c]available October, 1998.

CONTINUOUS FIELD FACILITIES

The continuous field facilities are located primarily at the NHMFL site in Tallahassee. The resistive and hybrid magnets are powered by state of the art 36 MW DC power supply with ripple and noise on the order of 10 ppm. It has an overload capacity of 40 MW for up to an hour and up to 68 MW for the order of minutes. The facility also has a very low vibration cooling system, sixteen magnet test cells, and extensive instrumentation.

The continuous fields magnets either extant or under development are summarized in Table 1.

The magnets are housed in six water cooled magnet stations. The 24.5 T magnet has 20 ppm homogeneity over a 10 mm dsv. The 45/50 T hybrid magnet is presently under development and should be operational by the end of 1998. The magnet consists of a 14 T, 610 mm, warm bore, cable-in-conduit, superconducting outsert, with a 20 MW, 31 T resistive insert. The magnet is designed for upgrade, in that it will accept inserts up to 40 MW, making possible a continuous field of 55 T in a 32 mm bore. The advanced cable-in-conduit design of the 14 T superconducting outsert is expected to provide a prototype of high field, large bore, magnets of the future.

In addition, there are four 20 T magnets at the NHMFL equipped with dilution refrigerators and variable temperature inserts for cryogenic work, and a new 12 T 150 mm bore, 30 x 70 mm split coil, superconducting magnet operating at 4.2 K, has recently been installed.

A facility for making measurements in high fields at low temperatures is now under development at the Gainesville campus of NHMFL in coöperation with the Microkelvin

Table 2. NHMFL Pulsed Magnets

Field (T), Bore (mm)	Rise, Total Time (ms)	Research Projects
50, 24	6, 30	Magneto-optics, UV through near IR,
60, 24		magnetization, specific heat, transport, high
63, 15	7, 35	pressure, NMR in highest fields – low
70, 10 – 14		resolution, mechanical properties, thermal
45, 24	9, 60[a]	expansion, high pressure
42, 24	100, 500	
60, 32	100 ms flat-top and variable pulse shape[b]	This quasi-continuous, controlled pulse magnet expands opportunities for all research conducted in pulsed magnet systems

[a]medium resolution.
[b]quasi-continuous magnet.

Laboratory there. This facility, known as the Ultra-High B/T facility will allow measurements that simultaneously require fields of the order of 20 T and temperatures as low as 500 μK. The resulting B/T ratio of 4×10^4 T/K will be the largest available anywhere. Many phenomena, which result from the establishment high spin polarizations or magnetizations, will become available for exploration in this facility. Examples include nuclear magnetism, magnetokinetics, polarized quantum fluids, quantum confined structures, and non-fermi liquids.

PULSED FIELD FACILITIES

The pulsed magnets are located on the Los Alamos campus of the NHMFL, as this site has unique capabilities for the production of pulsed electrical power and sites for flux compression experiments. The power for the long pulsed (quasi-continuous) magnets is supplied by a motor generator that delivers an energy pulse of 600 MJ, with capability to be upgraded to deliver a 2000 MJ pulse at a power level of 560 MW by addition of a flywheel and proper power supply modules. The various pulsed magnets available at the Los Alamos site are listed in Table 2.

These include 50 T and 63 T capacitive-driven magnets with 24 and 15 mm bores, respectively. A 60 T quasi-continuous (QC) magnet is expected to be commissioned in July, 1997. All the pulsed magnets are equipped for cryogenic operation with various types of dilution refrigerators, ^{3}He, variable temperature inserts and instrumentation that supports a variety of measurements such as transport, magnetization, high-pressure, and optical studies. Presently, a new non-destructive, 100T, 20 – 50 ms pulsed magnet is in the development stage, supported jointly by NSF and DOE.

Especially high fields are also available on a limited basis at the Los Alamos campus using explosive flux compression techniques. This technique employs chemical explosives to produce magnetic fields from 10^2 – 10^3 T for microsecond durations in 11 – 16 mm bores. Temperatures down to 2.3 K are available in these systems. A wide range of spectroscopic, electrical, and photographic instrumentation for fast time recording of data are available.

Table 3. NHMFL Magnetic Resonance Systems

Frequency	Field (T), Bore (mm)	Homogeniety	Measurements
	25,52	1ppm	Solid state NMR (1 GHz) EPR (700 GHz); ICR
850 MHz	18/20, 31	100 ppb	Solid state NMR
720 MHz	16.9, 50	1 ppb	Solid state NMR
600 MHz	14, 89	1 ppb	MRI; Solid state NMR
600 MHz	14, 50	1 ppb	Solution NMR, MRI/S
500 MHz	11.75, 50	1 ppb	Solution NMR
500 MHz	11.75, 50	1 ppb	Solid state NMR
300 MHz	7, 50	1 ppb	Solid state NMR
300 MHz	7, 89	1 ppb	Solid state NMR
200 MHz	4.7, 330	0.1 ppm	MRI/S of animals
125 MHz	3, 800	0.1 ppm	Whole body MRI/S
125 MHz	30, 32	100 ppm	ECR up to 7 THz
470 GHz	17, 61	3 ppm	EMR
400 GHz	14, 88	10 ppm	EMR
9 GHz	X-band		EPR
	20, 50	1000 ppm	ICR
	9.4, 210	10 ppm	ICR
	7, 150	10 ppm	ICR
	6, 150	10 ppm	ICR
	5.6, 90	10 ppm	ICR
	3, 150 (2)	10 ppm	ICR

MAGNETIC RESONANCE FACILITIES

Magnetic resonance facilities are located on both the Gainesville and Tallahassee campuses.

The NHMFL includes the Center for Interdisciplinary Magnetic Resonance (CIMAR), where nuclear magnetic resonance (NMR), magnet resonance imaging (MRI), electron magnetic resonance (EMR) including electron cyclotron resonance (ECR), and ion cyclotron resonance (ICR) mass spectroscopy are closely integrated. The extensive interaction between these fields is facilitated by the broad internal and external user program. The various facilities are located on various campuses, and the available instruments are summarized in Table 3.

In addition to these instruments, a new 1.1 GHz (25 T), high resolution NMR magnet is being developed. Initial work involves design and fabrication of a 140 mm cold bore, 110 mm warm bore, high resolution 900 MHz NMR magnet using conventional superconductor technology. Simultaneously, there is a development program of high T_c superconductors, along with the new magnet technology for these conductors. The 900 MHz magnet will serve as a platform for the incorporation of a high T_c central magnet, which will provide an additional field of *ca.* 5 T, to achieve the 1.1 GHz (25 T) goal.

The magnetic resonance facilities span all three NHMFL campuses, with the concentration of NMR, EMR, and ICR being in Tallahassee while MRI is concentrated in Gainesville.

The magnetic resonance imaging spectroscopy (MRI/S) facilities in Gainesville has concentrated on whole body and small animal MRI/S using large bore instruments. The NHMFL has established strong ties with the UF Center for Structural Biology and the UF College of Medicine, and will concentrate the MRI efforts at the UF Brain Institute, scheduled to open in 1998.

MATERIALS CHARACTERIZATION FACILITIES

There are also facilities at all three NHMFL campuses dedicated to the evaluation and characterization of magnet components such as conductors, insulators, reinforcement materials and composites in the full temperature range of magnet use; from the cryogenic to several hundred K.

The materials characterization facility has instruments specialized for measurements of mechanical, electrical, electro-mechanical, and thermal properties of materials, including a variety of scanning and transmission electron microscopes.

USERS' AND VISITORS' PROGRAMS

The NHMFL supports both active user and visitor programs. The user programs, presently implemented in the continuous field, pulsed field, CIMAR, and Ultra-High B/T facilities are open to any active researcher through a peer reviewed proposal process. There is no charge for magnet use, however a modest charge for consumables may be required. There is also a program for visitors to come to the NHMFL and participate in both experimental and theoretical programs as well as magnet development. Under this program, funds are available for sabbaticants and for supplemental travel for young investigators and first time users.

Further information about these programs may be obtained from the World Wide Web at *http://www.magnet.fsu.edu*, or from Dr. Bruce Brandt, Director of Operations and Instrumentation, NHMFL, 1800 E. Paul Dirac Drive, Tallahassee, FL 32310.

SELF-ADAPTIVE FINITE ELEMENT TECHNIQUES FOR STABLE BOUND MATTER-ANTIMATTER SYSTEMS IN CROSSED ELECTRIC AND MAGNETIC FIELDS

J. Ackermann

Institute of Molecular Biotechnology (IMB)
Department of Molecular Information Processing
Beutenbergstr. 11
07745 Jena, Germany

INTRODUCTION

In a recent paper the existence of stable bound matter-antimatter systems in crossed fields has been shown[1]. Lifetimes of the order of one year have been obtained for positronium ($e^- e^+$) for certain values of the field $\mathbf{B}$ and the pseudomomentum $\mathbf{K}$. The study includes a semi-classical calculation, a perturbational treatment as well as an accurate three-dimensional adaptive finite-element (FEM) treatment. In the present contribution we wish to focus on some technical details of the adapted FEM used.

Let us start with a brief description of the problem and the numerical challenge involved. We consider two particles with charges $Z_1 = e$, $Z_2 = -e$ and equal masses $m_1 = m_2 = m_e$ (positronium) in crossed electric and magnetic fields: $\mathbf{E} = 0$, $\mathbf{B} = B\hat{\mathbf{z}}$, and $\mathbf{K} = K\hat{\mathbf{y}}$, see ref. 1 for more details. $\mathbf{K}$ denotes the eigenvalue of the conserved pseudomomentum $\mathbf{K} = \mathbf{P} + \frac{e}{2}\mathbf{B} \times \mathbf{r}$. The effective Hamiltonian for the internal motion (neglecting spin) is given in Cartesian coordinates by (in atomic units)[2, 3]

$$H = \mathbf{p}^2 + \frac{B^2}{4}(x^2 + y^2) + \frac{BKx}{2} - \frac{1}{\sqrt{x^2 + y^2 + z^2}} + \frac{K^2}{4} \tag{1}$$

For the purpose of illustration let us choose $K = 0.4$ and $B = 5 \times 10^{-5}$, as in ref. 1. Beside the Coulomb singularity at $x = 0$ the potential has a local minimum at $x = -7987.46$ a$_o$ and a barrier with a maximum at $x_s = -322.81$ a$_o$. A set of Coulomb states are localized at the origin and a set of delocalized states exist in the outer well. An important question concerns the stability of the (excited) outer well states against direct annihilation and against transition to the unstable Coulomb states. To give a definite answer to this question it is indispensable to achieve an accurate approximation of the wave function in the outer well with a typical length in the order of several thousands a$_o$ and in the Coulomb regime with a typical length smaller than one a$_o$. A separation of the outer well states and the Coulomb states is possible only for sufficient large barriers between them. A hybrid expansion of the wave function in a basis set

Atoms and Molecules in Strong External Fields
Edited by Schmelcher and Schweizer, Plenum Press, New York, 1998

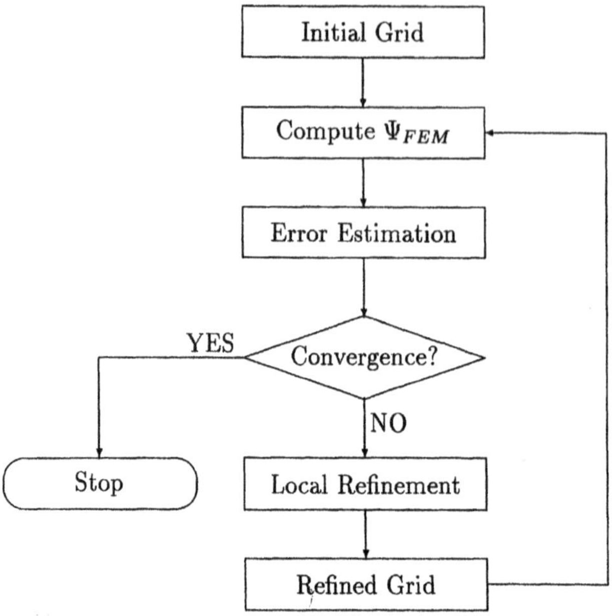

Figure 1. The flow diagram for the automatic refinement.

which includes global functions of the Coulomb type as well as appropriate outer well harmonic functions would fail to describe the correct tunneling through the barrier and would fail to approximate saddle states, which are neither of the Coulomb type nor typical outer well states. An attractive opportunity would be to solve the eigenvalue problem for the Hamiltonian (1) by a direct numerical method, for which no a priori knowledge about the wave function is necessary. This, however is a numerical task for several reasons.

The domain of the discretization has to be large enough to avoid an effect on the numerical results. An appropriate choice would be

$$\Omega := \{(x,y,z) \mid -r_c \leq x < r_c,\ 0 \leq y < r_c,\ 0 \leq y < r_c,\} \quad , \tag{2}$$

with $r_c = 10^5$ a_0. The Coulomb states, on the other hand, are functions showing peaks at the Coulomb singularity and a rather fine grid is needed to resolve these peaks. The stability against direct annihilation is determined by the probability of particle and anti-particle to be at the same space point. Thus, the approximation of the wave function should be accurate not only in the energy norm, but it should also indicate the correct value of the wave function locally at the two–body coalescence point. Moreover, no exact numerical Gauss integration formula exists for the integration of the Coulomb singularity at the origin.

ADAPTIVE FINITE ELEMENT METHOD

In this section we wish to give a brief description of the adaptive finite element method for stationary Schrödinger equations in three dimensions [4, 5]. The method is

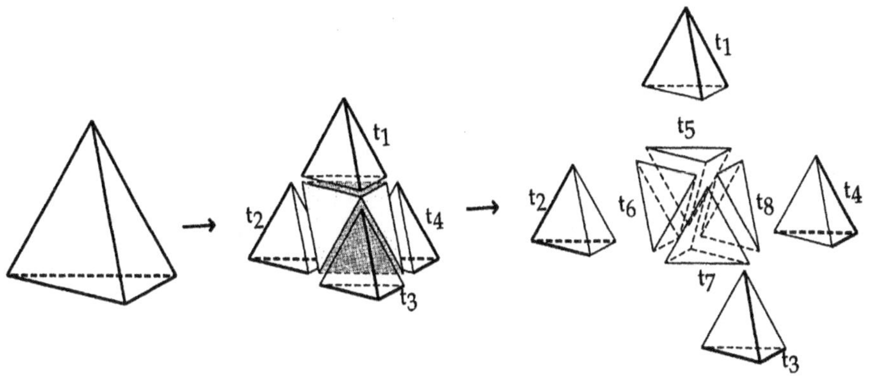

Figure 2. Numerically stable regular "red" refinement of a tetrahedron.

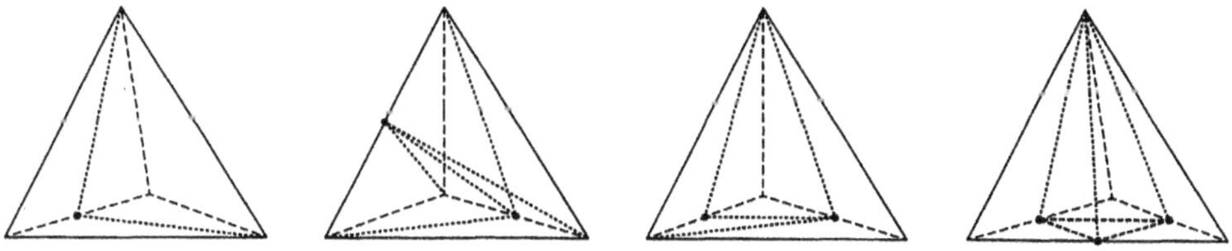

Figure 3. Green closures are used to complete the transition between refined and non-refined tetrahedra.

implemented by using the KASKADE toolbox[7]; a flow diagram is sketched in figure 1. We start with an initial coarse grid and compute the FEM approximation $\tilde{\Psi}$ on this grid. In general, this approximation will be a rough estimation for the real wave function. Now a local error estimator, see below, indicates which of the tetrahedra have to be refined to improve the FEM approximation. A refinement of the selected tetrahedra leads to a new grid. Computing a new FEM approximation, error estimation and refinement are repeated iteratively to construct a grading in the grid adapted to the exact wave function.

The refinement of the selected tetrahedron has to be performed in such a way a way that the numerical stability of the FEM is preserved. This can be done by a – so called – regular (red) refinement, see figure 2. By connecting the midpoints of the edges of a tetrahedron t, we obtain four new tetrahedra t_1, t_2, t_3, t_4, each of which corresponds to a vertex of t, and a remaining octahedron. The splitting of the octahedron in four tetrahedra t_5, t_6, t_7, t_8 is not unique, but depends on the selection of the interior diagonal edge which can be chosen in three ways. Each choice provides a regular (red) refinement of the tetrahedron, but has to be selected properly. The transition between refined and non-refined tetrahedra must be completed by green closures shown in figure 3.

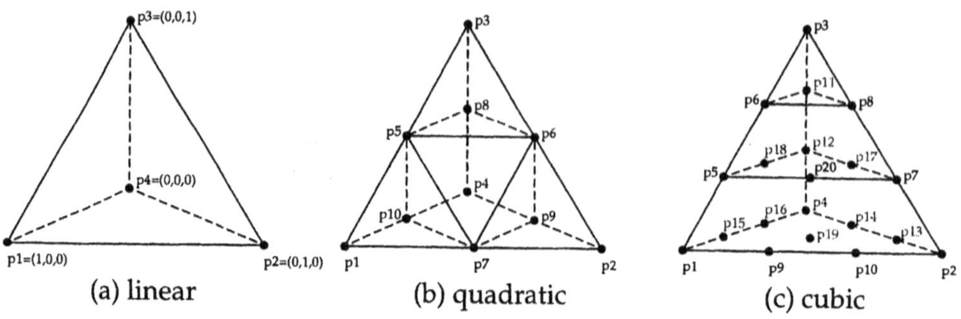

Figure 4. Nodes for linear, quadratic, and cubic Lagrange polynomials.

On each tetrahedron the wave function is approximated in terms of local polynomials. To improve the FEM approximation higher order Lagrange polynomials[6] up to quintic polynomials are used. Each Lagrange polynomial gives unity at one point and passes through N nodes on a tetrahedron. The 4, 8 and 20 nodes for linear, quadratic and cubic Lagrange functions (respectively) are shown in figure 4. In our computations the maximal number of basis functions on each tetrahedron is 56 for quintic polynomials.

The local error estimator used to select the tetrahedra to be refined is based on FEM expansion in these local polynomials. An exact expression for the discretization error on the ith tetrahedron is given by

$$\delta E_i^{(p)} = \langle (\Psi^{(p)} + \Psi), (H - \gamma)\delta\Psi_i^{(p)} \rangle \quad , \tag{3}$$

where $\Psi^{(p)}$ denotes the FEM approximation for a given grid and Lagrange polynomials of the order p. The difference between the exact wave function Ψ and the FEM approximation $\Psi^{(p)}$ in the tetrahedron i is given by $\delta\Psi_i^{(p)}$. The real shift parameter γ is arbitrary and can be used to replace the semi-bounded operator H by the positive definite (elliptic) operator $H - \gamma$ by setting $\gamma \leq E_0 := \sup\{e \in R \mid e \in \sigma(H)\}$. Note, that $H - \gamma$ can be replaced by the (positive definite) operator for any physical property. Since the exact wave function and E_0 are not known, they are replaced by the next order FEM approximation, i.e. $\Psi^{(p+1)}$ and $E^{(p+1)} = \langle \Psi^{(p+1)}, H\Psi^{(p+1)} \rangle$.

The precision required of the FEM approximation determines the number of grid points in the final grid constructed by the self–adaptive method. High precision results (relative precision better 10^{-10}) lead to dense grids and vice versa to general matrix eigenvalue problems for very large sparse matrices. Matrices with dimensions up to 5×10^5 have been used in adaptive FEM calculations and in these cases obviously iterative solvers have to be used. We implemented several iterative methods; most commonly used is the inverse vector iteration with an conjugated gradient method preconditioned with the successive over-relaxation method[8].

NUMERICAL DETAILS AND RESULTS

The most cumbersome term in the Hamiltonian (1) is the Coulomb potential for the particle–antiparticle attraction. The singularity at $x = y = z = 0$ produces a cusp of the

Table 1. The first eight states of positronium ($e^- e^+$) in crossed electric and magnetic fields for $K = 0.4$ and $B = 5 \times 10^{-5}$.

$n =$	state	Energy	$\langle x \rangle$	type of state
1	$1s$	-0.209999994	-0.00036	Coulomb
2	$2s + 2p_x$	-0.022560000	-6.013	Coulomb
3	$2p_z$	-0.022500027	-0.0125	Coulomb
4	$2p_y$	-0.022500002	-0.0125	Coulomb
5	$2s - 2p_x$	-0.022440000	5.987	Coulomb
6	$n_z = 0$	-0.000074135	-7987.60	outer well
7	$n_z = 1$	-0.000072188	-7987.89	outer well
8	$n_z = 2$	-0.000070275	-7988.18	outer well

wave function at the origin. This reduces the regularity of the solution and destroys the local convergence of the solution for an increasing order of the Lagrange polynomials. Note, that the local convergence for $p \to \infty$ is a condition for the applicability of the local error estimator. Moreover, the Coulomb term can not be exactly integrated by the numerical Gauß integration used to evaluate the matrix elements. In order to avoid these problems we have to isolate the Coulomb singularity in a small tetrahedron. The initial domain (2) can be described by two tetrahedra. Starting with this very coarse grid we successively refine only the tetrahedron containing the Coulomb singularity. The Coulomb ground state energy converged sufficiently after 23 such refinement steps and the resulting grid with 212 grid points is used for an further multi-level refinement according to the local error estimator. For the outer well states a similar convergence criterion is applied to achieve a starting grid for the automatic grid refinement. The first 8 states of the Hamiltonian (1) have been computed for $K = 0.4$ and $B = 5 \times 10^{-5}$, see table 1. Energy eigenvalues, expectation values for various operators and all dipole transition moments have been computed. The list of states include Coulomb as well as outer well states. The results are in good agreement with low order perturbation calculations and are consistent with an isolated potential picture (at least for this choice of parameters). The overlap between the Coulomb wave functions and the outer well wave functions is nearly zero and the transition between these states will be negligible. Since the wave function for the outer well states drops to zero for $x \to 0$, a direct annihilation is also suppressed. Lifetimes of the order of one year result for the outer well states.

CONCLUSION

The adaptive FEM turns out to be a general and powerful method for the numerical solution of Schrödinger eigenvalue problems in three space dimensions. For standard problems the method is competitive with standard global basis set expansions. In the case of new physical problems for which no global basis sets are established, such as the stable particle-antiparticle problem discussed above, a direct solution using the adaptive FEM is the most reliable method. The grid can be adapted to any shape of the wave function. Using special local error estimators the grid can be optimized according to any physical property of one or several wave functions. Local properties can be studied more systematically and errors can be controlled. Moreover, the adaptive FEM is based on firm mathematical grounds. No nonlinear parameters are used and no *a priori* knowledge is necessary.

A disadvantage of the adaptive finite element method is the rather non–compact description of the wave function. General eigenvalue problems for very large sparse matrices have to be solved. Iterative methods have to be used and to achieve convergence for the wave function is a numerical challange, in particular for the problem discussed above. The outer well and Coulomb states have very different spectra. The distances between the eigenvalues for both types of states differ by several orders of magnitude. Thus, a simultaneous description of outer well as well as Coulomb states lead to convergence problems for any iterative solver.

Finally, let us turn to the physical aspect of the result presented in ref 1. Crossed fields provide a unique way of stabilizing simple bound matter–antimatter systems. Stable bound matter–antimatter systems have never been proposed before and this result may have interesting consequences in many physical fields. For laboratory field strength the system need to be studied theoretically in detail, however, the first experimental production of such systems in laboratory may soon become feasible. Of special interest are saddle states which are neither of the Coulomb type nor outer well states. Work in this direction is in progress and will be published elsewhere.

REFERENCES

1. Ackermann, J., Shertzer, J., and Schmelcher P., 1997, Long–lived states of positronium in crossed electric and magnetic fields, *Phys. Rev. Lett.* 78:199
2. Avron J.E., Herbst I.W., and Simon B., 1978, Separation of center of mass in homogeneous magnetic fields, *Ann. Phys. (N.Y.)* 114:431
3. Dippel O., Schmelcher P., and Cederbaum L.S., 1994, Charged anisotropic harmonic oscillator and the hydrogen atom in crossed fields, *Phys. Rev. A* 49:4415
4. Ackermann, J., and Roitzsch, R., 1993, A two–dimensional multilevel adaptive finite element method for the time–independent Schrödinger–equation, *Chem. Phys. Lett.* 214:109
5. Ackermann J., Erdmann, B., and Roitzsch, R., 1994, A self–adaptive multilevel finite element mehod for the stationary Schrödinger equation in three space dimensions, *J. Chem. Phys.* 101:7643.
6. Zienkiewicz O.C., and Taylor R.L., 1989, The finite element method, 4th ed. (McGraw–Hill,London), Vol.1
7. Erdmann, B., Lang J., and Roitzsch, R., 1993, KASKADE Manual, Technical Report 93-5, Konrad–Zuse-Zentrum, Berlin
8. Golub, H.G., and Van Loan, Ch.F., 1989, Matrix computation, (John Hopkins University Press, Baltimore, MD)

A COMPUTATIONAL METHOD FOR QUANTUM DYNAMICS OF A THREE-DIMENSIONAL ATOM IN STRONG FIELDS

Vladimir S. Melezhik

Joint Institute for Nuclear Research,
Dubna, Moscow Region, 141980,
Russia

INTRODUCTION

In this report we give a brief consideration of a nonperturbative method, developed during the last few years, for the quantum dynamics of a three-dimensional atom in strong external fields.

So far the technique has been developed for both the stationary and the time-dependent Schrödinger equations and applied to the calculation of binding energies and transition probabilities (bound-bound and bound-continuum transitions have been considered). The essential elements of the approach, the method's attractive features and its origins are discussed in the next section. The efficiency of the method for different quantum mechanical problems, arising as a hydrogen-like atom interacts with strong external fields, is demonstrated in the subsequent section by using, as an example, a few unique results obtained with this approach. A short conclusion is given in the last section.

PRINCIPAL ELEMENTS OF THE METHOD

We take an interest in solving the three-dimensional ($X = (R, \theta, \phi)$) Schrödinger equation for a charged particle in the Coulomb $U_c(R)$ and the external (non-stationary in general case) $\Delta U(\vec{R}, t)$ fields, which do not permit the use of conventional methods of quantum mechanics such as variable separation or perturbation theory.

The key idea of the method[1,2] is an expansion of seeking wave function $\psi(\vec{R}, t)$ over the two-dimensional basis $\varphi_\nu(\Omega)$:

$$\psi(R,\Omega,t) = \frac{1}{R}\sum_{\nu j}^{N} \varphi_\nu(\Omega)\,\varphi_{\nu j}^{-1}\,\psi_j(R,t)\,,\; \varphi_\nu(\Omega) = P_l^m(\theta)\begin{cases} \cos m\phi, & m>0 \\ \frac{1}{\sqrt{2}}, & m=0 \\ \sin m\phi, & m<0, \end{cases} \tag{1}$$

where $P_l^m(\theta)$ are the associated Legendre polynomials, $\nu = \{l, m\}$ and $\varphi_{\nu j}^{-1}$ is the $N \times N$ matrix inverse to $\{\varphi_\nu(\Omega_j)\}$ defined on the grid $\Omega_j = \{\theta_j\, \phi_j\}_1^N$. The coordinates of the grid points are the nodes of the Gauss quadrature over the variables θ and ϕ. Thus, the problem is reduced to a system of Schrödinger-type equations with respect to the vector $\vec{\Psi}(R,t) = \{\sqrt{\lambda_j}\psi_j(R,t)\}_1^N = \{\sqrt{\lambda_j}\psi(R,\Omega_j,t)\}_1^N$ of unknown coefficients in(1)

$$i\frac{\partial}{\partial t}\vec{\Psi}(R,t) = \hat{H}(R,t)\,\vec{\Psi}(R,t), \qquad (2)$$

where

$$H_{kj}(R,t) = \{-\frac{1}{2M}\frac{\partial^2}{\partial R^2} + U_c(R) + \Delta U(R,\Omega_k,t)\}\delta_{kj} + \frac{1}{2MR^2}L^2_{kj}\,, \qquad (3)$$

$$L^2_{kj} = \frac{1}{\sqrt{\lambda_k\lambda_j}}\sum_{\nu=\{l,m\}}^{N} \varphi^{-1}_{\nu k}\, l(l+1)\,\varphi^{-1}_{\nu j}\,, \sum_{\nu=1}^{N} = \sum_{m=-(\sqrt{N}-1)/2}^{(\sqrt{N}-1)/2}\ \sum_{l=|m|}^{|m|+(\sqrt{N}+1)/2}$$

and λ_j are the weights of the Gauss quadrature.

In this approach the matrix $(U_c(R) + \Delta U(R,\Omega_k,t))\delta_{kj}$ of the interaction potential is diagonal and the only non-diagonal part is the operator L^2_{kj} of the angular momentum square.

First, the technique was developed for the stationary Schrödinger equation with the potential $\Delta U(R,\Omega_k) = \{\frac{1}{2M}(\vec{B}\cdot\vec{R}) + \frac{1}{8M}[\vec{B}\vec{R}]^2 + (\vec{F}\cdot\vec{R})\}\,|_{\Omega=\Omega_k}$, describing a hydrogen atom in external magnetic $\vec{B}$ and electric $\vec{F}$ fields of arbitrary mutual orientation. In this case the problem is reduced to solving of the eigenvalue problem[3]

$$\hat{H}(R)\vec{\Psi}(R) = \varepsilon\vec{\Psi}(R) \qquad (4)$$

for the system of N ordinary differential equations with the matrix elements $H_{kj}(R)$ defined by formulae(3). Note that the method of global approximation on a subspace grid Ω_k of the multidimensional wave function $\Psi(R,\Omega)$, first applied to a hydrogen atom in crossed fields[3], belongs to the class of the methods of discrete variable (a list of papers on the methods one can find in[4]) and is essentially close to "discrete variable representation"[5], "pseudospectral"[6] and Lagrange-mesh methods[7].

For the initial value problem (2),(3), an economical algorithm with the computational time approximately proportional to the number of unknowns N has been developed recently[2]. It is based on the splitting-up method suggested by G. Marchuk[8] for propagation in time $t_n \to t_{n+1} = t_n + \Delta t$

$$\vec{\Psi}(R,t_n+\Delta t) = (1+i\frac{\Delta t}{2}\hat{h})^{-1}(1-i\frac{\Delta t}{2}\hat{h})(1+i\frac{\Delta t}{2}\hat{H}_0)^{-1}(1-i\frac{\Delta t}{2}\hat{H}_0)\vec{\Psi}(R,t_n), \qquad (5)$$

where the operators $\hat{H}_0$ and $\hat{h}$ are defined as

$$\hat{H}_0(R) = \hat{H}(R,t) - \hat{h}(R,t), h_{k,j}(R,t) = \Delta U(R,\Omega_k,t)\delta_{kj}.$$

Thus, the problem is split-up into the boundary value problem

$$(1+i\frac{\Delta t}{2}\hat{H}_0(R))\vec{\Psi}(R,t_n+\frac{\Delta t}{2}) = (1-i\frac{\Delta t}{2}\hat{H}_0(R))\vec{\Psi}(R,t_n), \qquad (6)$$

$$\vec{\Psi}(R=0,t_n+\frac{\Delta t}{2}) = \vec{\Psi}(R=R_m,t_n+\frac{\Delta t}{2}) = 0\,, R_m \to \infty\,,$$

which may be diagonalised by the simple unitary transformation $S_{k\nu} = \varphi_\nu(\Omega_k)\sqrt{\lambda_k}$, and the system of N uncoupled algebraic equations

$$(1+i\frac{\Delta t}{2}\hat{h}(R,t_n))\vec{\Psi}(R,t_n+\Delta t) = (1-i\frac{\Delta t}{2}\hat{h}(R,t_n))\vec{\Psi}(R,t_n+\frac{\Delta t}{2}) \qquad (7)$$

with the diagonal matrix $h_{kj}(R, t_n)$. Applying the split-operator method(6),(7) to problem(2),(3) demands that the two-dimensional basis $\varphi_\nu(\Omega)$, used in (1), should be orthogonal on the grid Ω_k. The total CPU time, spent for the step $t_n \rightarrow t_{n+1}$ (5), is a sum of the times for computation of boundary value problem (6) $\sim \alpha N^2 + \beta N$ (in this problem the number of the operations for diagonalization of $\hat{H}_0$ is $\sim \alpha N^2$ and the number of the remaining operations is $\sim \beta N$) and uncoupled algebraic equations (7) $\sim \gamma N$. Due to the simplicity of the diagonalising procedure for the $\hat{H}_0(R)$ operator (which leads to $\alpha \ll \beta, \gamma$), the CPU time for solving problem (6),(7) is proportional to

$$(\alpha N^2 + \beta N) + \gamma N \approx (\beta + \gamma)N$$

for not very large N.

It is proven[8,9] that the above scheme is unconditionally stable, saves unitarity and, for the commuting operators $\hat{H}_0$ and $\hat{h}$, has the same order of accuracy $\sim O(\Delta t^2)$ as the conventional Crank-Nickolson algorithm. For non-commuting operators the accuracy becomes worse, $\sim O(\Delta t)$, but may be improved[9] to $\sim O(\Delta t^2)$ by alternating the sequence of the actions of the operators $\hat{H}_0$ and $\hat{h}$ (in time-evolution operator (5)) on two neighboring time-steps $t_{n-1} \rightarrow t_n = t_{n-1} + \Delta t$ and $t_n \rightarrow t_{n+1} = t_n + \Delta t$.

RESULTS AND DISCUSSION

With this approach a few unique results have been obtained for atoms in strong static and alternating fields. The research includes a rather broad set of quantum properties: energy spectrum of bound states, bound-bound and bound-continuum transitions.

Energy levels

First, the method was applied to low-lying energy levels of a hydrogen atom in strong external electric and magnetic fields arbitrarily oriented to one another[3]. The rearrangement of the low-lying part of the spectrum due to change in the mutual orientation of the fields was analyzed in the nonperturbative region for the first time[3]. In application to the classical problem of a hydrogen atom in a strong magnetic field the results we obtained are in agreement with the more advanced variational calculations for ground states and partly improve the data of the Tübingen group for low-lying excited states[10].

In the paper[11] the method was extended to Rydberg states by using, as an example, highly excited states of a Coulomb particle in a van der Waals field.

Note two attractive features that manifested themselves in these calculations[3,11]: the rapid convergence over N and the simplicity of generation of the matrix of effective potentials(3) in the system of Schrödinger-type equations (4).

Bound-Bound Transitions

The method was independently extended for bound-bound transitions in papers[11,12]. Because formula (1) gives a smooth interpolation procedure over the calculated coefficients $\psi(R, \Omega_j)$ for the sought-for wave function $\psi(R, \Omega)$, the method was rather effective in computation of the oscillator strengths for a broad class of the problems of atomic interaction with strong external fields[11–13]. The oscillator strengths were analyzed in a nonperturbative region for a hydrogen atom in a van der Waals field[11]. For the systematic analysis of stationary lines of a hydrogen-like atom in arbitrary electric and magnetic fields (which includes computation of wavelengths and oscillator strengths), a computer program was created by the Tübingen group[12,13] with a finite element approximation over the radial variable R for system of equations (4) instead of four-order finite-difference approximation used in[3,11].

Bound-Continuum Transitions

Recent development of the method has been done in application to nonstationary effects in external fields.

A good test for the approach in this kind of the problems is a classical effect of hydrogen atom ionization by a constant electric field. In this case the interaction Hamiltonian $\Delta U(R,\Omega_k) = (\vec{F}\cdot\vec{R})\mid_{\Omega=\Omega_k}$ in(3) does not depend on time. However, it is well known that switching-on of the external DC field leads to propagation to the anode of a part of the electron wave-packet, tunneling the screened Coulomb barrier. The results presented in Fig.1 for the ground and Rydberg states demonstrate the applicability of the method to quantitative analysis of the time evolution of the states populations and ionization of the atom. Particularly, one can see in Fig.1 that the ionization $\Gamma_{1s} = -\frac{1}{P(t)}\frac{dP(t)}{dt} \simeq 0.0145$ from the ground state (for $F = 0.1$), evaluated by the decay rate of the norm $P(t) = \mid<\vec{\Psi}(R,t)\mid\vec{\Psi}(R,t)>\mid$ with absorbing boundary conditions at $R_m \simeq 100$, agrees quite well with the recent numerical calculations of the widths $\Gamma = 0.01453811352$ of the Stark resonances in hydrogen[14].

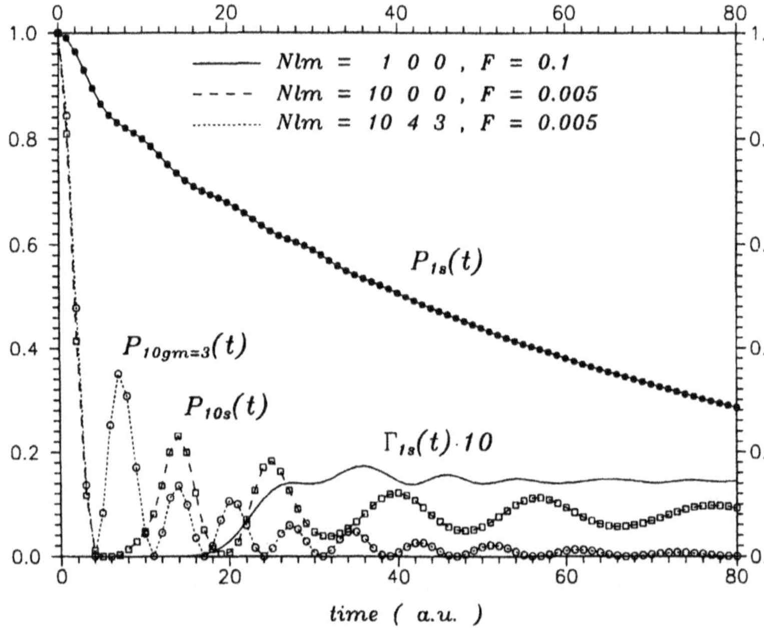

Figure 1. Calculated evolution of the populations $P_{Nlm}(t)$ of the ground and two excited ($N,l,m = 10,0,0$ and $10,4,3$) states stimulated by the switching-on of the external DC field F at $t = 0$. Computations are performed by (6),(7) with $N = 25$, $N_R = 2000$ ($R_m = 100$ for the ground and $R_m = 400$ for the excited states) and $\Delta t = 0.0125$.

With this method, a few unique results have been obtained[2] in the problem of harmonic generation ($n\hbar\omega$) by a hydrogen atom and its excitation in a strong laser field ($\hbar\omega$)

$$(H)_{1s} + \hbar\omega \rightarrow (H)_{Nlm} + n\hbar\omega.$$

A quantitative description of the atom in an elliptically polarized field

$$\Delta U(x,y,z,t) = \frac{E_0}{\sqrt{1+\varepsilon^2}} f(t)[z sin\omega t + \varepsilon y cos\omega t] \tag{8}$$

(where E_0 and ε are the amplitude and the ellipticity of the pulse, ω and $f(t)$ denote the laser frequency and the envelope of the pulse) required the nonperturbative approach(6),(7) for solving time-dependent four-dimensional Schrödinger equation(2). Most results were obtained in the attempt to solve the problem only with linearly polarized fields ($\varepsilon = 0$), i.e. for a three-dimensional case. In these calculations only the shape of the spectrum of the emitted harmonics ($n\hbar\omega$) and the cut-off position were analyzed[15]. Recently a few experimental groups[16] have begun to investigate more refined effects of the polarization of the emitted radiation in an elliptically polarized laser field. Treating of these effects has demanded development of more refined theoretical models[17]. In the paper[2] a quantitative analysis of the polarization properties of the hydrogen atom's radiation was done without additional model simplifications for an elliptically polarized laser field with laser intensity $I = E_0^2 c/(8\pi) = 2 \times 10^{13}\, W/cm^2$ and wavelengths $1064nm$ and $532nm$. The parameters E_0 and ω were chosen in the region $U_p = E_0^2/4\omega^2 \ll I_p = 1/2$ outside the "tunneling limit" (the region of the validity of the two-step model[17]) and the applicability region of perturbation theory.

The spectrum of the emitted radiation $\mid d_{z'}(\omega) \mid^2$ in the $\vec{z}'/\mid \vec{z}' \mid = (0, sin\theta', cos\theta')$ direction (the vector $\vec{z}'$ is rotated in the plane $x = 0$ of the polarization ellipse of the laser) was evaluated by the Fourier transform for the induced dipole of the atom

$$\mid d_{z'}(\omega) \mid^2 = \mid \int dt\, e^{-i\omega t} \int d\vec{R}\, \psi^*(\vec{R},t) \frac{(\vec{z}' \cdot \vec{R})}{\mid \vec{z}' \mid} \psi(\vec{R},t) \mid^2 . \tag{9}$$

The calculations revealed considerable alteration of the polarization (i.e. variation in the ellipticity and rotation of the polarization ellipse) of some of the radiated harmonics ($n\hbar\omega$) with respect to the laser polarization.

The rapid convergence of the method over N and the proportionality of the CPU time to N (see Fig.2) confirmed in this computation[2], indicates the possible use of the method in more complicated problems of atom-laser interaction.

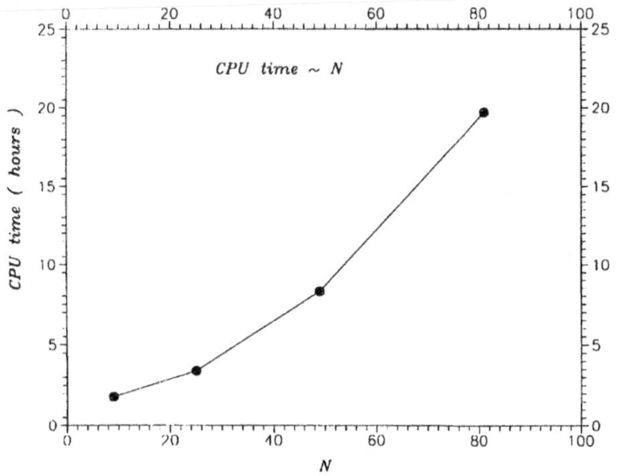

Figure 2. Dependence of the CPU time on the number of the terms N in (1) obtained in calculation of the harmonic generation by the hydrogen atom in field(8). The calculations were performed for $N_R \simeq 10^3$ and $N_T \simeq 10^5$ on DEC 3000-M600 work-station.

CONCLUSION

In this report we briefly reviewed a set of quantum problems solved with the nonperturbative approach developed for hydrogen-like atoms in strong external fields. The high efficiency and the flexibility of the method found in these computations, suggest the method development in application to other important problems of physics of "atoms and molecules in strong external fields". A natural step in this direction is an extension of the approach to the scattering problem[18] for a Coulomb particle in an external field (i.e. an extension to continuum-continuum transitions). Another problem, where the method may be fruitful, is a helium-like atom[19] (i.e. an extension to more complicated quantum systems).

Note also an interesting application of the approach to nonperturbative analysis of "shaking-off" reactions in muon physics[1] and in other few-body problems.

I am grateful to W. Schweizer and P. Schmelcher for fruitful suggestions and discussions which considerably stimulated the writing of the present report.

REFERENCES

1. V.S. Melezhik, New approach to the old problem of muon sticking in μCF, *Hyperfine Interaction* 101/102:365 (1996).
2. V.S. Melezhik, Polarization of harmonics generated from a hydrogen atom in a strong laser field, *Phys. Lett.* A230:203 (1997).
3. V.S. Melezhik, Three-dimensional hydrogen atom in crossed magnetic and electric fields, *Phys. Rev.* A48:4528 (1993).
4. D. Baye, Constant-step Lagrange meshes for central potentials, *J. Phys* B28:4399 (1995).
5. J.V. Lill, G.A. Parker, and J.C. Light, Discrete variable representations and sudden models in quantum scattering theory, *Chem. Phys. Lett.* 89:483 (1982).
6. R.A. Friesner, Solution of self-consistent field electronic structure equations by a pseudospectral method, *Chem. Phys. Lett.* 116:39 (1985).
7. D. Baye and P-H. Heenen, Generalised meshes for quantum mechanical problems, *J. Phys.* A19:2041 (1986).
8. G.I. Marchuk, On the theory of the splitting-up method *in* "Partial Differential Equations. II. SYNSPADE-1970," Academic Press, N.Y., L. (1971).
9. G.I. Marchuk."Methods of Computational Mathematics," Nauka, Moscow (1978) (*in Russian*).
10. W. Rösner, G. Wunner, H. Herold, and H. Ruder, Hydrogen atoms in arbitrary magnetic fields: I. Energy levels and wave functions, *J. Phys.* B17:29 (1984).
11. V.S. Melezhik, Nonperturbative behavior of a hydrogen atom in a van der Waals field, *Phys. Rev.* A52:R3393 (1995).
12. P. Fassbinder and W. Schweizer, Hydrogen atom in very strong magnetic and electric fields, *Phys. Rev.* A53:2135 (1996).
13. P. Fassbinder and W. Schweizer, Stationary hydrogen lines in magnetic and electric fields of white dwarf stars, *Astron. Astrophys.* 314:700 (1996).
14. F. M. Fernandez, Direct calculation of Stark resonances in hydrogen, *Phys. Rev.* A54:1206 (1996).
15. J.L. Krause, K.J. Schafer, and K.C. Kulander, Calculation of photoemission from atoms subject to intense laser field, *Phys. Rev.* A45:4998 (1992).
16. F.A. Weihe, S.K. Dutta, G. Korn, D. Du, P.H. Bucksbaum, and P.L. Shkolnikov, Polarization of high-intensity high-harmonic generation, *Phys. Rev.* A51:R3433 (1995).
17. P. Antoine, A. L'Huillier, M. Lewenstein, P. Salieres, and B. Carre, Theory of high-order harmonic generation by an elliptically polarized laser field, *Phys. Rev.* A53:1725 (1996).
18. V.S. Melezhik, New method for solving multidimensional scattering problem, *J. Comput. Phys.* 92:67 (1991).
19. V.S. Melezhik, New method for solving three-dimensional Schrödinger equation, *Nuovo Cimento* B106:537 (1991).

DISCRETIZATION TECHNIQUES APPLIED TO ATOMS UNDER EXTREME CONDITIONS

Wolfgang Schweizer[1,2], Matthias Stehle[1], Peter Faßbinder[1], Stefan Kulla[1], Ingo Seipp[1], and Rosario Gonzalez[1,3]

[1]Institut für Astronomie und Astrophysik, Theoretische Astrophysik Universität Tübingen, D-72076 Tübingen, Germany
[2]Fakultät für Physik und Astronomie, Theoretische Physik I, Ruhr-Universität Bochum, D-44780 Bochum, Germany
[3]Instituto 'Carlos I' de Fisica Teorica y Computacional Universidad de Granada, E-18071 Granada, Spain

INTRODUCTION

The availability of high-speed computers has opened the door to study non-integrable quantum systems. One of the fascinating aspects of non-integrable, low-dimensional systems is given by their relation to quantumchaology, but also to fundamental research in all areas of physics, e.g., in astrophysics, theoretical atomic physics or solid state physics.

We begin with the time-independent Schrödinger equation for a three-dimensional one-particle system,

$$\left[-\frac{\hbar^2}{2m}\Delta + V(\vec{r})\right]\psi_E(\vec{r}) = E\psi_E(\vec{r}) \quad , \tag{1}$$

where the wave function in a stationary state with energy E is denoted by $\psi_E(\vec{r})$. In this connection the question arises in what curvilinear coordinates the Schrödinger equation (1) can be separated. By the methods of group theory (Miller 1968) it can be shown, that in three dimensions the Laplace-Beltrami operator Δ can be separated in exactly 11 different curvilinear coordinates. For each of this coordinates the potential $V(\vec{r})$ have to fulfill certain properties, that the Schrödinger equation becomes separable.

For all orthogonal curvilinear coordinates q_j in three dimensions the following conditions hold:

$$\begin{aligned} ds^2 &= dx^2 + dy^2 + dz^2, \text{ with } (x,y,z) \text{ cartesian coordinates} \\ ds^2 &= \sum_{i,j=1}^{3} g_{ij}\,dq_i dq_j \quad \text{and} \quad g_{ij} = \delta_{i,j} g_{ii} \\ \Delta_{q1,q2,q3} &= \frac{1}{\sqrt{g}} \sum_{i=1}^{3} \frac{\partial}{\partial q_i}\left(\frac{\sqrt{g}}{g_{ii}}\frac{\partial}{\partial q_i}\right) \quad (\text{Laplace} - \text{Beltrami Operator}) \end{aligned}$$

$$dV = \sqrt{g}dq_1dq_2dq_3 \quad \text{(volume element)}$$

with $g = g_{11}g_{22}g_{33}$, and the quantum system becomes separable, if the potential can be written as

$$V(q_1, q_2, q_3) = \sum_{i=1}^{3} \frac{1}{g_{ii}} V(q_i) \tag{2}$$

For separable systems the eigenvalues (separation constants) of each of the 3 unidimensional (in general n dimensional) differential equations can be used to label the eigenfunctions $\psi(\vec{r})$ and hence serve as quantum numbers. Integrability of a n-dimensional Hamiltonian system requires the existence of n commuting observables O_i, $1 \le i \le n$. Each separable system is integrable, but not vice versa. Non-integrability occurs if the number of commuting observables is smaller than the number of degrees of freedom of the system and hence requires solving at least a two-dimensional partial differential equation or a coupled system of unidimensional differential equations.

For non-integrable systems there are mainly two textbook methods to compute eigensolutions. By rewriting the potential into an integrabel V_{int} and into a nonintegrable part V_{non}, with V_{non} sufficiently weak, solutions can be obtained either by pertubation theory or by solving the eigenvalue problem of the Hamiltonian matrix calculated in a suitable basis. If V_{non} becomes comparable with V_{int} both methods hardly converge and even before becomes very costly in terms of computational time. Discretization methods are more efficient to compute eigensolutions of non-integrable systems. Methods based on basis expansions, like pertubational or direct diagonalization treatments, suffer from the use of the globally defined basis, which is only correct for systems of certain symmetry. In contrast to that, a locally defined basis can be optimized with respect to the "true" system under consideration.

Non-integrability is rather the rule than the exception. Prominent examples are the hydrogen atom in strong external fields under astrophysical aspects (see Friedrich et al. and P. Faßbinder et al. this proceedings), the helium atom in strong external fields (see Braun et al. this proceedings), but also alkali atoms in external fields in quest of quantum chaos (Jans et al. 1993) or the hydrogen atom in front of a metal surface (Ganesan et al. 1996).

METHODS

Recent years have seen tremendous progress in studies of the properties of atoms in strong magnetic fields. Decisive stimulus came from the discovery of huge magnetic fields in the vicinity of white dwarf stars ($B \approx 10^2 - 10^5$ T) and neutron stars ($B \approx 10^7 - 10^9$ T). At these field strengths the magnetic forces acting on an atomic electron outweigh the Coulomb binding forces, even for low-lying energy values of the hydrogen atom. Thus atomic structure is completely changed and it is evident that pertubation theory is no longer applicable and a recalculation of the atomic structure with more advanced numerical methods is necessary.

BOUND STATES

The Hamiltonian for a hydrogen atom in an external uniform magnetic field along the z-axis and an additional electric field reads as follows (atomic units and spherical coordinates):

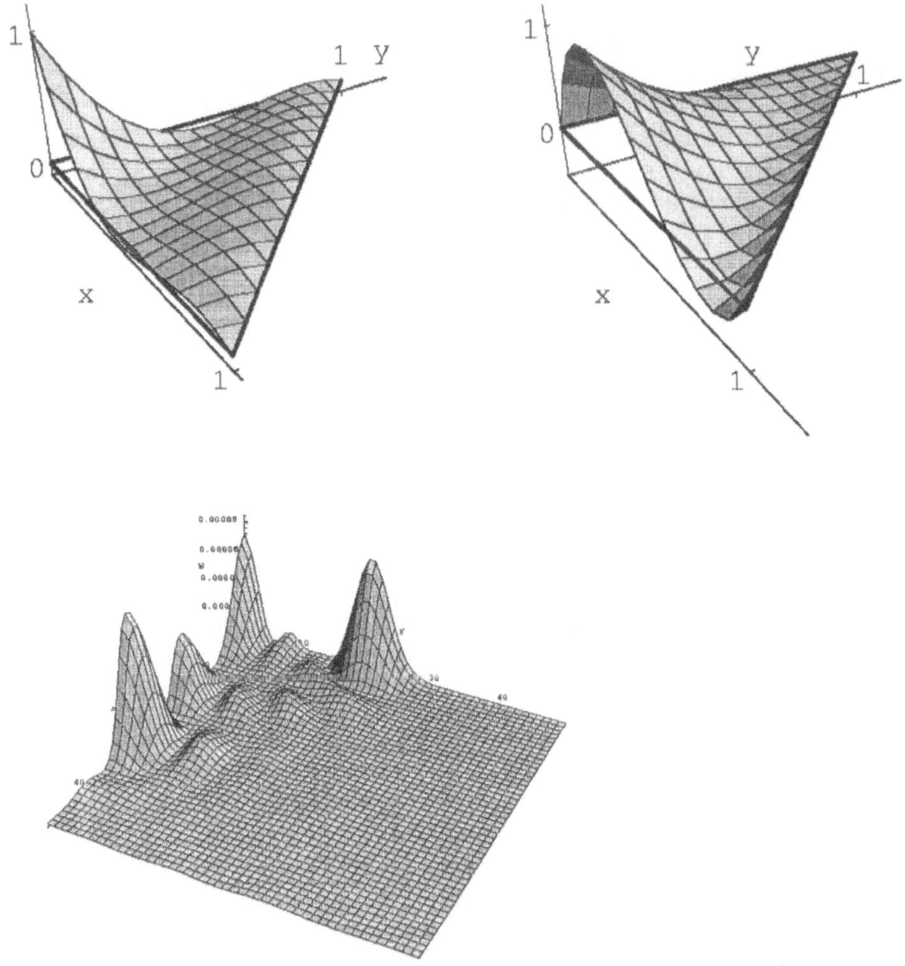

Figure 1. This figure shows as an example on top cubic form-functions used for the finite element calculations and the 25th m=0 eigenstate of the hydrogen atom localized in front of a metal surface in a distance of 50Å.

$$H(r,\vartheta,\varphi) = \frac{p^2}{2} - \frac{1}{r} - i\frac{\beta}{2}\frac{\partial}{\partial\varphi} + \frac{\beta^2}{2}r^2\sin^2\vartheta + \Phi r(\cos\vartheta\cos\delta + \sin\vartheta\cos\varphi\sin\delta) \quad (3)$$

Where the magnetic field axis points into the z-direction and is measured in units of $B_0 = 4.7 \cdot 10^5 T$, $\beta = \frac{B}{B_0}$, $\Phi = |\vec{F}|/F_0$ is the electric field strength measured in units of $F_0 \approx 5.14 \cdot 10^{11}$ and δ is the relative angle between the electric field and the magnetic field axis.

This system has the following qualitative properties: For parallel electric and magnetic fields the magnetic quantum number m is conserved and the problem can be reduced to a two degrees of freedom system. For perpendicular electric and magnetic fields three non-integrable degrees of freedom remain, but z-parity is conserved; whereas for arbitrary oriented fields no discrete symmetry is left. Hence all field free hydrogen eigenstates are coupled, which leads to additional allowed dipol transitions compared to the diamagnetic system.

For an effective numerical treatment we reduced the three-dimensional Schrödinger equation to a system of uni- or two-dimensional differential equations by discretization of the angular momentum operator (Melezhik 1997 and references therein) with respect to the spherical coordinate θ. For parallel electric and magnetic fields the magnetic quantum number m is conserved and hence this procedure results in a one-dimensional system. These differential equations for the radial coordinate are solved by the Finite-Element method (Faßbinder et al. 1996a,b).

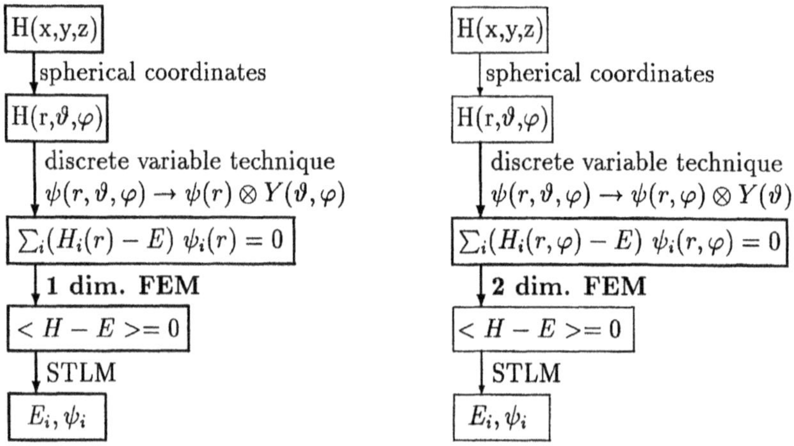

In the figure above we show graphically how we solve the resulted Hamiltonian differential equations. The procedure on the left can be used for any two-dimensional problem and on the right for three-dimensional non-integrable quantum systems, like the hydrogen atom in non-parallel electric and magnetic fields. Finally we compute eigensolutions by solving the Hamiltonian matrix with the spectrum transformed Lanczos method (STLM) of Ruhe and Ericsson (1980). The advantage of this discretization technique described above is its quick convergency which allows calculations on workstations. In contrast, by using Sturmian functions (Wunner et al. 1989) we failed solving the problems mentioned above even with extremely huge basis sizes of up to 325,000, which necessitates huge computer facilities.

ATOMIC RESONANCES

Due to the external electric fields the ionization threshold of the atom is lowered from the field-free value. In fact, all states become quasi-bound, as the electron can ionize by tunneling through the potential states. This process becomes important for states close to the classical ionization energy or above. Combining the complex coordinate method with an R-matrix quantum-defect method we have obtained results for hydrogen under astrophysical quest (Seipp et al. 1997) and for sodium atoms in parallel fields (Seipp et al. 1996).

In the complex coordinate method the real configuration space coordinates are transformed by a complex dilatation. The Hamiltonian of the system is thus continued into the complex plane. This has the effect that, according to the boundaries of the representation, complex resonances are uncovered with square-integrable wavefunctions. This square integrability is achieved through an additional exponentially decreasing term

$$r \to r \exp(i\Theta)\,. \tag{4}$$

After the coordinates entering the Hamiltonian have been transformed, the Hamiltonian is no longer Hermitian and thus can support complex eigenenergies associated with decaying states. The spectrum of a complex-rotated Hamiltonian has the following features:

- Its bound spectrum remains unchanged
- Continuous spectra are rotated about their thresholds into the complex plane by an angle of -2Θ
- Resonances are uncovered by the rotated continuum spectra with complex eigenvalues and square-integrable (complex rotated) eigenfunctions.

In the Stark effect the whole real energy axis is the continuum spectrum and no threshold exists in the unrotated Hamiltonian for the continuous spectrum to rotate about. The Hamiltonian for the hydrogen atom in parallel electric and magnetic fields, after the above complex transformation has been applied, becomes

$$\begin{aligned} H(\Theta) = & - \exp(-2i\Theta)\frac{\nabla^2}{2} - \exp(-i\Theta)\frac{1}{r} \\ & + \frac{1}{2}\gamma(l_z + 2s_z) + \exp(2i\Theta)\frac{1}{8}\gamma^2 r^2 \sin^2\theta + \exp(i\Theta)Fr\cos\theta \end{aligned} \tag{5}$$

using atomic units and spherical coordinates.

By this calculations we found, that the electric field has a strong influence on the spectra of the pure magnetic field. A rich resonance structure is created by the electric field starting from the ionization thresholds of the pure magnetic field. These resonances behave smoothly when the electric field is changed. For astrophysical situations the spectra appear as little modulated lines because the resonances move too quickly with change in the electric field and hence all resonance structures are smeared out due to the statistical origin of the electric fields in the atmosphere of white dwarf stars.

In future calculations we plan to merge the method of complex coordinate rotation with the discretization procedure (Gonzalez et al. 1997) mentioned above and to approximate alkali atoms via phenomenological (Schweizer et al. 1997) potentials. We expect by this combination to have a powerful and quick numerical procedure to calculate resonances under laboratory as well as astrophysical conditions.

THE HYDROGEN ATOM IN FRONT OF A METAL SURFACE

The Hamiltonian of the hydrogen atom in front of a metal surface reads in semi-parabolic coordinates ($\sqrt{(x^2+y^2)} = \mu\nu$, $z = \frac{1}{2}(\mu^2 - \nu^2)$)

$$H(\mu, \nu, p_\mu, p_\nu) = \frac{1}{2}\frac{p_\mu^2 + p_\nu^2}{(\mu^2 + \nu^2)} + V_{ms} \tag{6}$$

with

$$\begin{aligned} V_{ms} = & -\frac{1}{2d} - \frac{4}{\mu^2+\nu^2} + \frac{2}{\sqrt{\mu^2\nu^2 + (2d + \frac{1}{2}(\mu^2\nu^2))}} \\ & - \frac{1}{2(d + \frac{1}{2}(\mu^2 - \nu^2))} \end{aligned} \tag{7}$$

and d the distance from the metal surface. In radial coordinates the nodes of the wave function are quadratically widened, whereas in semiparabolic coordinates the nodes are approximately equally spaced. By expanding the Hamiltonian with respect to the distance d, Ganesan et al. (1996) calculated Rydberg eigenstates and eigenvalues in a Sturmian basis. Such an expansion is unnecessary by a two-dimensional finite element procedure. To solve this problem we used (Kulla et al. 1997) two-dimensional finite elements with quadratic and cubic form-functions. In Fig. 1 we show as an example the 25th eigenfunction of a hydrogen atom placed in a distance of 50 Å in front of a metal surface. Basis set calculations would necessitate huge Taylor-expansions in the Hamiltonian potential leading to serious convergency problems. Hence in conclusion discretization techniques are especially useful for computations in such extrem situations, where the hamiltonian is far from integrable limits, but the wave functions are still not too much structured.

ACKNOWLEDGEMENTS

This work was supported by the Deutsche Forschungsgemeinschaft.

REFERENCES

Ericsson, T., and Ruhe, A., 1980, The spectral transformation Lanczos method for the numerical solution of large sparse generalized symmetric eigenvalue problems, *Math. Comp.* 35:1251

Faßbinder, P., and Schweizer, W., 1996a, Hydrogen atom in very strong magnetic and electric fields, *Phys. Rev. A* 53:2135.

Faßbinder, P., and Schweizer, W., 1996b, Stationary hydrogen lines in magnetic and electric fields of white dwarf stars, *Astron. Astrophys.* 314:700.

Ganesan, K. and Taylor, K. T., 1996, Rydberg states of the hydrogen atom near a metal surface, *J. Phys.* B29: 1293

Gonzalez, R., and Schweizer, W. 1997, Direct calculation of resonances by finite element methods, in preparation.

Jans,W., Monteiro, T. S., Schweizer, W. and Dando, P. A., 1993, Phase-space distribution and spectral properties for non-hydrogenic atoms in magnetic fields, *J. Phys.* A 26: 3187

Kulla, S. and Schweizer, W. 1997, Finite elements for the hydrogen atom near a metal surface, in preparation

Melezhik, V. S., 1997, A computational method for quantum dynamics of a three-dimensional atom in strong fields, these proceedings.

Miller, W., 1968, Lie Theory and Special Functions, Academic Press, NewYork

Schweizer, W., and Faßbinder, P., 1997, Numerical studies of model potentials for alkali atoms, preprint.

Seipp, I., Taylor, K.T., and Schweizer, W., 1996, Atomic resonaces in parallel electric and magnetic fields, *J. Phys.* B29:1

Seipp, I., and Schweizer, W., 1997, Electric fields for hydrogen bound-free transistions in magnetic white dwarfs, *Astron Astrophys.* 318:990

Wunner, G., Schweizer, W., and Ruder, H., 1989, Hydrogen atoms in strong magnetic fields – in the laboratory and in the cosmos, *in*: "The Hydrogen Atom," F. Bassini et al., ed., Springer, Berlin.

COMPUTER–ALGEBRAIC DERIVATION OF ATOMIC FEYNMAN–GOLDSTONE EXPANSIONS

S. Fritzsche, B. Fricke and W.-D. Sepp

Fachbereich Physik, Universität Kassel
Heinrich–Plett–Str. 40, D–34132 Kassel, Germany

INTRODUCTION

Many–body perturbation theory (MBPT) has a long tradition in studying atomic and molecular properties. Beside of the traditional order–by–order approach, the Coupled–Cluster approximation in particular allows for accurate theoretical predictions for even (nearly) neutral systems[1,2]. In both schemes, one may also include additional perturbations like external fields which are treated similar to the interaction among the electrons. Apart from nonrelativistic studies, a number of relativistic perturbation calculations have also been caried out in recent years. These investigations include a variety of rather different properties, for instance, electron correlation energies[3], hyperfine structures and isotope shifts[4], radiative decay rates[5], g–factors[6] as well as the influence of external electric and magnetic fields[7].

During the past, however, most MBPT studies concerned atoms with a simple shell–structure only. This applies, for instance, for the rare gases and for atoms and ions with a single electron or hole apart from closed shells otherwise. Very few computations exist for atoms with two open shells or, more generally, with more than two valence electrons and/or holes, respectively. The main limitation to extend the many–body procedure beyond such simple shell–structures arise from the complexity of any perturbation expansion if more than just one or two *effective* particles appear in the derivation. To overcome this difficulty an efficient treatment of complex expansions is needed. — In this contribution, we will explain the basic steps in the derivation of atomic perturbation expansions. We will demonstrate how computer algebra (CA) can easily be used to facilitate such derivations. The kernel of the present work is a package of Maple procedures for the simplification of operator products in second quantization. A few important features of this CA tool will be shown below.

BASIC FORMULAS

Atomic perturbation expansions are often expressed in terms of Feynman–Goldstone diagrams in Rayleigh–Schrödinger (RS) perturbation theory. In such a graphical representation of some given atomic property, each diagram corresponds uniquely to

Atoms and Molecules in Strong External Fields
Edited by Schmelcher and Schweizer, Plenum Press, New York, 1998

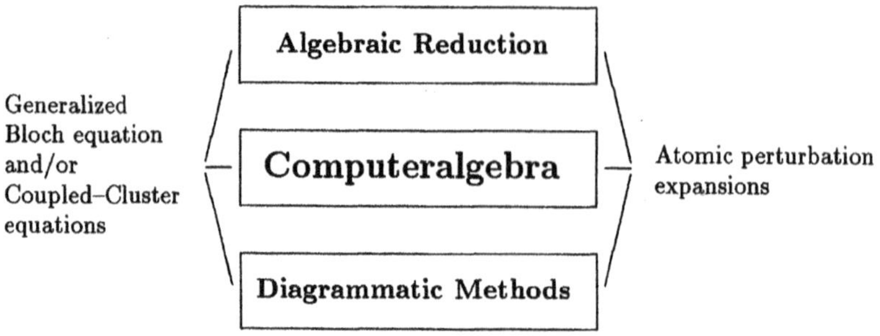

Figure 1. Different paths in the derivation of atomic perturbation expansions.

a certain algebraic expression. To derive perturbation series for open–shell atoms one may start directly from the generalized Bloch equation in intermediate normalization[8]

$$[\Omega,\, H_o]\, P = (V\,\Omega \;-\; \Omega\, P\, V\, \Omega)\, P\,. \tag{1}$$

defined within a multidimensional model space $\mathcal{M}$. This equation for the wave operator Ω is obtained from an *(a priori)* partitioning of the full Hamiltonian $H = H_o + V$ into the model operator H_o and the remaining perturbation V. The operator P projects any state onto the model space $\mathcal{M}$. Usually, H_o is taken to be a sum of *effective* one–electron operators whereby the interaction among the electrons is approximated by a suitable chosen central potential $U(r)$. The model operator must be a reasonable approximation to the full Hamiltonian and should also have eigenvalues and eigenfunctions which are simple to determine. For a given number of atomic states (with a definite projection inside $\mathcal{M}$), eq.(1) is equivalent to Schrödinger's equation of the many–electron atom. The model space $\mathcal{M}$ may in general also include non–degenerate states; a detailed account of the theory has been presented in the textbook of Lindgren and Morrison[9]. By using the wave operator Ω, one can define an *effective* Hamiltonian

$$H_{\text{eff}} \;=\; P\, H\, \Omega\, P \tag{2}$$

which has the same eigenvalues as the true Hamiltonian H but which is not hermitian in intermediate normalization.

We will not further discuss the standard graphical representation of the Bloch equation (1) in a given order and how graphical rules can be derived. The algebraic expression for each diagram in such Feynman–Goldstone expansions immediately reflects the second quantization of all operators. Similar graphs can be applied in the *all–order* and Coupled–Cluster (CC) approximations[1, 2] where the wave operator Ω is partitioned into one–, two–, ..., n–particle exitations. In this paper, we will focus our interest instead onto the computer–algebraic derivation of such expansions. Even though we restrict our discussion below to the order–by–order scheme, our computer–algebraic tools can be applied easily also to the CC approximation for which all basic operators have been implemented as well.

There are three different paths to derive perturbation expansions in second quantization as displayed in Figure 1. Apart from the (manual) stepwise commutation of the creation and annihilation operators by using the well–known fermion anticommutator rules different graphical methods have been developed during the past[10, 9]. A modern alternative is nowadays offered by computer algebra. All three paths serve the same purpose, i.e. the evaluation of operator products in second quantization like they appear in the Bloch equation (1).

Today, the main challenge in the application of RS perturbation theory for open-shell atoms arise from the complexity of the perturbation series for any property if higher orders and/or open shells are considered. In both cases, the number of operator terms in the perturbation expansions often increases drastically. Here, CA may demonstrate its power in the derivation of the formulas. Our aim therefore has been to develop a computer algebra program to facilitate these derivations.

EVALUATION OF PERTURBATION EXPANSIONS

The formal derivation of perturbation series starting from eq.(1) and the definition of the effective Hamiltonian (2) can be summarized by three individual tasks:

- The representation and evaluation of the wave operator Ω in second quantization (within a given approximation scheme),
- the evaluation of the matrix elements of H_{eff}, and (if required)
- the derivation of perturbation expansions for properties other than energy corrections. These expansions usually depend on the wave operator and on the representation of the zero-order wave functions $|\psi_\alpha^o\rangle$ within the model space $\mathcal{M}$ (see below).

These three tasks must be carried out rather independently. Here, we will give only a few remarks on them. A more detailed outline of these steps have been given elsewhere[9, 11]. In general, each of these steps results in a series of Feynman–Goldstone diagrams* where either some operator or matrix element is written as a sum of products of different one- and two-particle matrix elements. These sums of diagrams are the perturbation expansions we are interested in. They can later be calculated explicitly by using any appropriate one-particle spectrum. In most applications, some type of Hartree–(Dirac–)Fock spectrum is used due to its simplicity and due to the property that all one-particle excitations vanish identically. In this work, however, we will not consider this final step of numerical computation.

To provide some deeper insight into the algebraic derivation, it is helpful to give the operator P also a representation in second quantization

$$P = \sum_{\alpha \in \mathcal{M}} |\phi_\alpha\rangle \langle\phi_\alpha| = \sum_{\alpha \in \mathcal{M}} (\hat{\boldsymbol{a}}^\dagger)_\alpha |o\rangle \langle o| (\hat{\boldsymbol{a}})_\alpha , \tag{3}$$

where $(\hat{\boldsymbol{a}}^\dagger)_\alpha$ denotes a *string* of creation and/or annihilation operators in normal order. Such a string *creates* some model state $|\phi_\alpha\rangle = (\hat{\boldsymbol{a}}^\dagger)_\alpha |o\rangle$ out of the (closed-shell) vacuum. $(\hat{\boldsymbol{a}})_\alpha$ is the corresponding adjoint operator string.

The wave operator from the first step can be used to evaluate the effective Hamiltonian as well as any other atomic property. Usually, we need not an explicit representation of H_{eff} in terms of diagrams but want the matrix elements of $H_{\text{eff}}^{(n)} = P\, H\, \Omega^{(n)}\, P$ to a certain order n, evaluated with the basis functions of the model space $|\phi_\alpha\rangle \in \mathcal{M}$. ¿From the diagonalization of this Hamiltonian matrix (up to order n) we namely obtain the energy correction up to the order $(n+1)$ as well as the zero-order wavefunctions

$$|\psi_\alpha^o\rangle = \sum_{\beta \in \mathcal{M}} c_{\alpha\beta}\, |\phi_\beta\rangle \; . \tag{4}$$

Similar, we may evaluate the matrix elements of any (symmetric) one-particle operator $F = \sum_i^N f(\boldsymbol{r}_i)$ or two-particle operator $G = \sum_{i<j}^N g(\boldsymbol{r}_i, \boldsymbol{r}_j)$. In the study of transition

*For convenience, we use the term ***diagram*** also in the (computer–)algebraic derivation even though ***operator term*** would be a more appropriate one.

properties, for instance, they give rise to the effective operators $A_{\text{eff}} = \Omega_f^\dagger F \Omega_i$ and $A_{\text{eff}} = \Omega_f^\dagger G \Omega_i$ where the indices f and i denote the corresponding wave operators for the final and initial atomic states, respectively. By using the representation (4) for the zero-order wave functions, all these matrix elements can eventually be reduced to vacuum amplitudes

$$\langle \phi_\alpha \,|\, A_{\text{eff}} \,|\, \phi_\beta \rangle = \left\langle o \left| (\hat{\mathbf{a}})_\alpha \, A_{\text{eff}} \, (\hat{\mathbf{a}}^\dagger)_\beta \right| o \right\rangle \; . \tag{5}$$

After normal ordering of the full operator string in (5) only those terms will survive where all creation and annihilation operators are completely contracted.

BASIC STEPS IN THE COMPUTER–ALGEBRAIC DERIVATION

In the last section we saw that both, the wave operator Ω as well as the matrix elements of any effective operator can be expressed by a sum of Feynman–Goldstone diagrams. In most practical applications, we are even interested only in vacuum amplitudes [see eq.(5)]. We will now explain the basic steps which have to be carried out during any derivation. For this we use terms wich are particularly appropriate for CA manipulations of all operators. In the next section we will then describe a program package for the derivation of such perturbation expansions, appropriate also for open–shell atoms.

We assume that all operators in the derivation have a valid representation in second quantization and that all matrix elements will be calculated within some orthogonal basis $|\phi_\alpha\rangle \in \mathcal{M}$ of the model space. For the study of transition properties, this implies that both, the initial and final atomic states should have a zero–order representation (4) within $\mathcal{M}$. Then, the basic steps in the computer–algebraic derivation can be summarized as follows:

(i) Evaluation of all operator products on the rhs of the Bloch equation (1) or in the definition of the effective operator A_{eff} (5). This requires to bring all creation and annihilation operators into normal order. The result is a sum of normal–ordered operator terms.

(ii) Definition and representation of the model space $\mathcal{M}$. The character of the model space is specified by the number of occupied valence orbitals and the number of unoccupied valence–core orbitals in (3). For two and more *effective* particles and/or holes it might also be useful to couple the corresponding creation and annihilation operators in (3) to give rise to a definite angular momentum of the model states[12].

(iii) Calculation of the vacuum amplitudes on the rhs of (5). We obtain the effective operator A_{eff} from step (i) and the operator strings $(\hat{\mathbf{a}})_\alpha$, $(\hat{\mathbf{a}}^\dagger)_\beta$ from the definition of the model space. In contrast to the standard graphical representation of the Feynman–Goldstone diagrams there is no implicit summation involved in the definition of the model functions, i.e. for the strings $(\hat{\mathbf{a}})_\alpha$ and $(\hat{\mathbf{a}}^\dagger)_\beta$. This step (iii) yields a (completely contracted) sum of products of one– and two–electron matrix elements. In some cases, it might be appropriate to combine the two steps (ii) and (iii).

(iv) Angular reduction of all operator terms. By using a standard spherical representation of the one–particle spectrum (and a closed–shell reference state $|o\rangle$), the integration over all angular variables in the matrix elements can be performed analytically. The values of these integrals are often called *angular coefficients.*

After these four steps have been carried out, only integrals which depend on either one or two radial variables are left to be computed. The values of these *radial integrals* have to be multiplied by the corresponding angular coefficients and have to be summed over all allowed combinations of the particle–hole indices. The computation of the radial integrals is the most time–consuming part in practise and need to be performed with some algorithmic language like FORTRAN. This, however, is outside of the scope of the present work. We will now introduce the program APEX for the derivation of general *atomic perturbation expansions*.

APEX — A TOOL TO DERIVE PERTURBATION EXPANSIONS

The four steps in the derivation of perturbation series can be carried out by the help of computer algebra. To facilitate this procedure we developed the package APEX within the Maple† environment. The derivation of perturbation expansions using CA tools has some obvious advantages in comparison to diagrammatic methods. This concerns in particular the treatment of phases and weight factors. Moreover, CA also allows a syntax for the derivation which closely resembles the formal mathematical language. Once the work of a CA tool has been establihed, it also garanties a higher reliability compared with a manual derivation. In the interactive program mode, such tools even provide more transparency since all individual steps of the derivation can easily be followed.

The program APEX allows the evaluation of general operator products in second quantization. It is entirely written for interactive work. Some of the tasks which need often to be done during the evaluation can be seen immediately from the steps (i–iv) above:

- Definition of primitive operators like V, Ω, F, G, ... in second quantization.
- Evaluation of adjoint operators.
- Specification of the model space occupation.
- Evaluation of normal products as well as of operator products which also include all contracted terms.
- Classification and selection of different operator terms (diagrams).
- Transformation between different representations (orbital representation, radial–spherical representation, concept of pair functions, etc.).
- Task–adapted printout of the results.

By contrast, many other elementary tasks in a computer algebra approach are hidden to the user. For example, Wick's theorem has been implemented on such a hidden level. In our Maple program, each operator term is represented internally by means of list structures. Different data types are known to the program. The two most important types are `opterm` to represent a single diagram unit and `opsum` for a set of diagrams, i.e. a full perturbation expansion or just some part of it. Simpler data types represent, for example, the creation and annihilation operators or different one and two–particle matrix elements. There are no other limitations on the number of operator terms (diagrams) than those due to processing time and memory of the computer.

At present, the APEX package consists of about 140 individual procedures which support a variety of different tasks. This includes the algebraic solution of the Bloch equation (1) up to third and fourth order. All commands are built up at a hierarchy of four different levels[13]. At the user's level, however, about 20 of these commands are sufficient for all of the interactive work. These commands facilitate the derivation of rather general perturbation expansions in the orbital representation for open–shell atoms as well as in cases of non–trivial model spaces. A detailed documentation of the program in the style of *The Maple Handbook* by D. Redfern[15] is also available[13].

†Maple V is a widely used symbolic program and a trademark of Waterloo Maple Software, Inc.

The results of the symbolic derivation are represented by algebraic sums which are equivalent to the diagrammatic Feynman–Goldstone representation. A short example of an interactive session is shown below in Figure 2 for the second– and third-order energy corrections of a closed–shell system. The only real difference in dealing with open–shell atoms and/or more additional perturbations is the size of the final expansions.

We first define the perturbation operator V and assign it to the variable **VHF**.

```
> v0 := vs_setoperator(V0):
> v2 := vs_setoperator(V2):
> VHF := vs_addoperators(v0,v2):
```

Additionally, we need to define the shell–structure of the atom, i.e. the occupation of the model state functions $|\phi_\alpha\rangle$.

```
> vs_setmodelspaceoccupation(0,0);
Occupation numbers nvp = 0 and nvc = 0
```

The evaluation is performed stepwise for the energy corrections ($\mathbf{e}_n$) and the higher orders of the wave operator $\Omega^{(n)}$ ($\mathbf{w}_n$). To obtain the correlation correction $E^{(n)}$ we need Ω up to the order $(n-1)$.

```
> w1 := vs_solve_nextorder_Omega(VHF):
> e2 := vs_solve_nextorder_energy(VHF):
> w2 := vs_solve_nextorder_Omega(VHF):
> e3 := vs_solve_nextorder_energy(VHF):
```

We may print the result for **e2** by using the command

```
> vs_print(e2):

===> (-1)^0 1 V0 V0 X0
-------------------------------------------------------------------------- 1
===> (-1)^0 -1 V0 V0 X0
-------------------------------------------------------------------------- 2
===> (-1)^3 1/4 v2(c1;c2|r3;r4) v2(r4;r3|c2;c1) X0 1 / ( + Er4 + Er3 - Ec1 - Ec2 )
-------------------------------------------------------------------------- 3
===> (-1)^2 1/4 v2(c1;c2|r3;r4) v2(r4;r3|c1;c2) X0 1 / ( + Er4 + Er3 - Ec2 - Ec1 )
-------------------------------------------------------------------------- 4
===> (-1)^2 1/4 v2(c1;c2|r3;r4) v2(r3;r4|c2;c1) X0 1 / ( + Er3 + Er4 - Ec1 - Ec2 )
-------------------------------------------------------------------------- 5
===> (-1)^1 1/4 v2(c1;c2|r3;r4) v2(r3;r4|c1;c2) X0 1 / ( + Er3 + Er4 - Ec2 - Ec1 )
-------------------------------------------------------------------------- 6
```

Each of these terms may easily be represented by a Feynman–Goldstone diagram. However, we recognize that some of these diagrams are equivalent and, thus, can further be simplified.

```
> e2 := vs_simplifyoperator_bypermutation(e2):
> e2 := vs_reorder_indices(e2):
> vs_print(e2):

===> (-1)^3 1/2 v2(c1;c2|r3;r4) v2(r4;r3|c2;c1) X0 1 / ( + Er4 + Er3 - Ec1 - Ec2 )
-------------------------------------------------------------------------- 1
===> (-1)^2 1/2 v2(c1;c2|r3;r4) v2(r4;r3|c1;c2) X0 1 / ( + Er4 + Er3 - Ec2 - Ec1 )
-------------------------------------------------------------------------- 2

> e3 := vs_simplifyoperator_bypermutation(e3):
> e3 := vs_reorder_indices(e3):
```

Before we will print the final results, we directly convert the output to LaTeX. This is achieved by assigning the output to an external .tex file.

```
> vs_print_latex(e2):
> vs_print_latex(e3):
```

Figure 2. Interactive dialog for the derivation of the second– and third–order correlation energies of a closed–shell atom using a Hartree–Fock basis and including short comments. The output of the LaTeX file is shown in Figure 3.

ENERGY CORRECTIONS FOR CLOSED-SHELL ATOMS

The program APEX has almost no limitations with respect to the atomic shell-structure. For the sake of simplicity, however, in this example we restrict ourselve to the closed-shell case. This example will illustrate the handling of the program without that it produces a large amount of output. The application of the program can easily be extended to the open-shell case and to the study of additional perturbations like internal or external fields.

For a closed-shell atom we may start from a one-dimensional model space which is identical to the (many-electron) reference state, $|o_c\rangle \equiv |o\rangle$. The total energy in first order is then obtained from the expectation value of the Hamiltonian $H = H_o + V$ for the model state[14]

$$E^{(0)} + E^{(1)} = \langle o_c \,|\, H \,|\, o_c \rangle = \sum_a \epsilon_a + \sum_a \left[\frac{1}{2} (U_{\rm HF})_{aa} + \langle a \,|\, -U(r) \,|\, a \rangle \right]$$

$$= \sum_a \left\langle a \left| \frac{\boldsymbol{p}^2}{2} + U_{\rm nuc}(r) \right| a \right\rangle + \frac{1}{2} \sum_{ab} \left(g_{abab} - g_{abba} \right) \qquad (6)$$

where g_{abab} and g_{abba} represent the direct and exchange matrix elements of the Coulomb repulsion. As seen from eq.(6), the total energy is independent from the central potential $U(r)$ for the averaged electron-electron interaction used to create the one-particle spectrum in the computation. We may apply a Hartree-Fock potential $U(r) = U_{\rm HF}$ and, thus, obtain (in first order) the total energy

$$E_{\rm HF} = \sum_a \epsilon_a - \frac{1}{2} \sum_a (U_{\rm HF})_{aa} \; .$$

The correlation correction in second order follows from the expectation value of the perturbation V multiplied with the first-order wave operator $\Omega^{(1)}$, i.e. $\left\langle o_c \left| V \, \Omega^{(1)} \right| o_c \right\rangle$

$$E^{(2)} = -\frac{1}{2} \sum_{abrs} \frac{g_{abrs}\, g_{rsab}}{\epsilon_r + \epsilon_s - \epsilon_a - \epsilon_b} + \frac{1}{2} \sum_{abrs} \frac{g_{abrs}\, g_{srab}}{\epsilon_r + \epsilon_s - \epsilon_a - \epsilon_b} \, . \qquad (7)$$

In our notation, this is the perturbation expansion for the second-order correlation energy of a closed-shell atom using a Hartree-Fock basis. Here, we will not further consider the analogue expressions in third-order perturbation theory. Eq.(7) as well as the third-order expressions namely can be obtained from APEX within a few steps as shown in Figure 2. The results of the derivation (converted to LaTeX) are displayed in Figure 3 which also indicates that the size of the perturbation series rapidly increases if we extend the expansions to the next order. Only a few selected third-order terms are shown in this Figure.

SUMMARY

Even though (relativistic) MBPT has been proven to be a powerful and efficient method to predict a large variety of atomic properties, not so many accurate studies for open-shell atoms could be performed in the past. Often, the size and complexity of the corresponding perturbation expansions caused problems in any manual derivation and forced to develop alternative paths. Such an alternative is offered by computer algebra which, in the future, will become an even more important tool in studying many-particle systems.

$$-\frac{1}{2}\frac{\langle c_1c_2\,|v_{12}|\,r_3r_4\rangle\ \langle r_4r_3\,|v_{12}|\,c_2c_1\rangle\ \Omega_0\ \Omega_0}{(\epsilon_{r_4}+\epsilon_{r_3}-\epsilon_{c_1}-\epsilon_{c_2})} \tag{i}$$

$$\frac{1}{2}\frac{\langle c_1c_2\,|v_{12}|\,r_3r_4\rangle\ \langle r_4r_3\,|v_{12}|\,c_1c_2\rangle\ \Omega_0\ \Omega_0}{(\epsilon_{r_4}+\epsilon_{r_3}-\epsilon_{c_2}-\epsilon_{c_1})} \tag{ii}$$

$$-\frac{1}{2}\frac{\langle c_1c_2\,|v_{12}|\,r_3r_4\rangle\ \langle r_4r_3\,|v_{12}|\,r_5r_6\rangle\ \langle r_6r_5\,|v_{12}|\,c_2c_1\rangle\ \Omega_0\ \Omega_0}{(\epsilon_{r_6}+\epsilon_{r_5}-\epsilon_{c_1}-\epsilon_{c_2})\,(\epsilon_{r_4}+\epsilon_{r_3}-\epsilon_{c_1}-\epsilon_{c_2})} \tag{iii}$$

$$\frac{1}{2}\frac{\langle c_1c_2\,|v_{12}|\,r_3r_4\rangle\ \langle r_4r_3\,|v_{12}|\,r_5r_6\rangle\ \langle r_6r_5\,|v_{12}|\,c_1c_2\rangle\ \Omega_0\ \Omega_0}{(\epsilon_{r_6}+\epsilon_{r_5}-\epsilon_{c_2}-\epsilon_{c_1})\,(\epsilon_{r_4}+\epsilon_{r_3}-\epsilon_{c_2}-\epsilon_{c_1})} \tag{iv}$$

$$\vdots$$

Figure 3. LaTeX output of the calculated perturbation expansion terms for the second-order [eqs. (i) and (ii)] and third-order [eqs. (iii) and (iv)] correlation energies of a closed-shell atom. The full third-order expansion includes a total of 12 terms from which, however, only the first two are shown. The results could be compared with those found in the literature, for instance, the compilation for closed-shell and effective one-particle atoms by Blundell *et al.*[14] ($\Omega_0 \equiv 1$ in this case).

The program APEX is a first significant step to extend the application of MBPT towards open-shell atoms (and molecules). It might also allow that these methods become applicable to atomic physics groups which are not specialized in this field. In order to make more efficient use of this program, however, it has to be extended to perform the angular reduction (see step iv above) automatically. This step will require a close combination with known Racah algebra techniques. But our experience in dealing with perturbation expansions for open-shell atoms so far already demonstrate that symbolic computations allows for a very efficient treatment. In particular, it provides a valuable alternative and supplement to established diagrammatic methods.

REFERENCES

1. I. Lindgren, *Int. J. Quant. Chem.* S12:1273 (1978).
2. S.A. Blundell, W.R. Johnson, and J. Sapirstein, *Phys. Rev.* A43:3407 (1991); *Phys. Rev. Lett.* 65:1411 (1992)
3. W.R. Johnson, S.A. Blundell, and J. Sapirstein, *Phys. Rev.* A37:2764 (1988); *Phys. Rev.* A38: 2699 (1988); *Phys. Rev.* A42:1087 (1990).
4. A.-M. Mårtensson-Pendrill and A. Ynnerman, *Phys. Scr.* 41:329; *J. Phys.* B25:L551 (1992).
5. W.R. Johnson, D.R. Plante, and J. Sapirstein, *Phys. Rev.* A52:1976 (1995).
6. E. Lindroth and A. Ynnerman, *Phys. Rev.* A47:961 (1993).
7. R. Marrus and P. Mohr, in "Adv. in Atomic and Molecular Physics," vol. 14, 181 (1974).
8. I. Lindgren, *J. Phys.* B7:2441 (1974).
9. I. Lindgren and J. Morrison, "Atomic Many-Body Theory," 2nd edition, Springer, Berlin a.o., chapters 9–12 (1986).
10. J. Goldstone, *Proc. Roy. Soc.* A239:267 (1957).
11. S. Fritzsche, Habilitationsschrift, University of Kassel, to be published (1997).
12. E. Avgoustoglou, W.R. Johnson, D.R. Plante, J. Sapirstein, J. Sheinerman, and S.A. Blundell, *Phys. Rev.* A46:5478 (1992).
13. S. Fritzsche, Evaluation of operator products in second quantization, User manual, unpublihed (1996).
14. S.A. Blundell, D.S. Guo, W.R. Johnson, and J. Sapirstein, *At. D. Nucl. D. Tables* 37:103 (1987).
15. D. Redfern, "The Maple Handbook," Springer, New York, Berlin, a.o. (1994).

SCALED-ENERGY SPECTROSCOPY OF HELIUM AND BARIUM RYDBERG ATOMS IN EXTERNAL FIELDS

W. Hogervorst, A. Kips, K. Karremans,
T. van der Veldt, G.J. Kuik and W. Vassen

Laser Centre Vrije Universiteit, Department of Physics and Astronomy,
De Boelelaan 1081, 1081 HV Amsterdam, The Netherlands

INTRODUCTION

Rydberg atoms in many aspects strongly resemble the hydrogen atom and are readily accessible for detailed investigation of e.g. the influence of external magnetic or electric fields on the level structure. The hydrogen atom in a sufficiently strong magnetic field B is a prototype system that classically exhibits chaotic behaviour. This is a consequence of the diamagnetic effect, which is proportional to $(B \times r)^2$ and which at low field strength is responsible for the mixing of angular momentum states (r is the radius of the electron orbit). When diamagnetism dominates only two conserved quantities (energy W and z-component of angular momentum l) exist for this system with three spatial degrees of freedom, a prerequisite for chaos in classical physics. The diamagnetic effect not only becomes important for large values of B, but also for large values of r. As the radius r scales as n^2 (n is principal quantum number) it follows that the diamagnetic effect grows with n^4, so that in highly-excited states it may be studied at relatively moderate magnetic field strengths. The classically chaotic regime in the hydrogen atom is reached when the Lorentz force exerted by the field on the electron about equals the Coulomb force binding the electron. In the hydrogen ground state (n=1) this condition can only be fulfilled for the huge field strength of 2.35×10^5 Tesla, whereas in a Rydberg state with n=100 a field of 0.25 Tesla suffices. This, together with the fact that the hydrogen atom is experimentally not easily accessible, constitutes the major argument to investigate this effect also in Rydberg states of more complex atoms. An additional point of interest then relates to the influence of an extended atomic core, represented by a quantum defect in regular Rydberg sequences, on the observed spectra in the presence of a magnetic field.

Similar arguments hold for the investigation of Rydberg states in the presence of an external electric field E. Now it is the linear Stark effect, that at moderate field strengths induces mixing of angular momentum l-states, reflected in the appearance of angular momentum manifolds in the spectra, and at high field strengths also mixing of n-states. Finally field ionization will occur in sufficiently strong fields. In Rydberg states the sensitivity for electric fields is again strongly enhanced. In contrast to the

Atoms and Molecules in Strong External Fields
Edited by Schmelcher and Schweizer, Plenum Press, New York, 1998

magnetic field case the Hamiltonian of the hydrogen atom in an external electric field can be fully separated and chaos in the classical problem does not occur. It is a problem of significant interest in atomic physics to investigate Rydberg states of non-hydrogen atoms in the presence of strong electric fields as well, in particular with respect to core-induced effects.

In this contribution experimental results of a study of Rydberg states in the He atom in the presence of a magnetic field as well as of Rydberg states in the Ba atom, including autoionizing states, in the presence of an electric field will be presented and discussed. Spectra are recorded under conditions where the classically important scaled-energy parameter is kept constant. This allows for a Fourier transform of the experimental spectra to so-called scaled-action spectra, which reflect the closed periodic orbits of the system. A direct comparison with semi-classical periodic-orbit theory then is feasible, resulting in a beautiful interpretation of many of the observations.

This contribution is organized as follows. First the Hamiltonians of the hydrogen atom in the presence of external fields and the appropriate scale transformations will be discussed. Next the semi-classical periodic-orbit theory will be briefly summarized. Then the He experiment in a magnetic field will be presented, followed, finally, by a discussion of the Ba experiment in electric fields.

HAMILTONIAN OF H-ATOM IN MAGNETIC AND ELECTRIC FIELDS

The Hamiltonian of a spinless H-atom in a magnetic field B along the z-axis is:

$$H = H_0 + H_p + H_d = \frac{p^2}{2} - \frac{1}{r} + \frac{\gamma}{2}L_z + \frac{1}{8}\gamma^2\rho^2 \tag{1}$$

Here H_0 is the zero-field Hamiltonian, H_p the paramagnetic effect (Zeeman effect, H_p=0 for L_z=0) and H_d the diamagnetic effect ($\rho^2 = x^2 + y^2$); $\gamma = B/B_c$ with $B_c = 2.35 \times 10^5$ Tesla (critical field). This Hamiltonian (for H_p=0) can be transformed using scaling parameters

$$\tilde{r} = \frac{1}{2}\gamma^{-2/3}r; \tilde{p} = \gamma^{-1/3}p \tag{2}$$

resulting in the *scaled Hamiltonian* $\tilde{H}$:

$$\tilde{H} = \gamma^{-2/3}H = \frac{\tilde{p}^2}{2} - \frac{1}{\tilde{r}} + \tilde{\rho}^2 \tag{3}$$

This transformation shows that the classical motion of the electron is not determined by two independent parameters (energy W and magnetic field strength B) but only by a single parameter (*scaled energy*):

$$\tilde{\varepsilon} = \gamma^{-2/3}W \tag{4}$$

This scaled energy $\tilde{\varepsilon}$ is a measure for the onset of chaos ($\tilde{\varepsilon}$=-1 corresponds to the situation where Coulomb force and diamagnetic force on the electron are equal). It is of interest to note that also the commutator $[p_z, z] = i\hbar$ scales, resulting in:

$$[\tilde{p_z}, \tilde{z}] = i\gamma^{1/3}\hbar = i\hbar_{eff} \tag{5}$$

The semi-classical quantization condition for the action S along a given electron orbit n also transforms:

$$S_n = \frac{1}{2\pi}\int pdq = n\hbar \tag{6}$$

$$\tilde{S}_n = \frac{1}{2\pi}\int \tilde{p}d\tilde{q} = n\gamma^{1/3}\hbar = n\hbar_{eff} \tag{7}$$

The *scaled action* $\tilde{S}_n = \gamma^{1/3}S_n$ only depends on $\tilde{\varepsilon}$. This implies that the Fourier transform of a spectrum recorded at constant scaled energy, which is linear in the variable $\gamma^{-1/3}$, shows resonances at values of the scaled action $\tilde{S}_n = n\gamma^{1/3}\hbar; n\gamma^{1/3}$ is the conjugated variable.

The Hamiltonian of the H-atom in an electric field E along the z-axis is:

$$H = H_0 + H_e = \frac{1}{2}p^2 - \frac{1}{r} + fz \tag{8}$$

Here H_0 is the zero-field term, H_e the field term; $f = E/E_c$ with $E_c = 5.14\times10^9$ V/cm. Also in this case a scale transformation is possible:

$$\tilde{r} = f^{1/2}r; \tilde{p} = f^{-1/4}p \tag{9}$$

resulting in the *scaled Hamiltonian* $\tilde{H}$:

$$\tilde{H} = f^{-1/2}H = \frac{1}{2}\tilde{p}^2 - \frac{1}{\tilde{r}} + \tilde{z} \tag{10}$$

This again shows that the classical motion of the electron is governed by the single *scaled-energy parameter* $\tilde{\varepsilon}$:

$$\tilde{\varepsilon} = f^{-1/2}W \tag{11}$$

The value $\tilde{\varepsilon} = -2$ corresponds to the saddle-point energy, related to the maximum in the combined Coulomb and electric field potential determining the classical field ionization limit.

The recognition that scaling transformations do exist for the H-atom in external fields [1] now forms the basis of *scaled-energy spectroscopy.* Complex excitation spectra of Rydberg atoms in strong external fields, when recorded at constant scaled energy, show remarkable regular Fourier transforms. A limited set of resonances shows up in the scaled-action spectra, each resonance corresponding to a classical periodic orbit. This type of spectroscopy in the presence of a magnetic field was first applied to the H-atom itself, using pulsed laser excitations, by Holle et al [2]. Pioneering scaled-energy experiments in an electric field were performed by Eichmann et al [3] in Na Rydberg states and in Kleppners group on Li Rydberg states [4].

SEMI-CLASSICAL PERIODIC-ORBIT THEORY

In semi-classical periodic-orbit theory (see [5] and references therein) it is assumed that a frequency spectrum contains numerous sinusoidal oscillations, each related to a classical periodic orbit. These orbits may be regular or chaotic, depending on the parameters of the system. When an orbit is regular, it may be traversed many times. These multiple traverses give rise to higher harmonics in the frequency spectrum and interferences between all harmonics result in sharp resonances, corresponding to the quantum states. For chaotic orbits, however, the probability of repeated traversals is exponentially small. The signature of a chaotic orbit in the frequency spectrum is a deformed sine wave.

Du and Delos [6] and Gao and Delos [7] give a physical picture explaining the spectral features of hydrogenlike atoms in the presence of external fields using closed-orbit theory. In their model incident electromagnetic radiation excites the electron

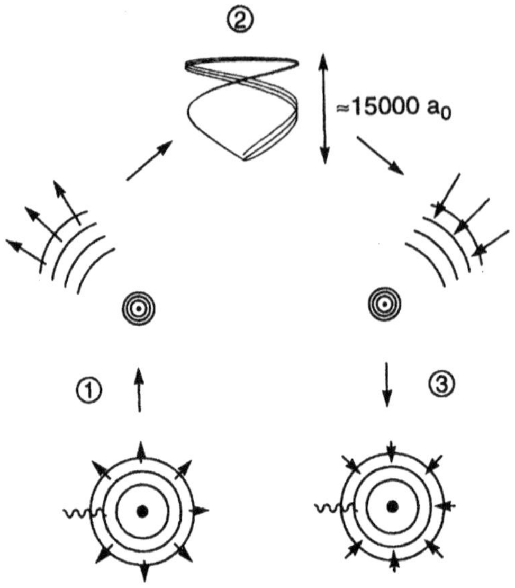

Figure 1. Schematic representation of closed-orbit theory [6,7]. (1) Outgoing Coulomb waves generated in the excitation process. (2) In the classical region the wave fronts follow classical trajectories. (3) Returning waves interfere with outgoing Coulomb waves producing oscillations in the frequency spectrum.

close to the nucleus in an outgoing, zero-energy Coulomb wave as near the core effects of the external field are negligible small (see Fig. 1). At large distances ($\sim 50a_0$, with a_0 the Bohr radius) the wave is assumed to propagate semi-classically. The outgoing wave fronts then follow classical trajectories and propagate far out in the field. As the influence of the magnetic or electric field grows with distance some of the trajectories will eventually curve back to the nucleus. Near the nucleus the incoming waves have to be treated quantum mechanically again. Incoming Coulomb waves interfere with the outgoing waves, resulting in oscillations in the absorption spectrum. These oscillations are described by the well known Gutzwiller trace formula [5,6], involving a summation over all closed periodic orbits and their traversals. A Fourier transform of the absorption spectrum results in an action spectrum, where each peak corresponds to a classical orbit. The intensity of each action peak then is determined by the stability of the orbit and by the external parameters (excitation process and light polarization). The effect of the core in a non-hydrogenic system may be incorporated by introducing a phase shift in the scattered wave function as in standard quantum defect theory.

Closed periodic orbits may be found assuming that the electron leaves the core region with a momentum directed perpendicular to the sphere with radius $r \sim 50a_0$ and with a value determined from eq. (1) for given value of the energy W:

$$p_r = \sqrt{W - \frac{1}{r} + \frac{1}{8}\gamma^2\rho^2} \tag{12}$$

Then the classical Hamilton equations are integrated. A closed orbit is found when the final momentum upon reaching the core region is again perpendicular to this sphere. The orbit may be traversed again in the same or reversed direction after scattering on the nucleus. The stability of the orbit follows from its sensitivity to small variations in

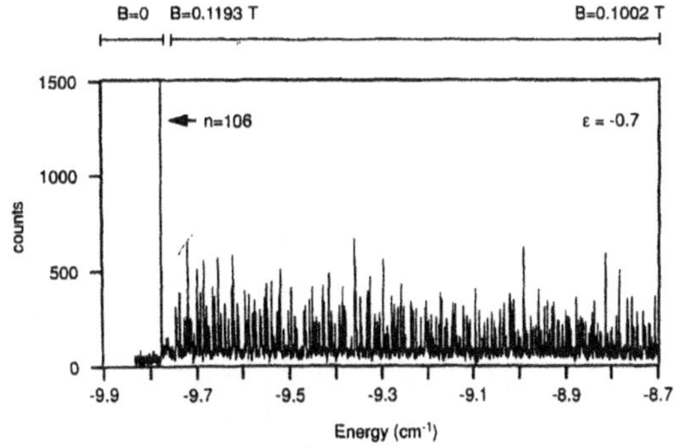

Figure 2. Experimental constant-scaled-energy spectrum for $\tilde{\varepsilon} = -0.700$. Laser frequency and magnetic field were varied simultaneously to keep the scaled energy constant.

the initial conditions, which becomes extreme in the high-field regime. For each closed orbit the (scaled) classical action can be calculated by straightforward integration along the closed trajectory and its value compared with experimental data.

Frequency spectra in the presence of external fields can also be calculated quantum mechanically. For highly-excited states this involves the diagonalization of huge energy matrices, for which special numerical techniques have been developed [8,9]. Such calculations are also possible under conditions of constant-scaled energy, so that a direct comparison with experimental data and Fourier transforms can be made.

HELIUM RYDBERG ATOMS IN A MAGNETIC FIELD

Scaled-energy experiments in a magnetic field were performed in helium Rydberg states at two values of the scaled energy: $\tilde{\varepsilon} = -0.700$ and $\tilde{\varepsilon} = -0.400$. Helium was excited to $1snp$ Rydberg states (up to n = 200) in a collimated beam of discharge-populated 2 $^3S_1$ metastable atoms with 260 nm light from a frequency-doubled CW ring dye laser pumped by an Ar-ion laser. Atomic beam and laser beam perpendicularly intersected (in a well-shielded interaction region) to eliminate Doppler effects to a large extend, resulting in a residual linewidth of 25 MHz. The magnetic field was produced with conventional current-driven coils (maximum value about 0.2 Tesla) and aligned very carefully parallel to the atomic beam to avoid motional Stark effects. The He Rydberg atoms were field-ionized after leaving the interaction chamber, mass selected with a quadrupole filter and counted with an electron multiplier. The visible CW laser was scanned continuously over 30 GHz (60 GHz at 260 nm), monitoring the change in frequency ν on-line with a Fabry-Perot etalon. This calibration spectrum in turn was used to control the current through the magnetic field coils in such a way that the scaled energy $\tilde{\varepsilon}$ is kept constant. In extended spectra several of these 30 GHz scans were overlapped, using the zero-field Rydberg resonances as references. More details of the experimental setup can be found in [10] and [11]. As an example in Fig. 2 a recorded spectrum at $\tilde{\varepsilon} = -0.700$ is shown, where the excitation energy was varied from -9.7 cm^{-1} to -8.7 cm^{-1} and the magnetic field simultaneously from 1193 Gauss to 1002 Gauss. ¿From this dense spectrum not much information can be extracted directly.

However, after Fourier transformation to a scaled-action spectrum a much more insightful picture results, revealing the closed periodic orbits of the system. In Fig. 3 (upper part) such a Fourier transformation is shown for ε=-0.400 (mixed regular-

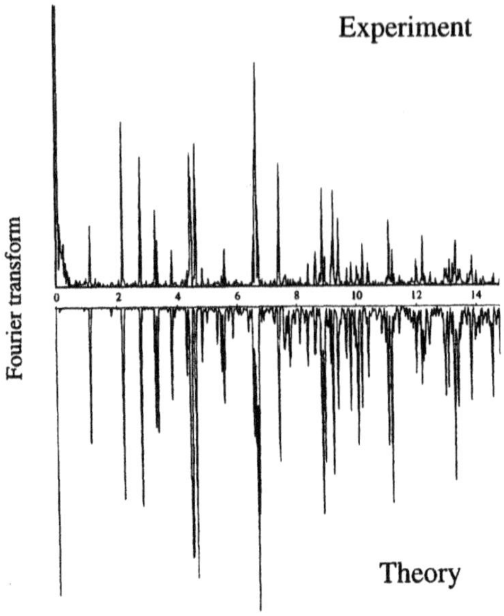

Figure 3. Squared Fourier transform of the constant-scaled energy spectrum at $\tilde{\varepsilon}$= -0.400. Upper part: transform of a spectrum in the range $112.3 < \gamma^{-1/3} < 161.1$. Lower part: transform of a quantum mechanically calculated energy spectrum.

chaotic regime). Although this scaled-action spectrum is less regular than the spectrum recorded at $\tilde{\varepsilon} = -0.700$ [11], many resonances still can be connected to two types of closed orbits and their traversals. The resonance at $\tilde{S}$=1.12 corresponds to the fundamental, so-called vibrator orbit V_1, directed parallel to the magnetic field, whereas there is a weak indication for the presence of the fundamental rotator orbit R_1, in the plane perpendicular to the magnetic field at $\tilde{S}$=0.92 (see [2] for this classification). The first bifurcation from the fundamental vibrator, traversed twice, appears at $\tilde{S}$=2.21 and is denoted V_2^1. Its action is about equal to the action of the two times traversed fundamental vibrator V_2. Higher traversals appear at $\tilde{S}$=3.27 (V_3^1) and 4.31 (V_4^1). The second bifurcation of the vibrator orbit, traversed three times, lies at $\tilde{S}$=3.35 (V_3^2). The first strong rotator orbit at $\tilde{S}$=2.80 corresponds to three traversals of the first bifurcation of the fundamental orbit (R_3^1), at $\tilde{S}$= 3.84 to R_4^1. The number of unstable orbits in the case of $\tilde{\varepsilon}$= -0.400 drastically increases compared to $\tilde{\varepsilon}$= -0.700, and also the number of resonances that does not fit the classification in terms of rotator - vibrator orbits, called exotics. Such exotic orbits may contribute to the resonances observed at e.g. $\tilde{S}$= 6.64, 8.85 and 9.15. In the action spectrum for $\tilde{\varepsilon}$= -0.400 there is clear evidence for electron scattering from vibrator into rotator orbits and vice versa. This is evidenced e.g. by the occurence of resonances at $\tilde{S}$= 5.01 and 9.03, which are the sum of actions for V_2^1 and R_3^1 orbits (2.21 and 2.80), respectively V_4^2 and R_5^2 orbits (4.42 and 4.61). The occurence of sum actions (not possible in the hydrogen atom) was for the first time observed in the $\tilde{\varepsilon}$=-0.700 spectrum [8]. The proliferation of unstable orbits as well as the more pronounced occurence of sum orbits results in less good agreement between the observed and calculated (with classical closed-orbit theory) action spectrum in the case of $\tilde{\varepsilon}$= -0.400 than in case of $\tilde{\varepsilon}$= -0.700 [11].

We also calculated a quantum mechanical scaled-energy spectrum for $\tilde{\varepsilon}$= -0.400 applying the same computer code as in the $\tilde{\varepsilon}$= -0.700 case [8]. The Fourier transform

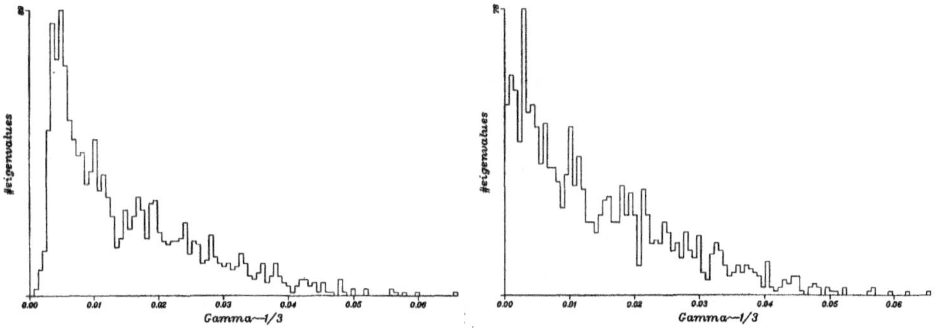

Figure 4. Calculated distribution of nearest neighbour separations for $\tilde{\varepsilon}$= -0.710 for the hydrogen (left) and helium (right) atom in a magnetic field. In helium the quantum defect of the 1snp Rydberg series has been taken into account. Along the horizontal axis the scaled separation $\gamma^{-1/3}$ is plotted.

of this quantum mechanically calculated spectrum is plotted in the lower part of Fig. 3. These calculations result in a much better agreement with observations, not only for the S-values where resonances occur but also with respect to their strengths, when compared with the closed-orbit calculations. Such quantum mechanical calculations may also be used to determine level statistics, i.e. the distribution of nearest neighbour separations (s). We calculated these distributions in the regular regime ($\tilde{\varepsilon}$= -0.700) for both the hydrogen and the helium atom. In hydrogen a clear Poissonian distribution $P(s) = P(0)exp(-s)$ is obtained, which is shown on the left in Fig. 4. However, in helium a Wigner type of distribution $P(s) = P(0)s \cdot exp(-\pi s^2/4)$ is found, as shown on the right in Fig. 4. To get good agreement between the experimental and calculated energy spectrum for the helium atom in a magnetic field for this case of $\tilde{\varepsilon}$= -0.700 it was necessary to include the quantum defect of 0.0684 for the 1snp Rydberg series [8]. Then the level statistics shown in Fig. 4 results. A change from Poisson to Wigner type of distributions as a function of the scaled-energy parameter is considered to be a signature of the transition from a classically regular system to a chaotic system. However, the effect calculated for the level statistics in the helium atom in the still regular regime does not relate to this type of transition. In helium it is the core-scattering effect (represented by the quantum defect) that is responsible for the change in level statistics as compared to hydrogen; it is therefore referred to as core-induced chaos [4]. The agreement between experimental and calculated energy spectrum at $\tilde{\varepsilon}$= -0.700 is excellent. This stimulates experiments to measure the level statistics under various circumstances. Such experiments are in progress.

BARIUM RYDBERG ATOMS IN AN ELECTRIC FIELD

Scaled-energy experiments in an electric field have been performed in three barium Rydberg series $6snf, 5d_{3/2}nf$ and $5d_{5/2}nf$ at two values of the scaled energy below the field-ionization limit ($\tilde{\varepsilon}$= -2.94 and -2.35) and one value above this limit ($\tilde{\varepsilon}$= -1.76). The $6snf$ states are bound and converge to the 6s-ionization limit of the atom, whereas the $5dnf$ states belong to weakly autoionizing series converging to the $5d_{3/2}$- and $5d_{5/2}$-excited states of the ion. So a study of these various series in principle allows for a comparison of differences induced by core effects. The states of interest were populated in a one-photon excitation process with narrowband, tunable CW laser radiation from the $6s5d^1D_2$ and $5d^{21}G_4$ metastable states. For the excitation of $6snfJ = 3$ states from $6s5d$ UV light from an intra-cavity frequency-doubled ring dye laser operating around 600 nm was used, whereas $5dnf$ states (dominantly J=5) were excited with a CW Stilbene ring dye laser operating in the blue spectral range around 450 nm. The laser beam perpendicularly intersected a well-collimated beam of Ba atoms between two capacitor plates in an excitation chamber, well shielded against stray electric fields. In the beam metastable states were populated by running a DC-discharge between the oven and a tungsten filament heating the oven. As in the helium case the frequency of the laser was scanned in the presence of an electric field and continuously monitored with a Fabry-Perot etalon. This etalon signal was used to adjust the electric field in such a way that spectra were recorded at constant-scaled-energy value. We quote an absolute uncertainty in the value of $\tilde{\varepsilon}$ of about 0.06, determined by the uncertainty in the measurement of the distance between the capacitor plates. Electrons directly produced by autoionization were, in case of the $5dnf$-series, detected using an electron multiplier positioned above one capacitor plate which contained a grounded fine wire mesh. In case of the $6snf$ states electrons were produced by field-ionizing the highly-excited atoms downstream from the excitation chamber in a second chamber. More details may be found in refs. [12,13].

Scaled-energy spectra were recorded for high-n Rydberg states (n=60-80) by overlapping several laser scans. Linewidths of $6snf$ states below the field-ionization limit did have the Doppler-limited value of about 10 MHz, whereas broadening was apparent above this limit. The linewidth of the $5dnfJ = 5$ autoionizing levels in zero field was of the order of 30 MHz, but broadened in the field by the admixture of states with a lower l-value with enhanced autoionization rates. Above the field-ionization limit autoionization into the 5d-continuum resulted in broad peaks with a width of about 100 MHz. In Fig. 5 scaled-action spectra for all three Rydberg series for $\tilde{\varepsilon}$= -2.35 are shown as obtained from a Fourier transform of the frequency spectra, whereas in Fig. 6 similar spectra for $\tilde{\varepsilon}$= -1.76 are reproduced. The typical grouping of lines, in particular in the spectra for $\tilde{\varepsilon}$= -2.94 and $\tilde{\varepsilon}$= -2.35, is similar to that observed in Na [3] and Li [4] experiments and can be interpreted directly for large negative $\tilde{\varepsilon}$. The energy difference between levels within a Stark manifold is responsible for the overall group structure, whereas the energy difference between adjacent Rydberg n-levels determines the splitting within each group. This latter difference also results in the resonance at low scaled action at about 0.44, which corresponds to the uphill / downhill orbits along the z-axis.

Comparing the Fourier spectra in Figs. 5 and 6 the resonances in the three series are found at nearly equal values of the scaled action, although for large scaled actions minor differences become apparent for the higher $\tilde{\varepsilon}$ values. This is as expected in periodic-orbit theory, where only the motion of the excited electron far outside the core is important. This motion is governed by classical equations, which do not involve the state of the core. However, the heights of the resonances show significant differences, in particular when comparing the Fourier spectrum of the $6snf$ series with those of the $5dnf$ series and at increased $\tilde{\varepsilon}$ values. The height of a resonance is determined by the the intensity of the recurring excitations in the oscillator strength distribution, which directlly relate to the transition probabilities at zero-field, and by core-scattering processes (sum orbits). The different probabilities for $6snf, 5d_{3/2}-$ and $5d_{5/2}nf$ excitations only partly explain the observed peak-height differences. Near and above the classical field-ionization limit ($\tilde{\varepsilon}$= -2.0) the frequency spectra for the 5dnf series show line broadening, resulting in high intensities for scaled actions close to zero as is clear in the $\tilde{\varepsilon}$= -1.76 spectrum when comparing the $5dnf$ results with those for $6snf$ (see Fig. 6). The

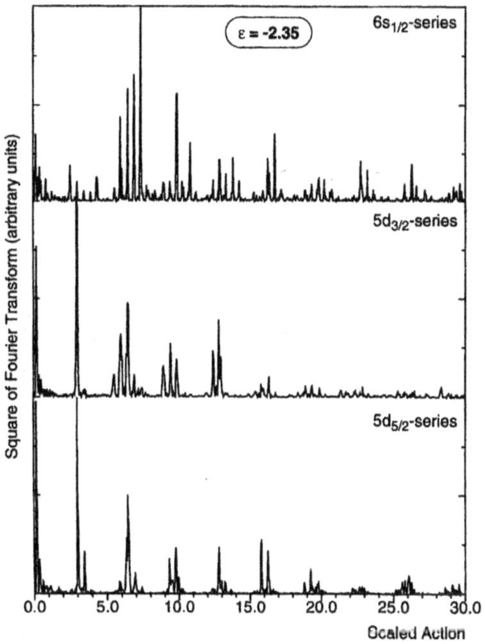

Figure 5. Squared Fourier transform of experimental spectra for $\tilde{\varepsilon}$= -2.35 (below the saddle point) for the 6snf and 5dnf Rydberg series.

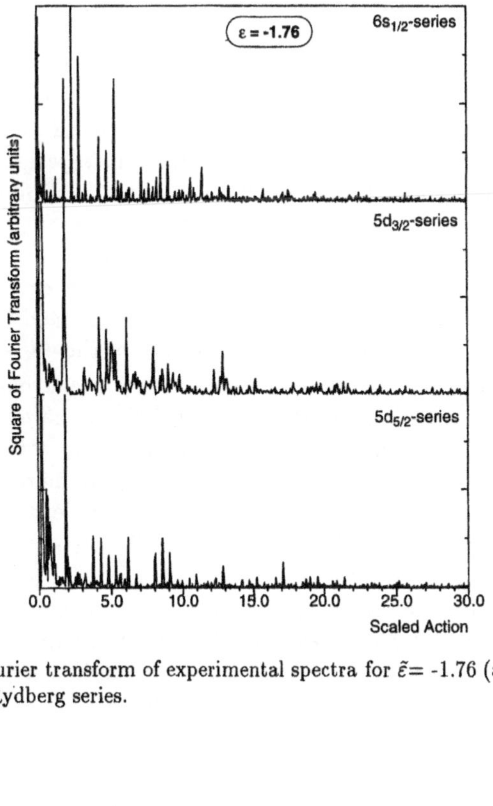

Figure 6. Squared Fourier transform of experimental spectra for $\tilde{\varepsilon}$= -1.76 (above the saddle point) for the 6snf and 5dnf Rydberg series.

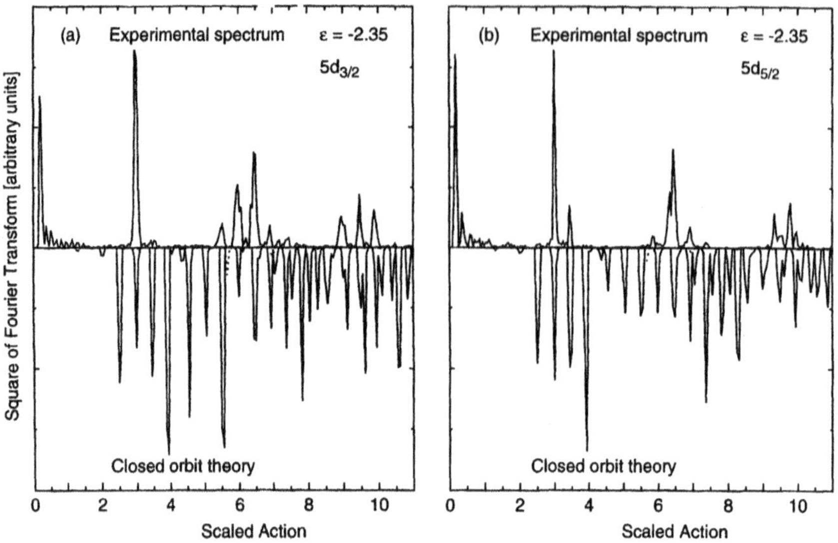

Figure 7. Comparison between experimental Fourier spectrum and closed-orbit calculculations for $\tilde{\varepsilon}$= -2.35 for the $5d_{3/2}$- (left) and $5d_{5/2}$- (right) series.

number of experimental peaks, in particular at high scaled action (recurrence time) strongly decreases with increasing value of $\tilde{\varepsilon}$ while also the typical grouping of peaks disappears. Above the saddle point only a limited number of closed orbits is observed. This can also be understood within the framework of closed-orbit theory. For $\tilde{\varepsilon} > 0$ only one unstable orbit, the uphill orbit parallel to the electric field, exists. For $\tilde{\varepsilon} < 0$ out of this orbit new orbits bifurcate, thus proliferating the number of peaks in the action spectrum. The downhill orbit, also parallel to the electric field axis, becomes possible at $\tilde{\varepsilon}$= -2. Closed-orbit calculations were performed for all experimental conditions, taking into account the respective quantum defects for the Rydberg electron: 0.17 for $6snfJ = 3$, 0.07 for $5d_{3/2}nfJ = 5$ and 0.14 for $5d_{5/2}J = 5$. Although in most cases the positions of the resonances could be reproduced well and can be assigned to classical orbits, it turned out to be extremely difficult to reproduce peak heights. Surprisingly, none of the strong peaks are due to a repetition of a parallel orbit, but many peaks are due to recurrences of other types of orbits, and in particular in the $6snf$ spectra, sum orbits strongly proliferate. This latter phenomenon makes peak-height calculations extremely difficult. Many more orbits are calculated than experimentally observed, whereas their theoretical intensities turn out to be of the same order of magnitude (see Figs. 7 and 8). Four peaks in the $\tilde{\varepsilon}$= -2.35 spectrum of the $5d_{5/2}$ -series could not be assigned. However, they show a remarkable connection: the strongest ones at $\tilde{S}$= 6.33 and $\tilde{S}$= 9.34 differ 3.01 in action, i.e. the action of the most intense experimental peak. The other two at $\tilde{S}$= 5.87 and $\tilde{S}$= 9.78 are shifted 0.44 in action from the stronger ones, which corresponds to the action of the uphill orbit. Remarkably the experimental spectra for $\tilde{\varepsilon}$= -2.35 and $\tilde{\varepsilon}$= -1.76 are dominated by a single strong peak, at $\tilde{S}$= 3.01 and at $\tilde{S}$= 1.98 respectively, and their recurrences. The corresponding orbits lie in the plane perpendicular to the electric field! Also the majority of other peaks are oriented in the same direction (θ= 90 degrees), for which no simple explanation is available. It was a surprise in itself that for increasing value of $\tilde{\varepsilon}$ the number of calculated orbits increases dramatically, in contradiction with observations. Especially for $\tilde{\varepsilon}$= -1.76 it might be expected that only a few orbits would remain that do not lead to escape of the Rydberg electron over the saddle point.

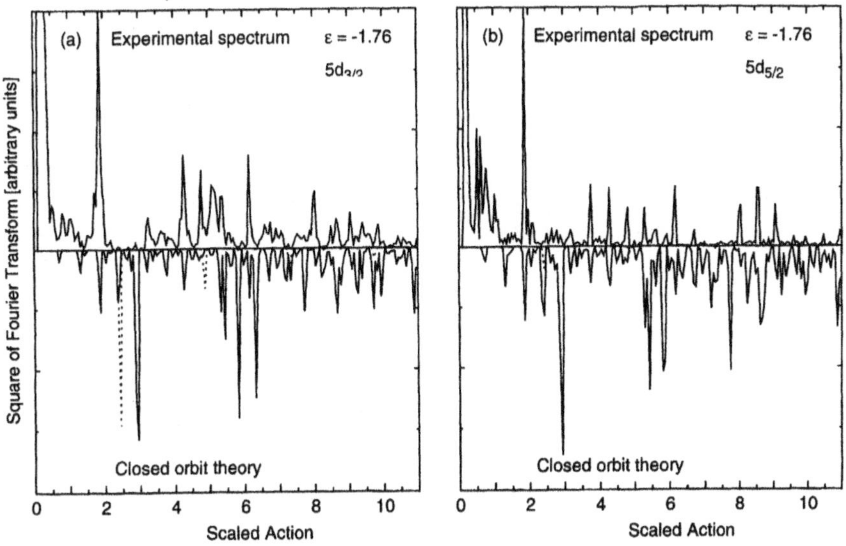

Figure 8. Comparison between experimental Fourier spectrum and closed-orbit calculculations for $\tilde{\varepsilon}$= -1.76 for the $5d_{3/2}$- (left) and $5d_{5/2}$- (right) series.

Acknowledgments

The authors are indebted to Sander Van Daatselaar for his calculations on the helium atom and gratefully acknowledge financial support from the Dutch Foundation for Fundamental research on Matter (FOM).

REFERENCES

[1]. D. Wintgen and H. Friedrich, Phys. Rev. A**35** (1987) 1464.

[2]. A. Holle, J. Main, G. Wiebusch, H. Rottke and K.H. Welge, Phys. Rev. Lett. **61** (1988) 161.

[3]. U. Eichmann, K. Richter, D. Wintgen and W. Sandner, Phys. Rev. Lett. **61** (1988) 2438.

[4]. M. Courtney et al., Phys. Rev. Lett. 73 (1994) 1340; Phys. Rev. Lett. **74** (1994) 2438; Phys. Rev. A**51** (1995) 1538.

[5]. M.C. Gutzwiller, Chaos in classical and quantum mechanics, (1991 Springer Verlag NY).

[6]. M.L. Du and J. Delos, Phys. Rev. A**38** (1988) 1896, 1913; Phys. Rev. Lett. **58** (1987) 1731; IEEE J. Quantum Elec. **24** (1988) 1445.

[7]. J. Gao, J. Delos, Phys. Rev. A**46** (1994) 869, 1455; Phys. Rev. A**42** (1992) 1449.

[8]. D. Delande, K.T. Taylor, M.H. Halley, T. van der Veldt, W. Vassen and W. Hogervorst, J. Phys. B**27** (1994) 2771.

[9]. P.A. Dando, T.S. Monteiro, D. Delande and K.T. Taylor, Phys. Rev. Lett. **74** (1995) 1099.

[10]. T. van der Veldt, Diamagnetism in helium Rydberg atoms (1993, thesis Vrije Universiteit Amsterdam).

[11]. T. van der Veldt, W. Vassen and W. Hogervorst, Europhys. Lett. **21** (1993) 903.

[12]. G.J. Kuik, High resolution studies of electric field effects on autoionising Rydberg states of barium (1997, thesis Vrije Universiteit Amsterdam).

[13]. G.J. Kuik, A. Kips, W. Vassen and W. Hogervorst, J. Phys. B**29** (1996) 2159.

ATOMS IN CROSSED FIELDS

J.-P. Connerade,[1] K. T. Taylor,[2] G. Droungas [1],
N. E. Karapanagiati,[1,3] M. S. Zhan,[1,4] and J. Rao[2,4]

[1]The Blackett Laboratory, Imperial College London SW7 2BZ UK
[2]DAMTP, Queen's University Belfast, Belfast BT7 1NN, UK
[3]L and A Division FORTH - IESL 71110 Heraklion Crete Greece
[4]CAS Wuhan Institute of Physics, Wuhan 430071, P.R. China

INTRODUCTION

We report in this article recent experimental work[1] and complementary theoretical work[2] carried out on Rydberg states of barium in crossed electric and magnetic fields. The most interesting experimental findings were the markedly different σ^+ and σ^- photoabsorption spectra. In the next two sections we first give overviews of the experiment and theory together with results from each. Before drawing conclusions and indicating possible future developments we report what understanding of the experimental findings theoretical analysis has produced.

THE EXPERIMENT

The earliest experimental studies of the quadratic Zeeman effect were the classic experiments of Garton and Tomkins[3] on the alkaline-earth elements. However, the agreement between theory and experiment, although broadly satisfactory as regards structure, was far from impressive as regards intensities. This was the starting point of our experimental investigation: it was clear to us that the experiments, first performed in 1969, were much in need of being refined and brought up to date.

The most obvious improvement required was in the method of detection. The early experiments were performed in photoabsorption, and the transmitted intensity was recorded photographically. Photoabsorption suffers from many disadvantages. First, because the method is inherently insensitive, one is led to increase the density of absorption above the optically thin regime in order to reveal fine detail in spectra. This leads to an optical saturation effect due to opacity in the centre of strong lines. Second, the noise increases with increasing signal. Lastly, atoms fly in all directions with respect to the magnetic field, which leads to varying amounts of quadratic Stark field. Photographic detection compounds the problems, since it is also subject to nonlinearity of response.

We therefore chose to work with an atomic beam, and to detect Rydberg atoms by field ionisation downstream from the interaction region. Our first idea was to direct the atomic beam parallel to the magnetic field, which was the geometry originally introduced at Imperial College to eliminate the motional Stark effect[4]. However, this was not possible with the magnet we had available. Since calculations using the standard formulae indicated that the motional Stark field should have a small effect for elements such as Ba or Sr, we decided to direct the direct the atomic beam perpendicular to the magnetic field, and the laser beam parallel to it. In this configuration, the quadratic Stark field is maximum but, by cofining the atomic beam to a narrow pencil, we hoped to restrict it more or less to a single value.

The geometry we had been forced to adopt had one advantage with respect to [4]: it allowed us to *separate* the σ^+ from the σ^- spectra. When we studied them separately, we had our first surprise: the structure in the two spectra was *not* the same 5, 6. This was contrary to the findings reported in the early papers, and also to the caculations on the motional Stark effect. The early papers had, in fact, only provided spectra for one polarisation. Fortunately, we had access to the original data of Garton and Tomkins[3]. By carefully digitising it, we discovered a similar discrepancy. Eventually, we were led to suspect that this might be due to the motional Stark field, and it is this realisation which led us into our detailed study of the crossed field problem.

Notice that experiments on crossed fields have evolved from experiments on the quadratic Zeeman effect, and have generally retained their original geometry. Thus, our present experiment appears to be the only one at the moment to have the laser beam parallel to the magnetic field, which confers the unique advantage of being able to separate σ^+ and σ^- spectra. This has turned out to be a crucial feature.

Another feature of our experiment arose from our decision to detect downstream. It turned out that the excited atoms had to fly 55mm from the interaction region to the point of detection. We found that we were able to gate our detector, and to set the time 'window' such that only a narrow range of velocities was accepted from within the Maxwellian distribution, thereby defining the motional Stark field rather accurately. We were also able to insert a pair of electric field plates above and below the horizontal atomic beam in Fig.1, and to apply an small, external electric field so as to cancel the motional Stark field. We thus had set up a crossed field configuration.

To cancel the motional Stark field, we began by studying the shifts of prominent spectral features in the diamagnetic spectrum as a function of applied electric field strength. This turned out to be quadratic, and so we could determine the zero field point by searching for the minimum shift on the parabola. To our relief, *the differences between the σ^+ and σ^- spectra disappeared at zero electric field.* We were now able to use this observation in order to set the zero electric field experimentally by 'balancing' the two spectra until their structure (when shifted by the linear Zeeman splitting to bring them into correspondence) became identical [7]. The results thus obtained exhibited much finer detail than the original spectra of 3, so that new calculations were required for comparison, but the results yielded extremely satisfactory agreement with experiment, both as regards the details of the structure and relative intensities of the lines [8].

Turning now to the crossed field problem, we found it very interesting that the σ^+ and σ^- spectra should differ from each other, a feature not accessible in earlier experiments. We therefore decided to record Stark maps of the diamagnetic spectrum under both polarisations. Since the excitation scheme we are using ($^1S_0 \rightarrow ^1P_1$) is the simplest one, the spectra we are using are ideal to explore this new situation. Also, we have chosen an element (Ba) for which the different ℓ-states of a given n are well

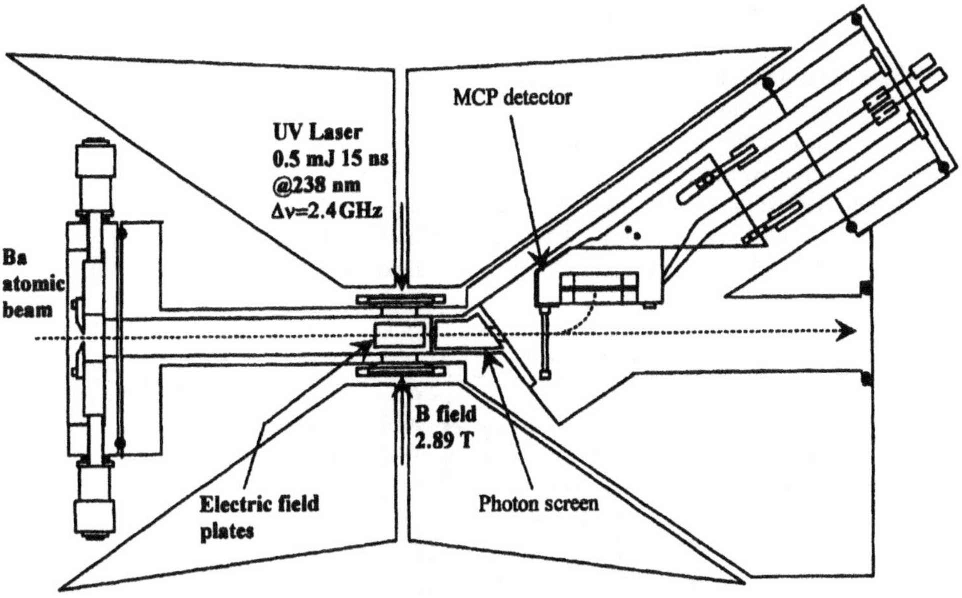

Figure 1. Layout of the interaction region, described in the text, showing how the laser and atomic beam are drossed inside the dipole magnet poles and the external field plates in a way allowing the separation of the σ^+ and σ^- spectra.

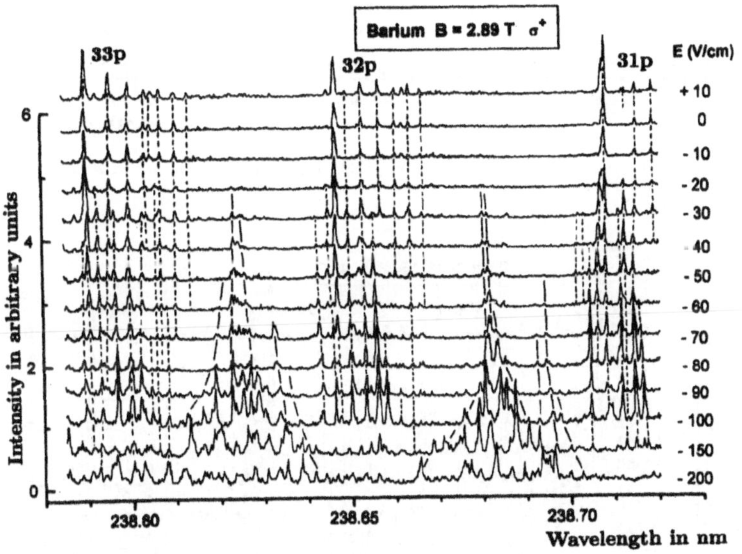

Figure 2. The Ba σ^+ experimental spectra near the 31p, 32p and 33p series members in crossed electric and magnetic fields.

spread out in energy. We have been able to follow most of the structure from one spectrum to the next as the electric field is varied in small steps 1, and our results show very clearly how important it is to be able to separate the polarisations: without this feature, the maps shown in Figs. 2-7 would have been impossible to make sense of, because two very different patterns, with different and complex evolution, would have been superimposed.

THE THEORETICAL METHOD

The availability of the new Imperial College experimental data encouraged the group in Belfast to devise a new theoretical treatment of non-hydrogenic atoms in crossed fields as an element of the collaboration with Professor Li's group at the Chinese Academy of Sciences Institute of Physics Wuhan. The new method is an extension of one previously developed for the 2-dimensional problem of non-hydrogenic atoms in parallel electric and magnetic fields. The basic method was reported by Halley *et al*[9, 10] (it will hence hereafter be referred to as the HDT method) and has already been used

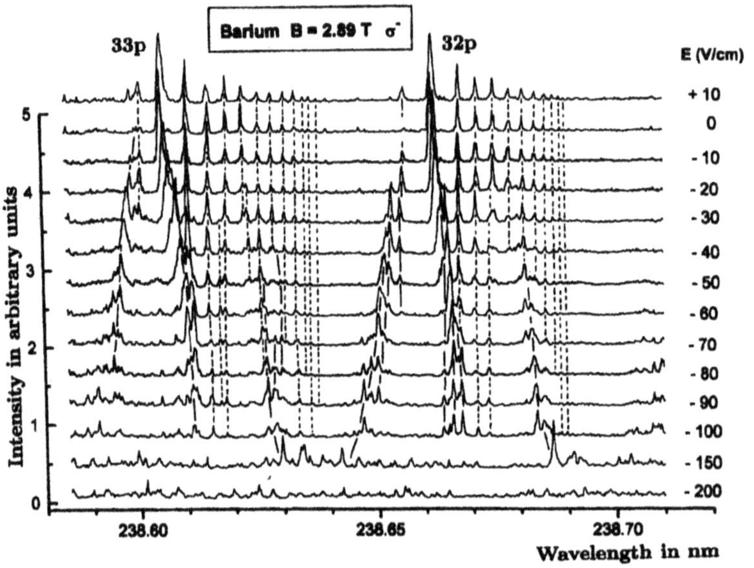

Figure 3. The Ba σ^- experimental spectra near the 31p, 32p and 33p series members in crossed electric and magnetic fields.

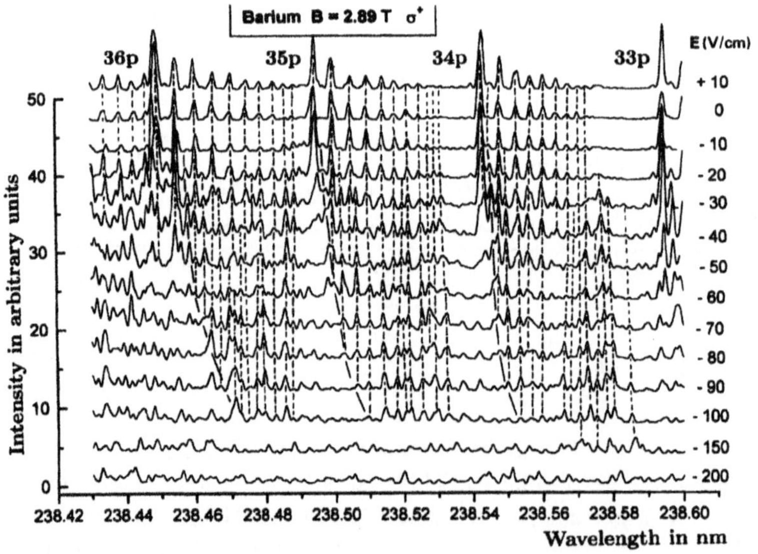

Figure 4. The Ba σ^+ experimental spectra near the 33p, 34p, 35p and 36p series members in crossed electric and magnetic fields.

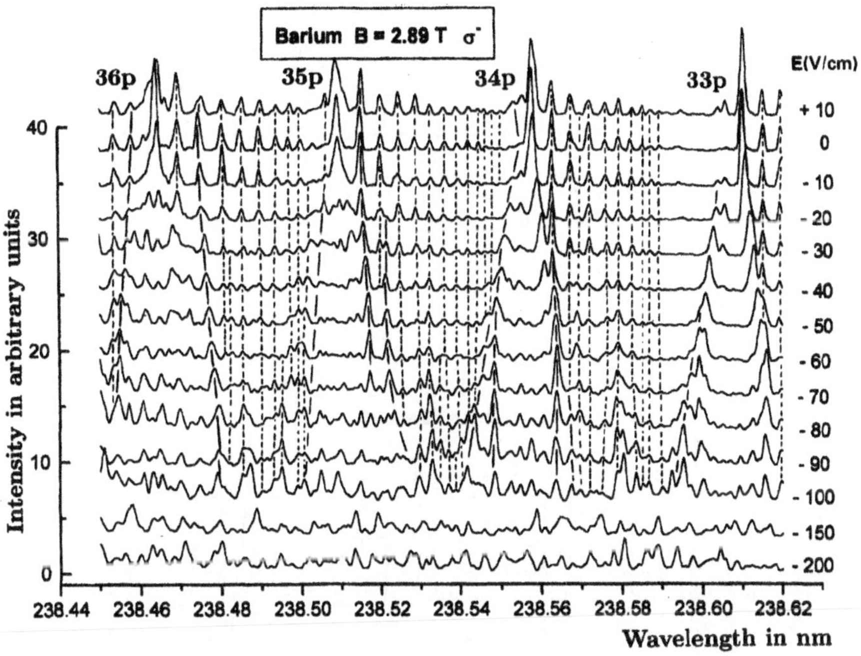

Figure 5. The Ba σ^- experimental spectra near the 33p, 34p, 35p and 36p series members in crossed electric and magnetic fields.

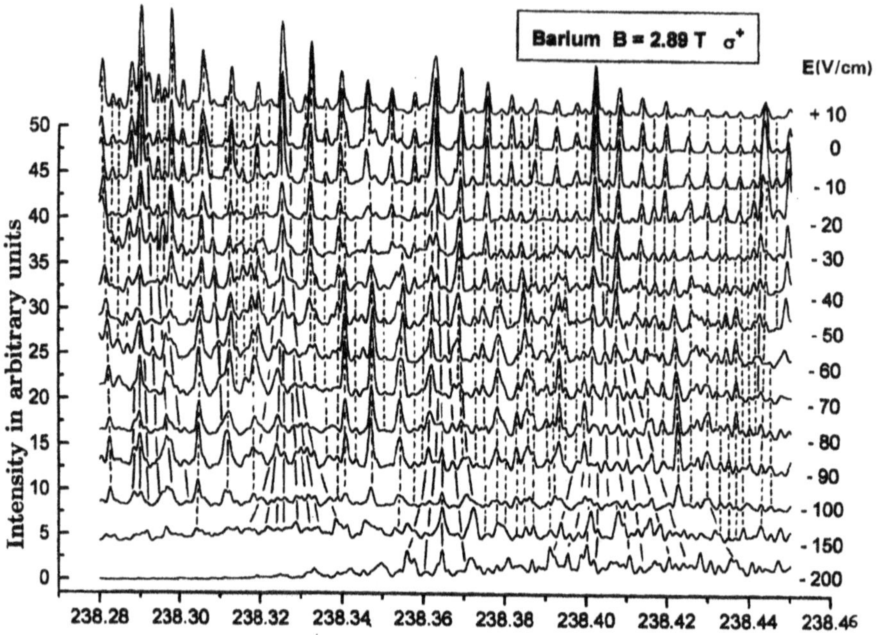

Figure 6. The Ba σ^+ experimental spectra near and above the 36p Rydberg member in crossed electric and magnetic fields.

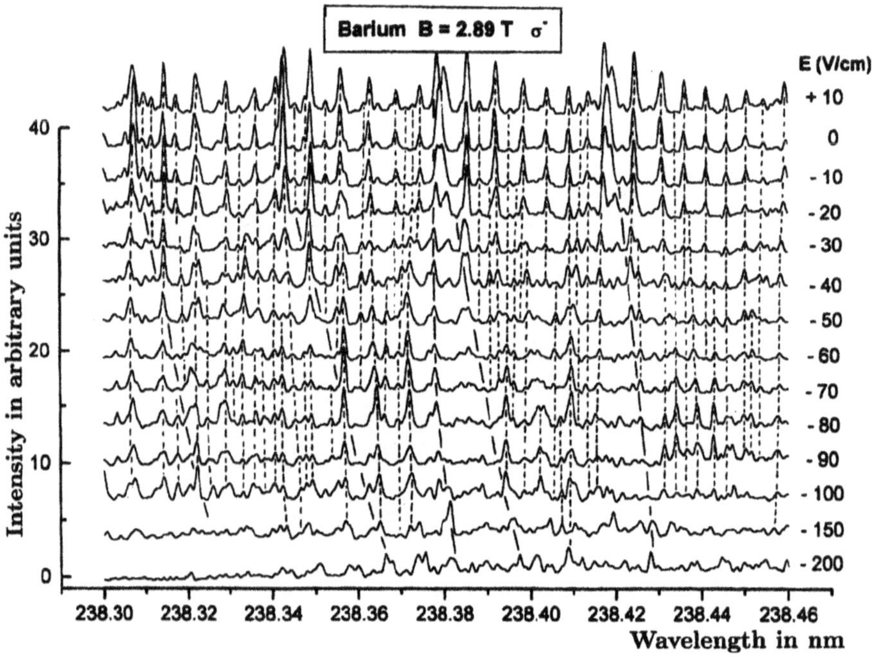

Figure 7. The Ba σ^- experimental spectra near and above the 36p Rydberg member in crossed electric and magnetic fields.

successfully in calculating the spectra of non-hydrogenic atoms in pure electric[11], pure magnetic[9, 10, 12] and parallel electric and magnetic fields[13] as well as the spectrum of the H_2 molecule in a magnetic field[14]. Since the experimental data relates to Rydberg states ($n \sim 30$) and is at a comparatively low level of resolution, we realised that a calculation to reproduce the experimental spectra would prove a useful test of the new method without requiring computing power beyond local workstations.

Before presenting an account of our new theoretical approach, we briefly review previous theoretical and computational work, within the quantum mechanical framework, on the crossed field problem.

Perturbation theory is appropriate when the influence of laboratory external fields is weak, most likely to be the case for low excited states. If the magnetic and electric perturbations are very small, then first order corrections to the Coulomb energy (with the quadratic Zeeman term dropped), are adequate. This was first established for hydrogen by Pauli[15]. As the external field strength increases, second order purturbation theory is useful, as demonstrated for energy-level positions by Solov'ev[16] and Braun and Solov'ev[17]. For non-hydrogenic atoms, Penent *el al*[18] combined first order perturbation theory with the resolvent formalism to analyse an experimental spectrum they obtained for rubidium in crossed weak fields. Marxer *et al*[19] used a basis of discrete Coulomb eigenfuctions from nearby manifolds to calculate the photoabsorption spectra of hydrogen and lithium in crossed fields in a low-energy region corresponding to principal quantum number n around 12. Main and Wunner[20, 21] calculated the spectrum of hydrogen in crossed fields going even above the ionization threshold by means of the complex co-ordinate rotation technique. Recently, Main *et al*[22] have calculated the quantum mechanical photoabsorption spectrum for hydrogen at constant scaled energy and at scaled electric field strength.

As far as we know however, there has hitherto been no effective quantum mechanical method for calculating the spectra of high-lying Rydberg states of non-hydrogenic atoms in crossed laboratory strength electric and magnetic fields. We now outline our new method.

Beyond a certain radial distance from the nucleus the Hamiltonian for a Rydberg electron of a multi-electron atom subject to an external magnetic field of strength B tesla along the z−axis and an electric field of strength F V/cm perpendicular to the magnetic field and along the x−axis can be written, in atomic units, as

$$H = -\frac{\nabla^2}{2} + V(r) + \frac{\gamma L_z}{2} + \frac{\gamma^2 r^2 \sin^2\theta}{8} + fr\sin\theta\cos\phi \qquad (1)$$

where $\gamma = B/B_0, B_0 = 2.35 \times 10^5$ T , $f = F/F_0, F_0 = 5.14 \times 10^9$ V/cm. $\gamma L_z/2$ and $\gamma^2 r^2 \sin^2\theta/8$ are the linear and quadratic Zeeman terms respectively and $V(r)$ is the potential (assumed radial) the electron experiences through interaction with the inner core formed by the other electrons and the nucleus. It is clear that the only remaining symmetry of this Hamiltonian is z - parity (reflection in the $x - y$ plane).

The HDT method combines a variant of the R-matrix method with quantum defect theory and also with complex coordinate rotation, if necessary. The latter allows the calculation of spectra above an ionization threshold[9, 10]. However the recent experiment on barium investigated a bound portion of the crossed field spectrum and a calculation over that spectral region does not require the complex coordinate rotation technique. Nevertheless even without this technique, the variant of the R-matrix method exploited by HDT brings about a considerable improvement in computational efficiency over previous applications of R-matrix techniques to bound states of non-hydrogenic atoms in external static fields. We here further exploit this aspect in order to deal with bound states of a non-hydrogenic atom in crossed fields.

The configuration space for the outermost electron of a non-hydrogenic atom in external static crossed fields of laboratory strength can be divided into an inner-region ($r \leq a$) and an outer-region ($r \geq a$). For a magnetic field less than say 5 tesla and an electric field less than say 1000 V/cm, a can be taken as about $10a_0$, where a_0 is the Bohr radius. Given such a value of a, then for $r < a$ the diamagnetic term $\gamma^2 r^2 \sin^2\theta/8$ and the Stark term $fr\sin\theta\cos\phi$ can be completely ignored in comparison with the potentials arising from the much stronger electron-nucleus and the electron-electron interactions. The description of this inner-region thus collapses to that of a conventional multi-electron atom completely free of any external fields. Moreover this size of inner-region is large enough to completely contain all but the outermost electron which can be taken to move in a pure Coulomb potential near $r = a$, i.e. in Eq. (1), near $r = a$ $V(r) = -1/r$. Hence quantum defect theory[23], allows the wavefunction for this electron to be written in each orbital angular momentum partial wave ℓ on the boundary $r = a$, as a linear combination of regular $f_\ell(a)$ and irregular $g_\ell(a)$ Coulomb functions,

$$f_\ell(a) + \tan\pi\mu_l\, g_\ell(a) \quad , \tag{2}$$

where the energy dependent quantum defect μ_ℓ -a parameter of the field-free atom-controls the linear combination.

In the outer-region $r > a$, the diamagnetic term, the linear Zeeman term and the Stark term cannot be ignored, but the problem reduces to that of one electron moving under the combined influence of the Coulomb potential and the external fields. In this region the eigenfunction ψ at energy E satisfies,

$$\left\{-\frac{\nabla^2}{2}-\frac{1}{r}+\frac{\gamma^2 r^2\sin^2\theta}{8}+\frac{\gamma L_z}{2}+fr\sin\theta\cos\phi\right\}\psi(r,\theta,\phi) = E\psi(r,\theta,\phi) \quad , \quad a\leq r\leq\infty \quad . \tag{3}$$

It is convenient here to use a spherical basis set expansion of ψ in terms of radial functions $F_{\ell,m}(r)$ for angular momentum ℓ and the magnetic quantum number m, multiplied by spherical harmonic functions $Y_{\ell,m}(\theta,\phi)$, viz:

$$\psi = \sum_{\ell,m}\frac{F_{\ell,m}(r)}{r}Y_{\ell,m}(\theta,\phi) \quad . \tag{4}$$

Substituting this expansion into Eq. (3) yields,

$$\sum_{\ell,m}\left\{-\frac{\partial^2}{\partial r^2}+\frac{\ell(\ell+1)}{r^2}-\frac{2}{r}+\frac{1}{4}\gamma^2 r^2\sin^2\theta+\gamma L_z+2fr\sin\theta\cos\phi\right\}F_{\ell,m}(r)Y_{\ell m}$$
$$= 2E\sum_{\ell,m}F_{\ell,m}(r)Y_{\ell m}, \quad . \quad a\leq r\leq\infty \quad . \tag{5}$$

In order to make the operators on the left hand side of Eq. (5) Hermitian over just the outer region we must follow Bloch[24], by replacing $\frac{\partial^2}{\partial r^2}$ for each partial wave ℓ by $\frac{\partial^2}{\partial r^2}+\delta(r-a)\frac{\partial}{\partial r}$. But added flexibility is obtained (and Hermiticity is preserved) if for each ℓ a non-derivative term $-\delta(r-a)B_{\ell,m}$ is also brought in. We thus arrive at,

$$\sum_{\ell,m}\left\{-\frac{\partial^2}{\partial r^2}-\delta(r-a)\left(\frac{\partial}{\partial r}-B_{\ell,m}\right)+\frac{\ell(\ell+1)}{r^2}-\frac{2}{r}+\frac{1}{4}\gamma^2 r^2\sin^2\theta\right.$$
$$\left.+\gamma L_z+2fr\sin\theta\cos\phi\right\}F^k_{\ell,m}(r)Y_{\ell m} = 2E_k\sum_{\ell,m}F^k_{\ell,m}(r)Y_{\ell m} \quad a\leq r\leq\infty \quad . \tag{6}$$

This is an eigenvalue equation which can be solved for (real) eigenvalues E_k and corresponding eigenfunction components $F^k_{\ell,m}(r)$. The crux of the method lies in the choice of $B_{\ell,m}$. If, as first pointed out by Schneider[25], we choose,

$$B_{\ell,m} = \frac{1}{F_{\ell,m}(r)} \frac{\mathrm{d}F_{\ell,m}(r)}{\mathrm{d}r}\bigg|_{r=a} \quad , \tag{7}$$

where $F_{\ell,m}(r)$ is the solution to Eq. (5) at energy E, it turns out that an eigenvalue E_k of Eq. (6) is co-incident with E and that the corresponding eigenfunction components $F^k_{\ell,m}(r)$ coincide with $F_{\ell,m}(r)$ over the range $a \leq r \leq \infty$.

It would seem from the form of $B_{\ell,m}$ that we actually need to solve Eq. (5) for $F_{\ell,m}(r)$ before such a $B_{\ell,m}$ can be constructed. But since we are considering Rydberg states where the density of states is very high, the value of $B_{\ell,m}$ changes very slowly from one level to the next at an a of typically 10 Bohr. In fact it can be evaluated at some chosen energy from the Coulomb function combination in Eq. (2) and the Lanczos algorithm used to find eigenvalues (and eigenfunctions) of Eq. (6) in the neighbourhood of this chosen energy. Typically, about 100 of these eigenvalues will be found to be eigenenergies of Eq. (3).

In practical calculations the $F^k_{\ell,m}(r)$ of Eq. (6) are expanded on a basis of Sturmian functions $S^{(\zeta)}_{n\ell}(r)$, $n = \ell+1, \ell+2, ..., n_{max}$ over the range $a \leq r \leq \infty$, giving rise to a generalised matrix eigenvalue equation which can be solved for eigenenergies E and the corresponding eigenvectors. For the definition of the Sturmian basis used see Clark and Taylor[26]. The number of Sturmian functions retained (i.e. the value of n_{max}) as well as the value chosen for the range parameter ζ are both governed by the convergence requirement of the computation. The use of this method allows us to calculate the energy levels and oscillator strengths of barium in crossed fields.

The major difference between the new crossed field implementation of the HDT method and previous applications is the need to handle a significant range of m values in a given calculation (only in one previous application of the HDT method, to diamagnetic H_2 by He et al (1995), was m not a good quantum number). This significant range in m inevitably leads to much larger basis sets, particularly as the strength of the electric field increases and larger ranges in m are required to gain convergence. We have ordered the basis functions so that the label n is run over most, and the label m least, rapidly.

We have calculated the barium photoabsorption spectra using the above method over the wavelength ranges explored experimentally and for electric field strengths up to -90 V/cm. The results were very sensitive to the magnetic field strength and we found best agreement between theory and experiment for a magnetic field strength of 2.86 tesla which is just within the uncertainty range of the magnetic field strength used in the experiment[1] reported[1] as 2.89 ± 0.03 tesla. In Figures 8 and 9 below we display our calculated σ^+ and σ^- photoabsorption spectra over the lowest spectral range where, to reproduce experimental conditions, we have built in a finite resolution of 3.6 GHz. We have also reported elsewhere[2] results for the higher spectral ranges.

THEORETICAL ANALYSIS OF EXPERIMENTAL RESULTS

The excellent agreement between calculation and experiment for the barium spectra is highlighted in Figures 10 and 11 where we compare directly sample results from experiment and calculation - in both σ^+ and σ^- polarization - from the lowest and highest energy spectral regions respectively.

We concentrate on Figures 2 and 3 and on Figures 8 and 9 which display respectively the experimental and theoretical results for the longest wavelength σ^+ and σ^-

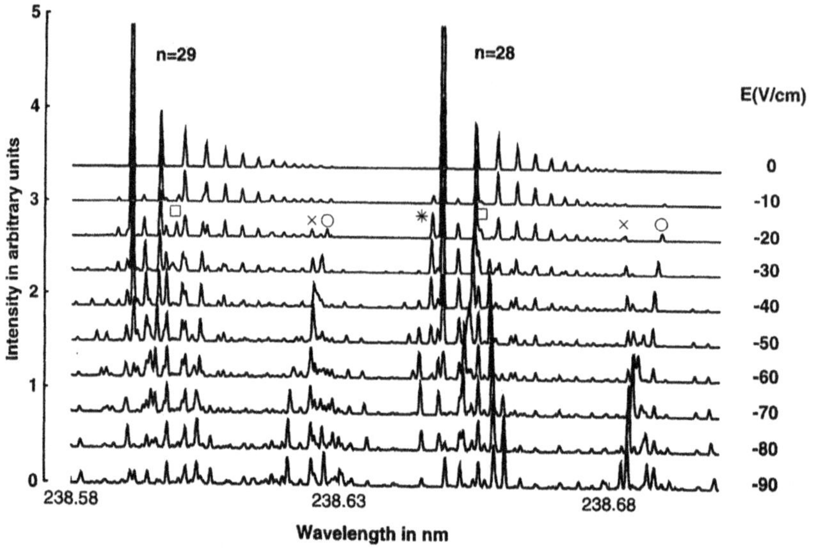

Figure 8. Calculated intensities in the σ^+ spectrum of barium in crossed electric and magnetic fields. For all spectra the magnetic field strength is 2.86 tesla. The wavelength region is that of Figure 2 above.

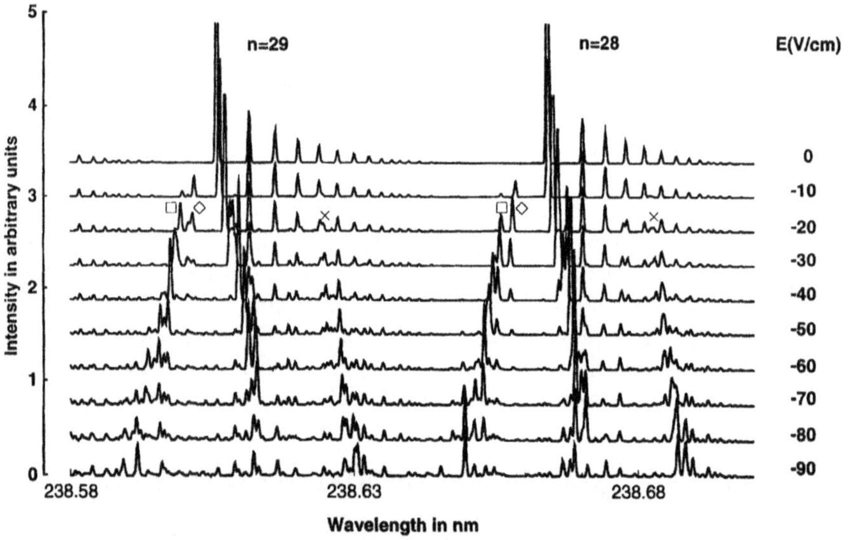

Figure 9. Calculated intensities in the σ^- spectrum of barium in crossed electric and magnetic fields. For all spectra the magnetic field strength is 2.86 tesla. The wavelength region is that of Figure 3 above.

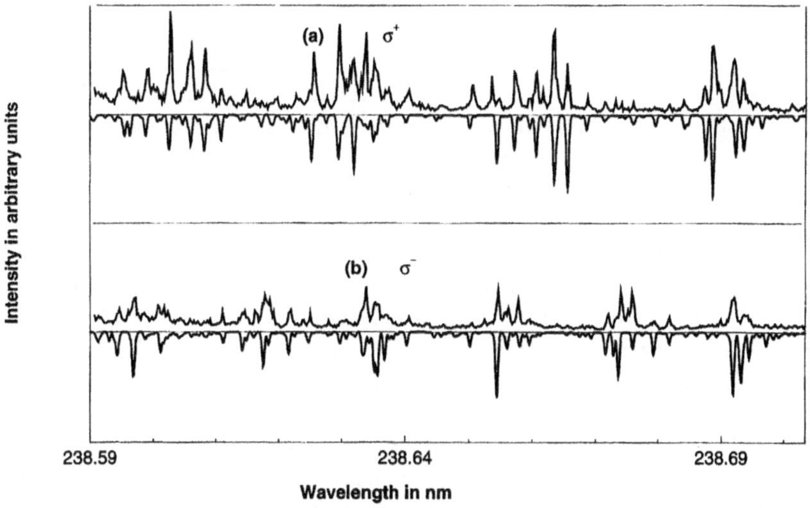

Figure 10. Comparison between experimental and calculated (mirror image) intensities in the barium crossed field photoabsorption spectra in a magnetic field of 2.86 tesla and an electric field strength of -90 V/cm, and over the wavelength range of Figures 2 and 3 above. The upper and lower frames correspond to σ^+ and σ^- polarization respectively.

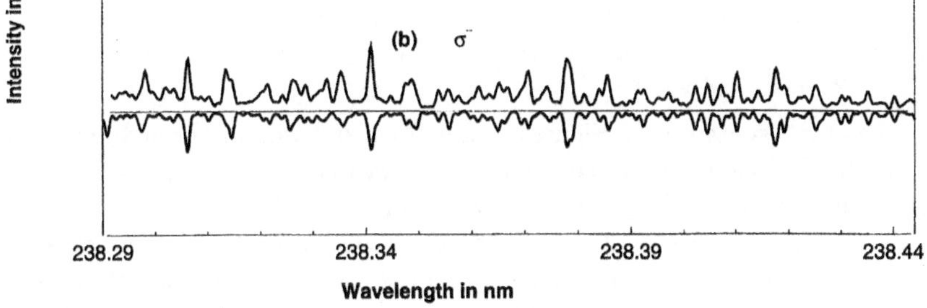

Figure 11. Comparison between experimental and calculated (mirror image) intensities in the barium crossed field photoabsorption spectra in a magnetic field of 2.86 tesla and an electric field strength of -30 V/cm and over the wavelength range of Figures 6 and 7 above. The upper and lower frames correspond to σ^+ and σ^- polarization respectively.

spectra. In both these σ^+ and σ^- spectra non-hydrogenic spectral patterns develop as the electric field increases, and these patterns are markedly different in the two cases. We find these patterns can be understood through identifying, at low electric field strengths, the quantum defect shifted locations of low m low ℓ states. These states effectively act as 'seed' states, bringing about significant oscillator strength concentrations at their non-hydrogenic locations as the electric field increases. If we first consider the σ^+ spectrum at -20 V/cm and work across from longer to shorter wavelengths, the first significant feature we encounter at about 238.69 nm is marked with an open circle. This feature we have identified, through analysis of the corresponding wavefunction, as a state which is predominantly $m = 2, \ell = 2$. The next feature to the left, marked with a cross, has been found through the same wavefunction analysis to be predominantly $m = 0, \ell = 0$. Both states are of course dipole excited by their admixture of adjacent $m = 1, \ell = 1$. Corresponding states near 238.62 nm pertaining to the next higher $n-$manifold are marked with identical symbols. (It is worth noting that the hydrogenic location of a predominantly $m = 2, \ell = 2$ state would instead be at about 238.64 nm (to the left of the line marked with a star which is predominantly $m = 2, \ell = 4$), and that of a predominantly $m = 0, \ell = 0$ state would be at about 238.66 nm, in the middle of the fundamental $n = 28$ cluster.) Although contributing only minor spectral features around 238.62 nm and 238.69 nm for an electric field of -20 V/cm, we note that as the electric field increases, $m = 1, \ell = 1$ character readily mixes in more and more -through the adjacent predominant values of m and ℓ in the states- and oscillator strength in the lines increases. By -50 V/cm there is secondary redistribution of $m = 1, \ell = 1$ character from these 'seed' states to others in the spectral neighbourhood and a multiline cluster of significant oscillator strength emerges. This explains the clusters of significant lines that emerge in Figure 4 at these wavelengths, and which are found[2], to be absent in hydrogen. There is one further m, ℓ combination, namely $m = 0, \ell = 2$, adjacent to the dipole driven $m = 1, \ell = 1$ combination, and which we can expect to mix strongly on increase of electric field. At -20 V/cm, states predominantly $m = 0, \ell = 2$ in character are marked with an open square at about 238.655 nm and at about 238.60 nm. In each case, because of the $d-$wave quantum defect shift they lie closer to the principal line of the fundamental cluster than is the case in hydrogen. Once again as the electric field strength increases, the consequence is a significant concentration of oscillator strength in the vicinity of the 'seed' state, to the disadvantage of states at somewhat longer wavelengths which, for hydrogen, gained oscillator strength.

To recap, we can understand the non-hydrogenic evolution of the barium σ^+ spectrum of Figures 2 and 8 under increasing electric field strength, to be controlled by the low electric field locations of predominantly $m = 0, \ell = 0$, $m = 0, \ell = 2$ and $m = 2, \ell = 2$ 'seed' states.

Can the σ^- spectrum of Figures 3 and 9 be understood through similar considerations? In that case the dipole driven combination is $m = -1, \ell = 1$ and adjacent combinations are $m = 0, \ell = 0$, $m = 0, \ell = 2$ and $m = -2, \ell = 2$. Thus as before let us begin by establishing the locations of states dominated by these symmetries in a low electric field strength spectrum. Again we choose the -20 V/cm case. Here in Figure 5, crosses near 238.69 nm and near 238.62 nm mark the locations of $m = 0, \ell = 0$ states, open squares near 238.66 nm and near 238.60 nm mark the locations of $m = 0, \ell = 2$ states and open diamonds (near the open squares) the locations of $m = -2, \ell = 2$ states. Once again, as in Figure 4, we see the role of these as 'seed' states: as the electric field strength increases, increasing oscillator strength enters these comparatively isolated spectral regions first through increased mixing of the 'seed' states with $m = -1, \ell = 1$ followed by more general, but localised, redistribution.

As we move to higher energy spectra in the two polarizations we find 'seed' states play a less and less commanding role. This is because even at zero electric field, the character of individual states is less and less dominated by a single m, l pair.

CONCLUSIONS AND FUTURE PROSPECTS

A number of important consequences have been found to ensue from the complementary nature of the laboratory measurements and theoretical calculations. The first consequence has been an ability to determine the laboratory magnetic field strength to a higher level of accuracy than possible in the experiment itself. A second consequence[2] has been the ability to determine quantum defects for the participating partial waves to a higher level of accuracy than previously possible through field-free measurements alone. The third and most important consequence has been the ability to elucidate in detail what specific electron dynamics underlie the observed (markedly non-hydrogenic) evolution of the crossed field barium spectra under weak, but increasing, electric field strength.

Prospects for further experiments include fixed scaled energy spectroscopy of Rydberg states of the alkaline earths in crossed fields. The present work has been useful in vindicating the accuracy of the new theoretical approach outlined above but in the near future, given appropriate computational resources, we would wish to extend this to handle other non-hydrogenic atoms and complement yet higher resolution experimental work.

Acknowledgements

On the experimental side MSZ and NEK are grateful to the Chinese Academy of Sciences (CAS) and the European Union respectively for financial support of short-term visits to Imperial College. The lasers used in the experiments were those of the Blackett Laboratory Laser Consortium. On the theoretical side the work at Queen's University Belfast was supported by the Royal Society of London and CAS under Joint Project Q692.

REFERENCES

1. J.-P. Connerade G. Droungas N.E. Karapanagioti and M.S. Zhan (1997) J. Phys. B: At. Mol. Opt. Phys. **30** 2047
2. J. Rao and K.T. Taylor (1997) J. Phys. B: At. Mol. Opt. Phys. **30** 3627
3. W.R.S. Garton and F.S. Tomkins (1969) Astrophys. J. **158** 1219
4. K.J. Drinkwater J. Hormes D.D. Burgess J.P. Connerade and R.C.M. Learner (1984) J.Phys.B: At. Mol. Opt. Phys. **17** L439
5. G. Droungas N.E. Karapanagioti and J.-P. Connerade (1995) Phys. Rev. **A51** 191
6. J.-P. Connerade G. Droungas R. Elliott X.H. He N. Karapanagioti W. A. Farooq H. Ma J.P. Marangos and M. Nawaz (1994) J. Phys. B: At. Mol. Opt. Phys. **27** 2753
7. R.J. Elliott G. Droungas and J.-P. Connerade (1995) J. Phys. B: At. Mol. Opt. Phys. **28** L537
8. R.J. Elliott G. Droungas J.-P. Connerade X.-H. He and K.T. Taylor (1996) J. Phys. B: At. Mol. Opt. Phys. **29** 3341
9. M.H. Halley, D. Delande and K.T. Taylor 1992 J. Phys. B: At. Mol. Opt. Phys. **25** L525
10. M.H. Halley, D. Delande and K.T. Taylor 1993 J. Phys. B: At. Mol. Opt. Phys. **26** 1775
11. I. Seipp and K.T. Taylor 1994 J. Phys. B: At. Mol. Opt. Phys. **27** 2785
12. D. Delande, K.T. Taylor, M. H. Halley, T. van der Velt, W. Vassen and W. Hogervorst 1994 J. Phys. B: At. Mol. Opt. Phys. **27** 2771
13. I. Seipp, K.T. Taylor and W. Schweizer 1996 J. Phys. B: At. Mol. Opt. Phys. **29** 1

14. X.-H. He, K.T. Taylor and T.S. Monteiro 1995 J. Phys. B: At. Mol. Opt. Phys. **28** 2621
15. W. Pauli W 1926 Z. Phys. **36** 336
16. E.A. Solov'ev 1983 Sov. Phys. - JETP **58** 63
17. P.A. Braun and E.A. Solov'ev 1984 Sov Phys - JETP **59** 38
18. F. Penent, D. Delande and J.-C. Gay 1988 Phys. Rev. **A37** 4707
19. H. Marxer, I. Moser, F. Mota-Furtado and P.F. O'Mahony 1994 J. Phys. B: At. Mol. Opt. Phys. **27** 4491
20. J. Main and G. Wunner 1992 Phys. Rev. Lett. **69** 586
21. J. Main and G. Wunner 1994 J. Phys. B: At. Mol. Opt. Phys. **27** 2835
22. J. Main, Mandelshtam and H.S. Taylor 1997 submitted to Phys. Rev. Lett.
23. M.J. Seaton 1983 Rep. Prog. Phys. **46** 167
24. C. Bloch 1957 Nuc. Phys. **4** 503
25. B.I. Schneider 1981 Phys. Rev. **A24** 1
26. C.W. Clark and K.T. Taylor 1982 J. Phys. B: At. Mol. Phys. **15** 1175

HYDROGEN-LIKE IONS MOVING IN A STRONG MAGNETIC FIELD

V. G. Bezchastnov,[1,3] G. G. Pavlov,[2] and J. Ventura[3]

[1]Ioffe Physical-Technical Institute, 194021, St.-Petersburg, Russia
[2]Pennsylvania State University,
525 Davey Lab, University Park, PA 16802, U.S.A.
[3]Department of Physics, University of Crete,
GR-711 09, Heraklion, Crete, Greece

INTRODUCTION

For a long time, the calculation of the energy states of an electrically charged two-body system (hydrogen-like ion) moving in a magnetic field has remained a challenging problem, and its full solution has not been obtained. It was addressed by many authors, starting from the paper of Avron et al. (1978). The main difficulty of this problem is that the center-of-mass (CM) motion across the magnetic field cannot be separated from the relative motion, so that the problem cannot be reduced to a one-particle Schrödinger equation. In this respect the charged two-body problem differs essentially from the neutral case (hydrogen atom), for which significant progress has been achieved in the investigation of the energy states with allowance for the effect of atomic motion (see, e.g., Vincke et al., 1992; Potekhin, 1994). To overcome this difficulty, Baye (1982) performed an approximate separation of the CM and relative motions of the charged system. This approach was further developed to obtain CM corrections for low-lying energy levels of hydrogenic ions in a magnetic field (Baye and Vincke, 1986). For highly excited states, the effect of CM coupling to the internal motion on the classical dynamics of moving ions was investigated by Schmelcher (1995).

A reliable numerical method to determine the eigenstates of a complicated quantum system invokes multi-configurational Hartree-Fock (MCHF) computations (see, e.g., Ruder et al., 1994). For the hydrogenic ion moving in a magnetic field, such a method has not yet been employed, mainly due to the absence of appropriate Hartree-Fock basis states. A convenient *two-particle* basis set was suggested by Bezchastnov, 1995 (hereafter B95). It is constructed from the Landau states of an electron and nucleus and corresponds to the *exact* integrals of motion in the presence of the Coulomb interaction. This set is most useful for strong magnetic fields, when $\gamma = a_0^2/a_{\rm m}^2 = B/B_0 \gg 1$, where a_0 is the Bohr radius, $a_{\rm m} = \sqrt{c\hbar/(eB)}$ is the magnetic length, and $B_0 = 2.35 \times 10^9$ G. We report here results of MCHF computations for the He^+ energy states obtained by exploiting the B95 basis states (Section 2). We also suggest a simple perturbation treatment of the finite nucleus mass effect (Section 3). We start with a brief description of the main quantum mechanical operators related to the problem (Section 1), and give the conclusions in Section 4.

1. BASIC OPERATORS

Let us consider the system of two particles with masses m_- and m_+ and charges $-e$ and Ze ($e > 0$, $Z > 1$), respectively, moving in a uniform magnetic field $\boldsymbol{B} = (0, 0, B)$. The particle longitudinal and transverse coordinates are denoted by $z_\mp$ and $\boldsymbol{r}_\mp$, and the relative coordinates are defined as $z = z_- - z_+$ and $\boldsymbol{r} = \boldsymbol{r}_- - \boldsymbol{r}_+$. The conventional notations, $M = m_- + m_+$ and $\mu = m_- m_+/M$, will stand for the total and reduced masses. We neglect the contribution of spin terms into the total two-particle Hamiltonian and adopt the cylindrical gauge $\boldsymbol{A}(\boldsymbol{r}_\mp) = (1/2)\boldsymbol{B} \times \boldsymbol{r}_\mp$ for the vector potential. The CM motion along $\boldsymbol{B}$ is free and can easily be separated. Then our problem reduces to searching for eigenstates of the Hamiltonian

$$H = \frac{p_z^2}{2\mu} + \frac{1}{2m_-}\left(\boldsymbol{p}_- + \frac{e}{2c}\boldsymbol{B} \times \boldsymbol{r}_-\right)^2 + \frac{1}{2m_+}\left(\boldsymbol{p}_+ - \frac{Ze}{2c}\boldsymbol{B} \times \boldsymbol{r}_+\right)^2 - \frac{Ze^2}{\sqrt{z^2 + r^2}}, \quad (1)$$

where $p_z = -i\hbar\partial/\partial z$ and $\boldsymbol{p}_\mp = -i\hbar\partial/\partial \boldsymbol{r}_\mp$. The commuting integrals of motion of the Hamiltonian (1) are determined by the pseudomomentum squared, P_0^2, where

$$\boldsymbol{P}_0 = \boldsymbol{p}_- - \frac{e}{2c}\boldsymbol{B} \times \boldsymbol{r}_- + \boldsymbol{p}_+ + \frac{Ze}{2c}\boldsymbol{B} \times \boldsymbol{r}_+, \quad (2)$$

and by the longitudinal component L_z of the total angular momentum (see, e.g., Avron et al., 1978). The corresponding eigenvalues are given by $P_0^2 = (\hbar/a_{\mathrm{m}})^2(Z-1)(2N_0+1)$ and $L_z = -\hbar L$, where $N_0 = 0, 1, 2, \ldots$ and $L = L_{\min}, L_{\min} + 1, L_{\min} + 2, \ldots$ It can be shown that $L_{\min} \geq -N_0$, and it depends on two additional quantum numbers to be introduced to describe the transverse motion of the system.

2. MCHF APPROACH

The two-particle basis wave functions introduced in B95 read

$$\langle \boldsymbol{r}_+, \boldsymbol{r}_- | N_0, L, n_-, n_+ \rangle = U \sum_{k=0}^{N+n_-} c_k^{(Z,N+n_-,n_+)} \Phi_{N_0,N+n_--k}^{(Z-1)}(\boldsymbol{r}_+) \Phi_{n_-,k}^{(1)}(\boldsymbol{r}), \quad (3)$$

where $U = \exp[(\mathrm{i}e/2\hbar c)(\boldsymbol{B} \times \boldsymbol{r}_-)\boldsymbol{r}_+]$, $N = N_0 + L$, $\Phi_{N_0,N+n_--k}^{(Z-1)}$ and $\Phi_{n_-,k}^{(1)}$ are the Landau functions, and the coefficients $c_k^{(Z,N+n_-,n_+)}$ are generated by a recurrence relation (see Eqs. (A2), (A3), and (32) of B95). The basis states describe the motion of two non-interacting particles across $\boldsymbol{B}$ with the energy spectrum $E_{n_-,n_+}^{\mathrm{Lan}} = E_{n_-}^{\mathrm{Lan}} + E_{n_+}^{\mathrm{Lan}}$ determined by the Landau level numbers $n_-, n_+ = 0, 1, 2, \ldots$ of the negative and positive charges,

$$E_{n_-}^{\mathrm{Lan}} = \frac{\hbar e B}{m_- c}\left(n_- + \frac{1}{2}\right), \quad E_{n_+}^{\mathrm{Lan}} = \frac{\hbar Z e B}{m_+ c}\left(n_+ + \frac{1}{2}\right). \quad (4)$$

They are also the common eigenstates of the operators P_0^2 and L_z that determine the integrals of motion for the Hamiltonian (1). Expanding the total two-particle wave function over the basis states, one arrives at a set of coupled differential equations for the expansion coefficients which are functions of the relative longitudinal coordinate (Eq. (41) of B95). We have developed a numerical code to solve these equations for the

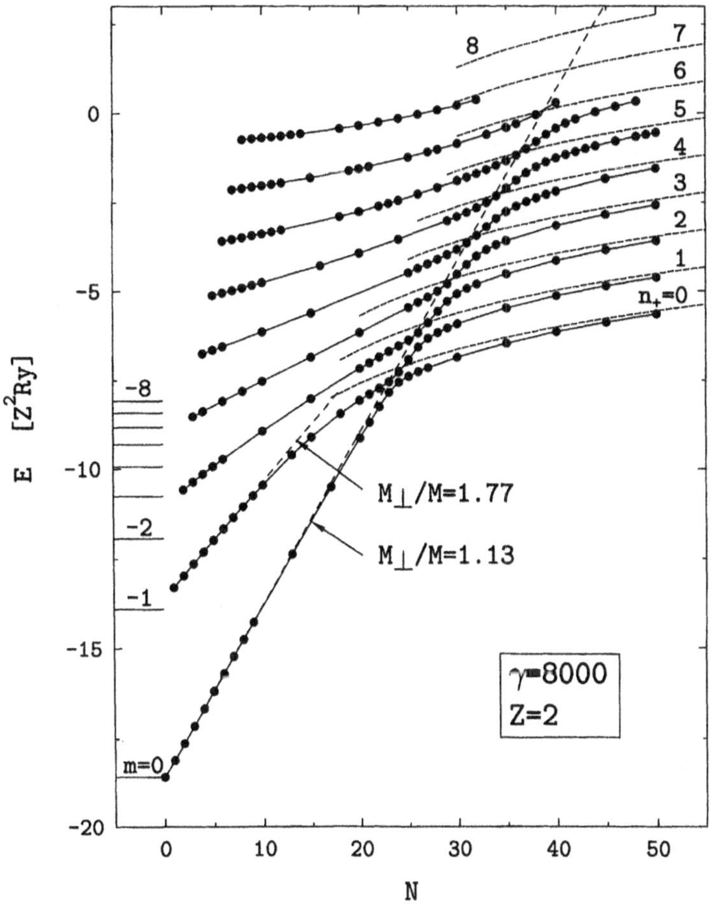

Figure 1. Discrete energy states (in units of Z^2Ry) for the nine lowest tightly-bound levels of the He^+ ion (nucleus charge number $Z = 2$, ion mass $M = 4.0032$ a.u.m.) moving across the magnetic field $B = 1.88 \times 10^{13}$ G.

discrete energy states and performed the computations for the He^+ ion moving in the magnetic field $B = 1.88 \times 10^{13}$ G corresponding to $\gamma = 8000$.

In Figure 1 we present the discrete energies for the nine lowest tightly-bound states plotted against N. We exclude from all the energies the additive term of kinetic energy of the CM motion along the magnetic field, as well as the zero-point Landau energies for both charges, so that the zero energy in Figure 1 corresponds to the continuum edge. The results of the multi-configurational computations are plotted by dots in this graph and connected by smooth solid lines for convenience. Short-dash lines show results obtained in the adiabatic (one-configurational) approximation when only one basis state is included in the wave function expansion (these basis states correspond to $n_- = 0$ and $n_+ = 0, 1, \ldots, 8$ indicated near the curves).

For low N, the discrete energies grow linearly with N. However, this oscillator-like behavior holds only for N below some critical value N_c (growing with the number of the energy level), at which the neighboring levels approach each other. For higher values of N, the neighboring levels diverge from each other (thus exhibiting anticrossings at $N \approx N_c$) and then approach smoothly the adiabatic asymptotes. This means that with further increase of N, the Coulomb binding becomes negligible, and the binding energy for the ground level tends to zero, whereas excited levels approach the threshold values, $E_{0,n_+}^{\rm Lan} - E_{0,0}^{\rm Lan}$, lying in the continuum. Therefore, each excited level in Figure 1 is subject to autoionization at sufficiently large N.

3. PERTURBATION APPROACH

Let us define the zero-order approximation to the Hamiltonian (1) as its infinite nucleus mass limit, $H^{(0)} = H(m_+ \to \infty)$. Then the zero-order states are determined by the electron motion around a static Coulomb center. For $\gamma \gg 1$, they can be conveniently labeled by the electron Landau level number $n_- = 0, 1, 2, \ldots$, magnetic quantum number $m = n_-, n_- - 1, n_- - 2, \ldots$ and "longitudinal" quantum number (number of nodes of the z-dependence of the bound electron wave function) $\nu = 0, 1, 2, \ldots$ Due to an arbitrary location of the Coulomb center ($H^{(0)}$ still depends on $\boldsymbol{r}_+$), two additional quantum numbers should be specified in order to define the corresponding eigenfunctions. The best option is to use N_0 and L that determine the integrals of motion. Then the zero-order wave functions read

$$\langle \boldsymbol{r}_+, \boldsymbol{r}_-, z | N_0, L, n_-, m, \nu \rangle = U\, \Phi^{(Z-1)}_{N_0, N+m}(\boldsymbol{r}_+)\, \Phi^{(1)}_{n_-, n_- - m}(\boldsymbol{r})\, g_{n_-, m, \nu}(z). \tag{5}$$

In this equation, $\Phi^{(Z-1)}_{N_0, N_0+L+m}$ and $\Phi^{(1)}_{n_-, n_- - m}$ are the Landau functions, the first one takes into account the integrals of common motion, and the second one describes the transverse confinement of the electron in a given cyclotron orbit. The longitudinal part of (5), $g_{n_-, m, \nu}$, describes the electron states bound by the one-dimensional potential well $V_{n_-, m}(z) = \langle n_-, m | - Ze^2/\sqrt{r^2 + z^2} | n_-, m \rangle$ (adiabatic approximation). We denote the corresponding binding energies as $\epsilon_{n_-, m, \nu}$. Using the nucleus charge scaling, $\epsilon_{n_-, m, \nu}(Z, B) = Z^2 \epsilon_{n_-, m, \nu}(1, B/Z^2)$, they can be easily obtained from the binding energies for the static hydrogen atom which were precisely calculated and tabulated for different values of the magnetic field (see Ruder et al., 1994). The zero-order energies are then given by $E^{(0)}_{n_-, m, \nu} = E^{\rm Lan}_{n_-} - \epsilon_{n_-, m, \nu}$. In Figure 1, we show the zero-order energy spectrum for $n_- = \nu = 0$ and $m = 0, -1, \ldots, -8$ by the horizontal solid bars. We have checked that the binding energies for $m = 0, 1$ and 2, available from Ruder et al. (1994), are reproduced by our MCHF computations if one inputs zero for the electron-to-nucleus mass ratio, and we have used our code in this way to compute the energies for the states with higher m.

Now, we can treat the difference $H - H^{(0)}$ as the perturbation and estimate the dependence of the energy on the quantum number $N = N_0 + L$ of the collective motion. After calculation of the first- and second-order energy corrections, we can present the result in the following form

$$E_{N, n_-, m, \nu} = E^{(0)}_{n_-, m, \nu} + \Delta E_\perp + \frac{\hbar (Z-1) eB}{M_\perp c}(N + m), \tag{6}$$

where $\Delta E_\perp = \Delta E_{n_-, m, \nu}$ is an energy shift, and $M_\perp = M_{n_-, m, \nu}$ is an effective transverse mass of the ion. Neglecting the difference between μ and m_- at $m_+/m_- \gg 1$ and taking into account that $\langle g_{n_-, m\pm1, \nu'} | g_{n_-, m, \nu} \rangle \simeq \delta_{\nu', \nu}$ at $\gamma \gg Z^2$, we can write for these quantities the following equations:

$$\Delta E_{n_-, m, \nu} = \frac{\hbar Z e B}{2 m_+ c} + \frac{\hbar e B}{m_+ c}(n_- - m)\left[1 - \frac{\hbar (Z-1) e B}{m_+ c}\frac{1}{\epsilon_{n_-, m, \nu} - \epsilon_{n_-, m+1, \nu}}\right],$$
$$\frac{m_+}{M_{n_-, m, \nu}} = 1 - \frac{\hbar e B}{m_+ c}\left[\frac{n_- - m + 1}{\epsilon_{n_-, m, \nu} - \epsilon_{n_-, m-1, \nu}} - \frac{n_- - m}{\epsilon_{n_-, m+1, \nu} - \epsilon_{n_-, m, \nu}}\right]. \tag{7}$$

As an example, let us estimate $M_\perp$ for the He^+ ion ($Z = 2$, $m_+ = 4.0026$ a.u.m.) at $B = 1.88 \times 10^{13}$ G ($B/Z^2 = 4.7 \times 10^{12}$ G which corresponds to $\beta_Z = \gamma/(2Z^2) = 1000$). From the Table A1.2 of Ruder et al. (1994) we have $\epsilon_{0,0,0} = 18.61\ Z^2$Ry, $\epsilon_{0,-1,0} = 13.90\ Z^2$Ry and $\epsilon_{0,-2,0} = 11.94\ Z^2$Ry for the zero-order binding energies. Using these three quantities, we obtain $M_\perp/M = 1.13$ for the ground level and $M_\perp/M = 1.77$ for the first excited level. The perturbation energies of these levels are shown in Figure 1 by two long-dash straight lines. We see that they agree fairly well with the results of our MCHF computations.

CONCLUSIONS

Generally, the dependences of the binding energies on N for the hydrogen-like ion are similar to the corresponding dependences for the hydrogen atom on $K_\perp^2$, the square of the (continuous) eigenvalue of the transverse generalized momentum (cf. Vincke et al., 1992; Potekhin, 1994). Similar to the case of neutral atom, the ion's motion across the magnetic field changes the selection rules so that transitions forbidden in the limit of infinite nucleus mass become permitted. Moreover, since the transverse motion of the charged two-body system is quantized, a fine structure of the ion spectral lines and photoionization edges should appear when the fine structure components are narrower than the spacing between them. We expect these results to be important for modeling of atmospheres of neutron stars with strong magnetic fields, $B \sim 10^{12} - 10^{13}$ G.

ACKNOWLEDGMENTS

This work was partially supported by INTAS Grant 94-3834, RBRF Grant 96-02-16870a, RBRF-DFG Grant 96-02-00177G, NASA Grant NAG5-2807 and NATO Grant CRG 931446.

REFERENCES

Avron, J. E., Herbst, I. W., and Simon B., 1978, Separation of center of mass in homogeneous magnetic fields, *Ann. Phys. NY* 114:431.

Baye, D., 1982, An approximate constant of motion for the problem of an atomic ion in a homogeneous magnetic field, *J. Phys. B: At. Mol. Opt. Phys.* 15:L795.

Baye, D., and Vincke, M., 1986, Centre-of-mass energy of hydrogenic ions in a magnetic field, *J. Phys. B: At. Mol. Opt. Phys.* 19:4051.

Bezchastnov, V. G., 1995, A new basis of the Landau states for a hydrogen-like ion moving in a strong magnetic field, *J. Phys. B: At. Mol. Opt. Phys.* 28:167 (B95).

Potekhin, A. Y., 1994, Structure and radiative transitions of the hydrogen atom moving in a strong magnetic field, *J. Phys. B: At. Mol. Opt. Phys.* 27:1073.

Ruder, H., Wunner., G., Herold, H., and Geyer, F., 1994, Atoms in Strong Magnetic Fields, Springer-Verlag, Berlin–Heidelberg.

Schmelcher, P., 1995, Interaction of the collective and electronic motion of atomic ions in magnetic fields, *Phys. Rev. A* 52:130.

Vincke, M., Le Dourneuf, M., and Baye, D., 1992, Hydrogen atom in crossed electric and magnetic fields: transition from weak to strong electron-proton decentring, *J. Phys. B: At. Mol. Opt. Phys.* 25:2787.

CENTER-OF-MASS EFFECTS ON ATOMS IN MAGNETIC FIELDS

D. Baye[1] and M. Vincke[2,1]

[1] Physique Nucléaire Théorique et Physique Mathématique, CP 229, Université Libre de Bruxelles, B 1050 Brussels, Belgium
[2] AIB-Vinçotte Nuclear, Avenue du Roi 157, B 1190 Brussels, Belgium

INTRODUCTION

The center of mass (c.m.) problem in a magnetic field had long remained a mathematically interesting but rather academic topic. The physical importance of the effects was too small in fields accessible to experiment. Striking properties appeared to occur in very strong fields but did not seem to take place in any realistic environment. The situation changed in recent years. Potentially observable effects entirely due to the c.m. are now predicted to occur even for the hydrogen atom: in crossed electric and magnetic fields, a new family of strongly decentered states is predicted. Unexpectedly, the very existence of such states qualitatively modifies the modelling of neutron-star atmospheres and a correct description of c.m. effects is now an essential ingredient of these models.

Let us first consider the c.m. problem in a chronological perspective. Although important steps were made before,[1, 2] the status of the c.m. problem completely changed in 1978. During that year, the existing mathematical knowledge on the problem was summarized and generalized by Avron, Herbst and Simon.[3] Almost simultaneously, Virtamo and Simola published a variational calculation on the hydrogen atom in a magnetic field showing that c.m. effects have an observable influence in very strong fields, typical of neutron stars.[4] Because they qualified the c.m. as "non separable", different authors published refutations based on a correct c.m. separation where they still concluded that c.m. effects are negligible.[5, 6] The controversy was clarified by Wunner, Ruder, and Herold.[7, 8] The explanation is as follows. The total energy of the hydrogen atom in very high fields is close to the electron cyclotron energy, proportional to the field strength. The c.m. correction is close to the proton cyclotron energy, which is indeed smaller by a factor m_e/m_p. However, physical effects do not rely on *total* energies but on *binding* energies, which are of the order of magnitude of the Rydberg unit and increase much more slowly with the field. At some field strength, the c.m. correction becomes comparable with the hydrogen binding energy and qualitatively affects physical properties. This remains true for other neutral atoms. Neutral systems behave as free particles whose collective motion is however affected by their internal state.[3] The c.m. separation is then called a pseudoseparation.

The c.m. problem for ions is very different. Ions are globally charged and undergo a collective cyclotron motion, also depending on the internal state. In that case, an exact separation of the c.m. motion is not possible.[3] However, the existence of an approximate constant of motion[9] allows an approximate pseudoseparation up to very high fields.[10] Accurate c.m. corrections were calculated for hydrogenic ions up to field strengths as large as 10^9 T.[11] A reinvestigation in the cases of neutral hydrogen and positronium provided previously unnoticed instability properties[12, 13] and a unified treatment of the charged and neutral cases was discovered.[14] General c.m. corrections for arbitrary atoms and ions were proposed.[15]

When neutral atoms undergo a transverse motion with respect to the field direction, their properties may be affected significantly. Such effects are also obtained in crossed electric and magnetic fields. For neutral hydrogen, a new type of states, where the electron motion is strongly decentered from the proton location was predicted in a gauge-independent way.[16] Such effects were anticipated in a simple potential model[17] but the prediction depended on a gauge choice. A potential approach[18] provided similar results independently of the gauge choice.[19] The transition from the centered regime to the decentered regime was studied in the strong field region.[20, 21] This allowed taking transverse motion into account in models of neutron-star atmospheres.[22, 23]

Here, we review the status of the quantum treatment of c.m. corrections for atomic systems. Other interesting aspects (classical treatment, molecules) are discussed in other contributions. We use the reduced field strength

$$\gamma = B/2.35 \times 10^5, \tag{1}$$

where B is expressed in Teslas.

CONSTANTS OF MOTION

Single Particle

Let $\boldsymbol{r}$ and $\boldsymbol{p}$ be the coordinate and momentum of a particle with mass m and charge q in a homogeneous magnetic field $\boldsymbol{B}$. Its kinetic (or mechanical) momentum reads

$$\boldsymbol{\pi} = \boldsymbol{p} - q\boldsymbol{A}(\boldsymbol{r}), \tag{2}$$

where $\boldsymbol{A}$ is the vector potential in an arbitrary gauge ($\boldsymbol{B} = \mathrm{rot}\boldsymbol{A}$). Here we shall not need the explicit form of this vector potential.

The non-relativistic spinless Hamiltonian

$$H = \boldsymbol{\pi}^2/2m \tag{3}$$

possesses several constants of motion: the pseudomomentum or generalized momentum[24]

$$\boldsymbol{k} = \boldsymbol{\pi} + q\boldsymbol{B} \times \boldsymbol{r} = q\boldsymbol{B} \times \boldsymbol{r}_c, \tag{4}$$

where, for $q \neq 0$, $\boldsymbol{r}_c$ is the guiding center, and a generalization of the parallel component of the orbital momentum[25]

$$\mathcal{L}_\| = \hat{\boldsymbol{B}} \cdot \boldsymbol{r} \times \tfrac{1}{2}(\boldsymbol{k} + \boldsymbol{\pi}) = (2qB)^{-1}(\boldsymbol{k}^2 - \boldsymbol{\pi}^2), \tag{5}$$

where $\hat{\boldsymbol{B}}$ is a unit vector in the field direction. The last expression is valid only for $q \neq 0$.

The operators $\boldsymbol{k}$ and $\boldsymbol{\pi}$ verify with $\mathcal{L}_{\|}$ the commutation relations

$$\begin{aligned}
&[\pi_\lambda, \pi_\mu] = \mathrm{i}\hbar q \epsilon_{\lambda\mu\nu} B_\nu, \quad [\pi_\lambda, k_\mu] = 0, \quad [k_\lambda, k_\mu] = -\mathrm{i}\hbar q \epsilon_{\lambda\mu\nu} B_\nu, \\
&[\boldsymbol{\pi}, \mathcal{L}_{\|}] = \mathrm{i}\hbar \hat{\boldsymbol{B}} \times \boldsymbol{\pi}, \qquad\qquad\qquad [\boldsymbol{k}, \mathcal{L}_{\|}] = \mathrm{i}\hbar \hat{\boldsymbol{B}} \times \boldsymbol{k},
\end{aligned} \tag{6}$$

where λ, μ, ν represent x, y, z and $\epsilon_{\lambda\mu\nu}$ is the antisymmetric tensor. Eqs. (6) correspond to the algebra of free motion for $q = 0$ or of cyclotron motion for $q \neq 0$. The components of $\boldsymbol{k}$ and $\boldsymbol{\pi}$ parallel to the field are identical and commute with the transverse components. They also commute with $\mathcal{L}_{\|}$.

We note here an important difference between the neutral and charged cases: if $q = 0$, all the operators belonging to the algebra (6) are constants of motion; if $q \neq 0$, $\boldsymbol{\pi}$ is not a constant of motion. This property has important consequences in the c.m. problem.

System of $N + 1$ Particles

Let us consider an atom or ion involving a nucleus with charge q_0 and mass m_0 and N electrons with charges $q_i = -e$ and masses $m_i = m_e$. The Hamiltonian reads

$$H = \sum_{i=0}^{N} \frac{\boldsymbol{\pi}_i^2}{2m_i} + \sum_{i>j=0}^{N} V_{ij}(|\boldsymbol{r}_i - \boldsymbol{r}_j|), \tag{7}$$

where V is rotation and translation invariant. Because of these invariances and of Eqs. (6), the constants of motion are the total pseudomomentum[3]

$$\boldsymbol{K} = \sum_{i=0}^{N} \boldsymbol{k}_i, \tag{8}$$

and the generalized parallel component of the total orbital momentum[13]

$$\mathcal{L}_{\|} = \sum_{i=0}^{N} \mathcal{L}_{i\|}. \tag{9}$$

The components of the total pseudomomentum $\boldsymbol{K}$ satisfy

$$[K_\lambda, K_\mu] = -\mathrm{i}\hbar Q \epsilon_{\lambda\mu\nu} B_\nu, \tag{10}$$

where Q is the total charge. Its transverse components commute with the parallel component but do not commute with each other if Q differs from zero. The commutator of $\boldsymbol{K}$ with $\mathcal{L}_{\|}$ is

$$[\boldsymbol{K}, \mathcal{L}_{\|}] = \mathrm{i}\hbar \hat{\boldsymbol{B}} \times \boldsymbol{K}. \tag{11}$$

In the neutral case ($Q = 0$), the transverse components of $\boldsymbol{K}$ and $\mathcal{L}_{\|}$ satisfy the Euclidean algebra. The c.m. behaves as a free particle with $\boldsymbol{K}$ playing the role of a momentum. Since all the operators appearing in (10) and (11) are then constants of motion, a c.m. separation is possible. In the charged case ($Q \neq 0$), the c.m. does not behave collectively as a charged particle because a constant-of-motion collective kinetic momentum is not available.

Generalized Kinetic Momentum

A c.m. kinetic-momentum operator can be defined as

$$\boldsymbol{C}(\alpha) = \boldsymbol{K} + Q\boldsymbol{B} \times \boldsymbol{R}(\alpha), \tag{12}$$

where α is a real parameter and

$$\boldsymbol{R}(\alpha) = \boldsymbol{r}_0 + \alpha \sum_{i=1}^{N} (\boldsymbol{r}_i - \boldsymbol{r}_0). \tag{13}$$

Indeed, it satisfies with $\boldsymbol{K}$ and $\mathcal{L}_{\|}$ the commutation relations

$$[C_\lambda, C_\mu] = \mathrm{i}\hbar Q \epsilon_{\lambda\mu\nu} B_\nu, \quad [C_\lambda, K_\mu] = 0, \quad [\boldsymbol{C}, \mathcal{L}_{\|}] = \mathrm{i}\hbar \hat{\boldsymbol{B}} \times \boldsymbol{C}, \tag{14}$$

forming with (10) and (11) an algebra similar to (6). For $Q = 0$, $\boldsymbol{C}$ reduces to $\boldsymbol{K}$.

The commutator of $\boldsymbol{C}(\alpha)$ with the Hamiltonian

$$[H, \boldsymbol{C}(\alpha)] = \mathrm{i}\hbar Q \boldsymbol{B} \times \left[\frac{\boldsymbol{\pi}_0}{m_0} + \alpha \sum_{i=1}^{N} \left(\frac{\boldsymbol{\pi}_i}{m_e} - \frac{\boldsymbol{\pi}_0}{m_0} \right) \right] \tag{15}$$

does not vanish if Q or B is not zero. Nevertheless, it is "small" for $\alpha = O(m_e/m_0)$ and $\boldsymbol{C}$ can then be considered as an approximate constant of motion.

CANONICAL TRANSFORMATIONS

Operators x_j and p_k form a canonical ensemble if

$$[x_j, x_k] = 0, \quad [x_j, p_k] = \mathrm{i}\hbar\delta_{jk}, \quad [p_j, p_k] = 0. \tag{16}$$

A canonical transformation[26, 27] is a transformation

$$x'_{j'} = f_{j'}(x_j, p_k), \quad p'_{k'} = g_{k'}(x_j, p_k), \tag{17}$$

where the functions f and g are chosen in such a way that the $x'_{j'}$ and $p'_{k'}$ also form a canonical ensemble, i.e. also satisfy (16).

Canonical transformations occur in well known situations, such as a change of variables where the functions f do not depend on momenta,

$$x'_{j'} = f_{j'}(x_j), \quad p'_{k'} = g_{k'}(x_j, p_k), \tag{18}$$

a gauge transformation where only momenta are modified,

$$x'_j = x_j, \quad p'_k = p_k + \varphi(x_j), \tag{19}$$

and the simple exchange of coordinates and momenta

$$x'_j = p_j, \quad p'_k = -x_k, \tag{20}$$

which is nothing but a Fourier transform, except for a phase factor.

Of particular importance are the linear canonical transformations (LCT)

$$\begin{pmatrix} x'_j \\ p'_k \end{pmatrix} = A \begin{pmatrix} x_j \\ p_k \end{pmatrix}, \tag{21}$$

where A is a real $2n \times 2n$ matrix. Relations (16) imply that A must satisfy

$$A\begin{pmatrix} 0 & I \\ -I & 0 \end{pmatrix}A^T = \begin{pmatrix} 0 & I \\ -I & 0 \end{pmatrix}, \tag{22}$$

where I is an $n \times n$ unit matrix. The example (20) is a LCT. A LCT is very easily performed with simple linear algebra. When the canonical relations (16) are proved for the new variables, LCT become simpler than changes of variables since momenta are treated as coordinates. This simplicity replaces more complicate mathematical manipulations, i.e. integral transforms[26, 27] or unitary transformations.[26, 3]

In the context of magnetic fields, a slight but useful generalization of the notion of LCT is obtained by working with the $\boldsymbol{\pi}_{i\perp}$ and $\boldsymbol{k}_{i\perp}$,[10, 14] which transform as

$$\begin{pmatrix} \boldsymbol{\pi}'_{i'\perp} \\ \boldsymbol{k}'_{j'\perp} \end{pmatrix} = M\begin{pmatrix} \boldsymbol{\pi}_{i\perp} \\ \boldsymbol{k}_{j\perp} \end{pmatrix}, \tag{23}$$

where M is a real $2n \times 2n$ matrix. Relations (6) imply that M must satisfy

$$M\begin{pmatrix} I & 0 \\ 0 & -I \end{pmatrix}M^T = \begin{pmatrix} I & 0 \\ 0 & -I \end{pmatrix}, \tag{24}$$

as is easily verified. In order to unify the treatment of transverse and parallel components, it is even more convenient to use only physical observables, i.e. the kinetic momenta $\boldsymbol{\pi}_i$ and coordinates $\boldsymbol{r}_i$,[12, 14] transforming as

$$\begin{pmatrix} \boldsymbol{\pi}'_{i} \\ \boldsymbol{r}'_{j} \end{pmatrix} = N\begin{pmatrix} \boldsymbol{\pi}_{i} \\ \boldsymbol{r}_{j} \end{pmatrix}, \tag{25}$$

where N is a real $2n \times 2n$ matrix. Here, N conserves the relations

$$[\pi_{i\lambda}, \pi_{j\mu}] = \mathrm{i}\hbar q_i \delta_{ij}\epsilon_{\lambda\mu\nu}B_\nu, \quad [r_{i\lambda}, \pi_{j\mu}] = \mathrm{i}\hbar\delta_{ij}\delta_{\lambda\mu}, \quad [r_{i\lambda}, r_{j\mu}] = 0. \tag{26}$$

UNIFIED TREATMENT OF CM SEPARATION

In both the neutral and charged cases, we perform a LCT of type (25) in which the generalized kinetic momentum and the total pseudomomentum respectively become the kinetic momentum and the pseudomomentum associated with the collective motion,[15] i.e.

$$\begin{aligned} \boldsymbol{\pi}'_0 &= \boldsymbol{C}(\alpha), & \boldsymbol{r}'_0 &= \boldsymbol{R}(\alpha), \\ \boldsymbol{\pi}'_i &= \boldsymbol{\pi}_i - \alpha\boldsymbol{C}(\tfrac{1}{2}\alpha), & \boldsymbol{r}'_i &= \boldsymbol{r}_i - \boldsymbol{r}_0. \end{aligned} \tag{27}$$

This transformation takes its simplest form when $\alpha = 0$.

For arbitrary α, the transformed Hamiltonian can be written as the sum

$$H' = H_0 + H_{\mathrm{int}} + H_{\mathrm{c}}. \tag{28}$$

For $\alpha = 0$, $\boldsymbol{r}'_0 = \boldsymbol{r}_0$ and the collective motion is described by the Hamiltonian

$$H_0 = \boldsymbol{\pi}'^2_0/2m_0. \tag{29}$$

The internal motion is governed by the internal Hamiltonian

$$H_{\mathrm{int}} = H_\infty + \frac{1}{2m_0}\left(\sum_{i=1}^{N}\boldsymbol{k}'_i\right)^2, \tag{30}$$

where the infinite-mass Hamiltonian is given by

$$H_\infty = \sum_{i=1}^{N} \frac{\boldsymbol{\pi}_i^2}{2m_e} + \sum_{i=1}^{N} V_{i0}(|\boldsymbol{r}_i - \boldsymbol{r}_0|) + \sum_{i>j=1}^{N} V_{ij}(|\boldsymbol{r}_i - \boldsymbol{r}_j|). \tag{31}$$

Both motions are linked by the coupling term

$$H_c = -\frac{1}{m_0}\boldsymbol{\pi}_0' \cdot \sum_{i=1}^{N} \boldsymbol{k}_i'. \tag{32}$$

A decoupling only occurs in the neutral case when the eigenvalue of the transverse part $\boldsymbol{\pi}_{0\perp}' = \boldsymbol{K}_\perp$ of the constant of motion $\boldsymbol{K}$ vanishes, i.e. when the neutral atom does not move transversally to the field. In the charged case, $\boldsymbol{\pi}_0'$ is not an exact constant of motion but the eigenstates of (29) are easy-to-use harmonic-oscillator states. Therefore, the charged case is not more complicated than the neutral case with a transverse motion.

A variant is obtained for $\alpha = m_e/M$ (where M is the total mass). The coordinate $\boldsymbol{R}(\alpha)$ is then the traditional c.m. coordinate and H_0 and H_c read

$$H_0 = \boldsymbol{\pi}_0'^2/2M, \tag{33}$$

and

$$H_c = \frac{1}{M}\left(Q\frac{m_e}{M} + e\right)\boldsymbol{\pi}_0' \cdot \boldsymbol{B} \times \sum_{i=1}^{N} \boldsymbol{r}_i'. \tag{34}$$

Eq. (34) is known as the "motional Stark effect".[1, 3] The internal Hamiltonian differs from (30) by Q times a term of order $(m_e/m_0)^2$ which is almost always negligible (the exact form is given in Ref. 14).

The internal Hamiltonian can also be expressed as

$$H_{int} = H_\mu + \frac{qB}{m_0}\mathcal{L}_\parallel' + \frac{1}{m_0}\sum_{i>j=1}^{N} \boldsymbol{k}_i' \cdot \boldsymbol{k}_j', \tag{35}$$

where H_μ is obtained by replacing m_e by the electron-nucleus reduced mass μ in (31). ¿From this expression, one sees that the internal energy of an atom or ion in a magnetic field is the sum of three terms: the eigenvalue of H_∞ conveniently scaled, an exact correction related to the constant of motion $\mathcal{L}_\parallel'$ of H_∞ and H_μ, and a generalization of the well-known mass polarization term. Since the last two terms of (35) arise from the same last term of (30), one expects them to be of comparable importance.

APPLICATIONS

Binding Energies

A first application of the formalism of the previous section is the calculation of binding energies. The precise definition of a binding energy in a magnetic field has been given in Ref. 15.

For an hydrogenic system with $q_0 = Ze$,[13] Eqs. (28), (35) and (31) provide when taking the coupling term (32) or (34) into account the binding energy (in Rydberg-energy units)

$$E_B(\gamma) = \frac{\mu}{m_e} E_{B\infty}\left(\frac{m_e^2}{\mu^2}\gamma\right) + 2\frac{m_e}{m_0}m\gamma - E_{CM}^\perp, \tag{36}$$

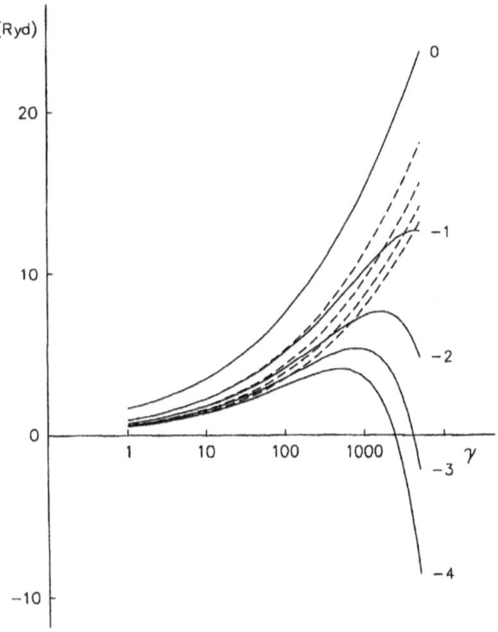

Figure 1. Hydrogen binding energies without (broken lines) and with (full lines) c.m. corrections.

where the eigenvalue m of $\mathcal{L}'_{\parallel}$ is negative. The infinite-mass binding energy $E_{B\infty}$ is obtained for $Z > 1$ by scaling energies of neutral hydrogen. The first and second terms of (36) respectively correspond to the first and second terms of (35). The collective energy $E^{\perp}_{\rm CM}$ arises from the eigenvalue of H_0, modified by the coupling effects due to H_c. For $Q = 0$, it is a function of the eigenvalue $\boldsymbol{K}_{\perp}$ of the transverse total pseudomomentum,

$$E^{\perp}_{\rm CM}(\gamma, K_{\perp}) = K^2_{\perp} m_e / M_{\rm eff}(\gamma, K^2_{\perp}). \tag{37}$$

For $Q \neq 0$, the oscillator quantum number ν_0 associated with $\boldsymbol{C}^2_{\perp}$ leads to

$$E^{\perp}_{\rm CM}(\gamma, \nu_0) = 2\nu_0 (Z-1)\gamma m_e / M_{\rm eff}(\gamma, \nu_0). \tag{38}$$

In both cases, the effective mass $M_{\rm eff}$ takes the coupling effects into account. Small higher-order corrections are neglected in (38). As an example, we display in Fig. 1 the c.m. corrections on binding energies of hydrogen.[13] States with $m < 0$ become unstable beyond some critical field, because of these c.m. corrections.

Binding energies for N-electron systems are more complicated because mass polarization corrections must be evaluated and because of the occurrence of three types of dissociations:
(i) neutral atom → ion + electron ($N = Z$),
(ii) ion → neutral atom + electron ($N = Z + 1$),
(iii) ion → ion + electron ($N < Z$ or $N > Z + 1$).
A simplified expression of the binding energy reads[15]

$$E_B(\gamma) = E_{B\infty}(\gamma) + 2(m_e/m_0)m\gamma + \Delta_{\rm MP} + \Delta_{\rm CM}, \tag{39}$$

where the eigenvalue m of $\mathcal{L}'_{\parallel}$ is negative and small scaling effects are neglected. The quantity $\Delta_{\rm MP}$ corresponds to the difference of mass-polarization corrections for both

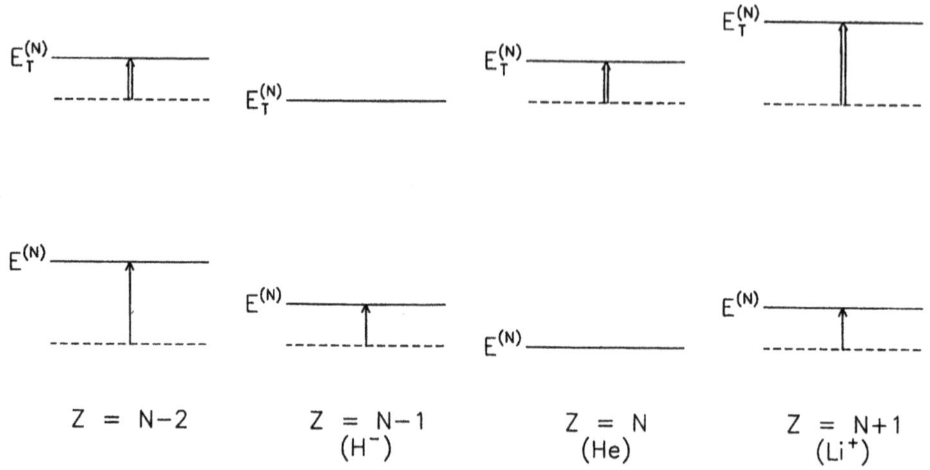

Figure 2. Zero-point energies before separation (single arrows) and after separation (double arrows), as a function of $Z - N$.

atoms and must be calculated from the eigenfunctions of H_∞. This term is *not* negligible. The difference Δ_{CM} is well approximated by

$$\Delta_{\text{CM}} = \Delta'_{\text{CM}} - \begin{cases} 2(m_e/m_0)\gamma & (N > Z), \\ 0 & (N \leq Z), \end{cases} \tag{40}$$

where Δ'_{CM} is a difference of terms from (38) and (37). The additional term appearing in (40) corresponds to the difference of zero-point energies which depends on the dissociation type (i), (ii), or (iii), as illustrated in Fig. 2. The zero-point energies reduce the binding energies for $N > Z$. This important c.m. effect can be observed in Fig. 3 where different H^- states become unstable at high fields.

Decentered States

The hydrogen atom in crossed electric and magnetic fields is equivalent to the same atom with a transverse motion in a pure magnetic field. Indeed, the electric field can be eliminated with the Galilean LCT[20] of type (25),

$$\boldsymbol{r}'_i = \boldsymbol{r}_i, \quad \boldsymbol{\pi}'_i = \boldsymbol{\pi}_i - m_i \boldsymbol{v}_d, \tag{41}$$

where $\boldsymbol{v}_d = B^{-1}\boldsymbol{F} \times \hat{\boldsymbol{B}}$ is the velocity of the drift frame, in which the electric field vanishes. In this frame, the pseudomomentum becomes $\boldsymbol{K}' = \boldsymbol{K} - M\boldsymbol{v}_d$.

The Hamiltonian of an hydrogen atom in a pure magnetic field reads

$$H = (2m_0)^{-1}\boldsymbol{\pi}_0^2 + (2m_e)^{-1}\boldsymbol{\pi}_1^2 - e^2/|\boldsymbol{r}_0 - \boldsymbol{r}_1|. \tag{42}$$

The simple LCT of transverse components[12, 16] of type (23),

$$\begin{aligned} \boldsymbol{\pi}'_{0\perp} &= \boldsymbol{k}_{1\perp}, & \boldsymbol{k}_{0\perp} &= \boldsymbol{k}_{0\perp}, \\ \boldsymbol{\pi}'_{1\perp} &= \boldsymbol{\pi}_{1\perp}, & \boldsymbol{k}'_{1\perp} &= -\boldsymbol{\pi}_{0\perp} \end{aligned} \tag{43}$$

is combined with the usual c.m. transformation for parallel components. The Hamiltonian becomes

$$H' = (2M)^{-1}K_\parallel^2 + (2m_e)^{-1}\boldsymbol{\pi}_1'^2 + (2m_0)^{-1}\boldsymbol{k}_1'^2 - e^2/|\boldsymbol{r}'_1 - \boldsymbol{r}_c|, \tag{44}$$

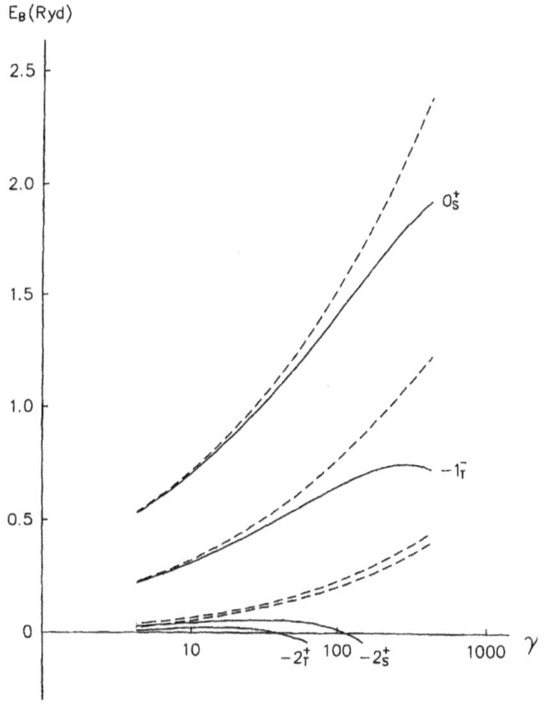

Figure 3. H⁻ binding energies of the main singlet and triplet states without (broken lines) and with (full lines) c.m. corrections.

where $\boldsymbol{r}_c$ is related to $\boldsymbol{K}'_\perp$ by (4). Eq. (44) displays an essential advantage of the transformation (43): the transformed potential term only depends on the relative coordinate $\boldsymbol{r}'_1$ and on the constant of motion $\boldsymbol{K}'_\perp$. For large r_c, an expansion of the potential term in (44) up to second order in $\boldsymbol{r}'_1$ provides a triaxal harmonic-oscillator Hamiltonian which can be solved exactly,[28] independently of the choice of the vector potential. The eigenvalues then read[16]

$$E_B \approx (2/r_c) - 2(m_e/m_0)\gamma n_- - r_c^{-3/2}(2n_\parallel + 1)/2, \tag{45}$$

where n_- and $n_\parallel$ are oscillator quantum numbers. The corresponding eigenstates present a strong decentering of the electron with respect to the proton, approximately equal to $r_c = K_\perp/\gamma$. Under realizable laboratory conditions ($B = 6$ T and $F = 3$ kV/cm), binding energies are very small ($E_B = 16\ \mu$eV) and the decentering r_c reaches 0.087 mm. Similar results are obtained in a gauge-independent potential model.[18, 19]

The observation of such states will require the understanding of the transition between the centered and decentered regimes. This transition has only been studied in the strong field range. An example is given for the lowest field studied until now ($\gamma = 10$) in Fig. 4.[20] The centered states appear in the left-hand-side while decentered states become dominant beyond $K_\perp \approx 70$.

Electromagnetic Transitions

The momentum of the emitted photon is given by the difference between the initial and final pseudomomenta.[3] The effect of c.m. corrections on electromagnetic transitions has been discussed in Ref. 29. For the dominant dipole transitions, the correction essentially consists in taking the c.m. corrected energy differences into account, when

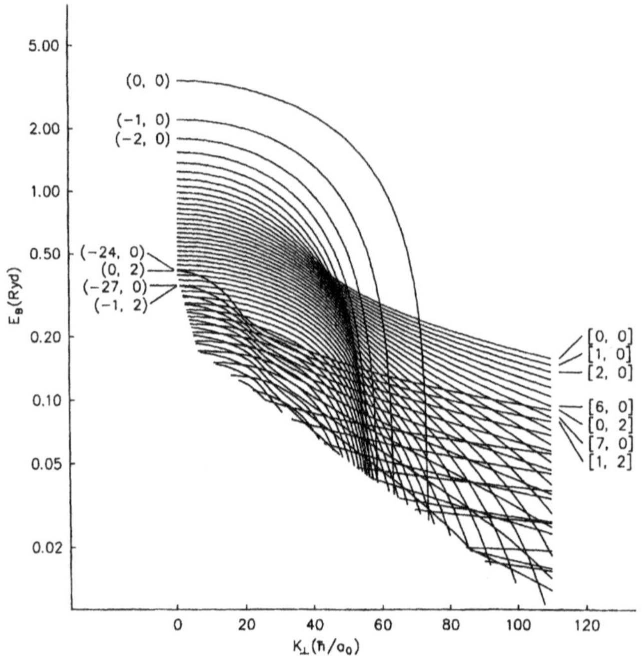

Figure 4. Hydrogen binding energies as a function of $K_\perp$ for $\gamma = 10$. The labels are $(m, n_\parallel)$ (left) and $[n_-, n_\parallel]$ (right).

the effects of transverse motion can be neglected. The situation is more complicated for other transitions where several types of energy differences (infinite-mass and c.m. corrected) occur simultaneously. When the transverse motion cannot be neglected, electric dipole transitions have been studied by Potekhin.[21]

CONCLUSIONS

By their simplicity of use, linear canonical transformations provide an ideal tool for studying the c.m. problem in a magnetic field. They simultaneously perform the changes of variables and gauge transformations necessary in other approaches. They allow a unified treatment of the very different neutral and charged systems. In fact, contrary to a common expectation, the c.m. problem is not more complicated for a charged system than for a neutral system in the general case, involving a transverse motion.

The c.m. problem in a magnetic field is now well understood when the collective transverse or cyclotron motion is weak. C.m. effects are not trivial but can be calculated reliably when the infinite-mass problem can be solved with sufficient accuracy.

Interesting physical phenomena related to strongly decentered states occur in crossed fields or, equivalently, for a large transverse motion. An important goal is now observing them in experiments but the planning of such experiments will require theoretical information on the preparation of such states i.e. on the transition between centered and decentered states with increasing electric field under laboratory-feasible magnetic fields. Additional studies of such states at higher fields, and in particular simple analytical approximations of binding energies and of the different types of transition probabilities, will also be useful for neutron-star models.

REFERENCES

1. Lamb, W.E. Jr, 1952, Fine structure of the hydrogen atom III, *Phys. Rev.* 85:259.
2. Carter, B.P., 1969, Partial separation of the Schrödinger equation for two charged particles in a magnetic field, *J. Math. Phys.* 10:788.
3. Avron, J.E., Herbst, I.W., and Simon, B., 1978, Separation of center of mass in homogeneous magnetic fields, *Ann. Phys. (NY)* 114:431.
4. Virtamo, J., and Simola, J., 1978, The spreading of the proton of a hydrogen atom in strong magnetic fields, *Phys. Lett.* A 66:371.
5. O'Connell, R.F., 1979, Effect of the proton mass on the spectrum of the hydrogen atom in a strong magnetic field, *Phys. Lett.* A 70:389.
6. Pavlov-Verevkin, V.B., and Zhilinskii, B.I., 1980, Neutral hydrogen-like system in a magnetic field, *Phys. Lett.* A 78:244.
7. Wunner, G., Ruder, H., and Herold, H., 1980, Comment on the effect of the proton mass on the spectrum of the hydrogen atom in a very strong magnetic field, *Phys. Lett.* A 79:159.
8. Herold, H., Ruder, H., and Wunner, G., 1981, The two-body problem in the presence of a homogeneous magnetic field, *J. Phys.* B 14:751.
9. Baye, D., 1982, An approximate constant of motion for the problem of an atomic ion in a homogeneous magnetic field, *J. Phys.* B 15:L795.
10. Baye, D., 1983, Separation of centre-of-mass motion for a charged two-body system in a homogeneous magnetic field, *J. Phys.* A 16:3207.
11. Baye, D., and Vincke, M., 1986, Centre-of-mass energy of hydrogenic ions in a magnetic field, *J. Phys.* B 19:4051.
12. Baye, D., and Vincke, M., 1988, Spontaneous dissociation of $m \neq 0$ positronium states in strong magnetic fields, *Phys. Lett.* A 129:109.
13. Vincke, M., and Baye, D., 1988, Centre-of-mass effects on the hydrogen atom in a magnetic field, *J. Phys.* B 21:2407.
14. Baye, D., and Vincke, M., 1990, Centre-of-mass problem in a magnetic field : unified treatment of charged and neutral systems, *Phys. Rev.* A 42:391.
15. Baye, D., and Vincke, M., 1990, Centre-of-mass corrections on atomic binding energies in a magnetic field, *J. Phys.* B 23:2467.
16. Baye, D., Clerbaux, N., and Vincke, M., 1992, Delocalized states of atomic hydrogen in crossed electric and magnetic fields, *Phys. Lett.* A 166:135.
17. Burkova, L.A., Dzyaloshinskii, I.E., Drukarev, G.F., and Monozon, B.S., 1976, Hydrogenlike system in crossed electric and magnetic fields, *Sov. Phys. JETP.* 44:276.
18. Schmelcher, P., and Cederbaum, L.S., 1993, Two-body effects of the hydrogen atom in crossed electric and magnetic fields, *Chem. Phys. Lett.* 208:548.
19. Dippel, O., Schmelcher, P., and Cederbaum, , L.S., 1994, Charged anisotropic harmonic oscillator and the hydrogen atom in crossed fields, *Phys. Rev.* A 49:4415.
20. Vincke, M., Le Dourneuf, M., and Baye, D., 1992, Hydrogen atom in crossed electric and magnetic fields: transition from weak to strong decentring, *J. Phys.* B 25:2787.
21. Potekhin, A.Y., 1994, Structure and radiative transitions of the hydrogen atom moving in a strong magnetic field, *J. Phys.* B 27:1073.
22. Lai, D., and Salpeter, E.E., 1995, Motion and ionization equilibrium of hydrogen atoms in a superstrong magnetic field, *Phys. Rev.* A 52:2611.
23. Pavlov, G.G., and Potekhin, A.Y., 1995, Bound-bound transitions in strongly magnetized hydrogen plasma, *Astrophys. J.* 450:883.
24. Johnson, M.H., and Lippmann, B.A., 1949, Motion in a constant magnetic field, *Phys. Rev.* 76:828.
25. Mitchell, T.P., 1981, Dynamical symmetries of the Schrödinger equation, *J. Math. Phys.* 22:1948.
26. Moshinsky, M., and Quesne, C., 1971, Linear canonical transformations and their unitary representations, *J. Math. Phys.* 12:1772.
27. Wolf, K.B., 1979, "Integral Transforms in Science and Engineering," Plenum, New York.
28. Baye, D., and Vincke, M., 1989, Systems of particles in harmonic interaction in a magnetic field, *Phys. Rev.* A 39:2806.
29. Cuvelliez, C., Baye, D., and Vincke, M., 1992, Center-of-mass corrections to the electromagnetic transitions of hydrogen atoms in strong magnetic fields, *Phys. Rev.* A 46:4055.

SCALING PROPERTIES FOR ATOMS IN EXTERNAL FIELDS

H. Friedrich

Physik-Department, Technische Universität München, 85747 Garching, Germany

1. INTRODUCTION

One important advantage of studying atoms or molecules in external fields is, that the field parameters are tunable variables, and investigation of the properties of the atom or molecule as function of these variables provides a much richer body of information than can be observed in the isolated specimen. Due to scaling properties, the classical dynamics of an atom or molecule in external fields depends on some combinations of field parameters in a trivial way. It is important to understand the rules behind such scaling properties in order to avoid redundancies in measurement and analysis of data and in order to extract genuine physical effects from the data.

The aim of this contribution is to summarize in a unified way the scaling properties of an atom (or molecule) in an external electric or magnetic constant or time dependent field, or in any superposition thereof. The results are not new, but I hope the following presentation will be helpful in clarifying the simple structure of the scaling rules applicable in the various situations. My starting point is the notion of mechanical similarity for a classical system described by a homogeneous potential or a superposition of homogeneous potentials. The section on quantum mechanics shows how the scaling properties of the corresponding classical system can be used to study quantum properties of a given system as function of an "effective Planck's constant" and defines the semiclassical limit of the system. It also illuminates the merits of a popular spectroscopic technique, which should most appropriately be called "scaled fields spectroscopy". A final section discusses the transfer of the scaling rules developed for atoms (or molecules), which are characterized by a Coulombic potential, to objects such as quantum dots, which can be thought of as atoms of a sort, but not necessarily with a Coulombic potential.

2. CLASSICAL MECHANICS

2.1 Mechanical Similarity for Homogeneous Potentials

Consider a conservative system S with a finite number of degrees of freedom described by a kinetic energy,

$$T = \frac{m}{2}\left(\frac{d\vec{r}}{dt}\right)^2 , \tag{1}$$

and a potential energy $U(\vec{r})$. [The mass m can be different for the various degrees of freedom, but this is irrelevant for the following.] The *similarity transformation*,

$$\vec{r}' = \alpha\vec{r} , \quad t' = \beta t , \quad \frac{d\vec{r}'}{dt'} = \frac{\alpha}{\beta}\frac{d\vec{r}}{dt} , \tag{2}$$

with the two positive constants α and β, transforms the system S into a system S', whose kinetic energy T' is related to the kinetic energy T in (1) by

$$T' = \left(\frac{\alpha}{\beta}\right)^2 T . \tag{3}$$

Suppose the potential energy in the system S is given by a homogeneous function $V(\vec{r})$ of degree d, i.e.

$$V(\alpha\vec{r}) = \alpha^d V(\vec{r}) , \tag{4}$$

multiplied by a parameter F, which gives us a handle on the potential strength, $U(\vec{r}) = FV(\vec{r})$. Let the potential energy U' in the system S' be given by the same (homogeneous) function V, multiplied by a strength parameter F', $U'(\vec{r}') = F'V(\vec{r}')$. Because of homogeneity (4), the potential energy U' is related to the potential energy U in S by

$$U'(\vec{r}') = \frac{F'}{F}\alpha^d U(\vec{r}) . \tag{5}$$

If and only if the field strengths fulfill the relation

$$\alpha^d F' = \left(\frac{\alpha}{\beta}\right)^2 F , \tag{6}$$

then the potential energies are related by the same multiplicative factor $(\alpha/\beta)^2$ as the kinetic energies (3). The Lagrangian $T' - U'$ in the system S' is then just a multiple of the Lagrangian $L = T - U$ in S, and the equations of motion in both systems are the same.[1] The trajectory $\vec{r}(t)$ is a solution of the equations of motion in S if and only if the trajectory $\vec{r}'(t')$, which is related to $\vec{r}(t)$ by the similarity transformation (2), is a solution of the equations of motion in S'. This is the property of mechanical similarity of the systems S and S', and the condition for mechanical similarity is, that the field strengths obey (6). The (conserved) energie $E = \frac{1}{2}m\dot{\vec{r}}(t)^2 + U(\vec{r}(t))$ of motion along the trajectory $\vec{r}(t)$ in S is related to the associated energy E' in S' via

$$E' = \left(\frac{\alpha}{\beta}\right)^2 E . \tag{7}$$

The condition (6) contains two parameters α and β and can always be fulfilled for any values of the field strengths F and F'. Together with the relation (7) we can, for any field strengths F and F' and energies E and E' uniquely determine the constants

α and β defining the similarity transformation (2) connecting the trajectory $\vec{r}(t)$ in S with the trajectory $\vec{r}'(t)$ in S',

$$\alpha = \left(\frac{F}{F'}\frac{E'}{E}\right)^{1/d} \quad , \quad \beta = \alpha\sqrt{\frac{E}{E'}} \quad . \tag{8}$$

[It shall be taken for granted throughout, that potential strengths and energies have the same sign in S' as in S.] From equation (8) we see e.g., that trajectories at different energies E, E' for one and the same potential strength, $F' = F$, are related by a stretching factor $\alpha = (E'/E)^{1/d}$ in coordinate space, whereas the traversal times are stretched by the factor $\alpha\sqrt{E/E'}$.

2.2 Potential Consisting of Several Homogeneous Contributions

The considerations above are readily generalized to a potential which can be written as a sum of n homogeneous terms of degree d_i, $i = 1, \ldots, n$. The potential U in the system S is now

$$U(\vec{r}) = \sum_{i=1}^{n} F_i V_i(\vec{r}) \; ; \quad V_i(\alpha\vec{r}) = \alpha^{d_i} V_i(\vec{r}) \; , \quad i = 1, \ldots, n \quad , \tag{9}$$

and the potential U' in the system S' differs only through different potential strengths,

$$U'(\vec{r}') = \sum_{i=1}^{n} F_i' V_i(\vec{r}') \quad . \tag{10}$$

The systems S and S' are mechanically similar, if U' is just U multiplied by $(\alpha/\beta)^2$, when $\vec{r}'$ and t' are related to $\vec{r}$ and t via (2). The condition (6) must now be fulfilled for each of the n terms independently, and the first equation (8) is replaced by the n equations,

$$\alpha = \left[\left(\frac{\alpha}{\beta}\right)^2 \frac{F_i}{F_i'}\right]^{1/d_i} = \left(\frac{E'}{E}\frac{F_i}{F_i'}\right)^{1/d_i} \quad , \quad i = 1, \ldots, n \quad . \tag{11}$$

The relation between the total energies E and E' is again given by (7).

Equating the right-hand sides of (11) for two different terms i and j and collecting unprimed and primed quantities on separate sides leads to the condition

$$\frac{|F_i/E|^{d_j}}{|F_j/E|^{d_i}} = \frac{|F_i'/E'|^{d_j}}{|F_j'/E'|^{d_i}} \quad . \tag{12}$$

If we consider an ensemble of systems S corresponding to different field strengths F_i and energies E (excluding changes of sign), then Eq. (12) shows that the classical dynamics within the ensemble is invariant within mechanical similarity if

$$\frac{|F_i/E|^{d_j}}{|F_j/E|^{d_i}} = \text{const.} \tag{13}$$

for each pair of labels i, j. For $n > 2$ these conditions are not independent. The $n+1$ parameters E, F_i $(i = 1, \ldots, n)$ are effectively subjected to $n-1$ independent conditions, because α and β generate a two-parameter manifold of mechanically similar systems.

2.3 Examples: Atom in External Static Fields

The Coulomb potential describing the forces in an atom (or molecule or ion) is homogeneous of degree $d_1 = -1$, and the corresponding strength parameter F_1 may be assumed to be constant for a given specimen. This fixes the scaling parameters,

$$\alpha = \frac{E}{E'} , \quad \beta = \left(\frac{E}{E'}\right)^{3/2} \quad , \tag{14}$$

according to (7), (11). In the presence of $n-1$ homogeneous external fields of degree d_j ($j = 2, \ldots, n$) the conditions (13) reduce to

$$F_j/|E|^{d_j+1} = \text{const.} , \quad j = 2, \ldots n \quad , \tag{15}$$

when inserting $d_1 = -1$, $F_1 = \text{const.}$ for $i = 1$. The $n-1$ conditions for mechanical similarity are thus, that the *scaled field strengths* $\tilde{F}_j$, defined by

$$\tilde{F}_j = F_j/|E|^{d_j+1} \quad , \tag{16}$$

be constant. The values of these $n-1$ scaled field strengths determine the properties of the classical dynamics which are invariant to within similarity transformations (2). For each set of values of the scaled field strengths there is now a one-parameter family of mechanically similar systems and not a two-parameter family, because the field strength F_1 is kept fixed.

For a homogeneous external electric field the potential V_2 is homogeneous of degree one, and F_2 is the electric field strength f. The scaled electric field strength is

$$\tilde{f} = f/E^2 \quad , \tag{17}$$

and all systems with the same value of $\tilde{f}$ (and the same sign of E) are mechanically similar.

A homogeneous external magnetic field is studied more conveniently by directly subjecting the equations of motion for a charged particle in such a magnetic field to the similarity transformation (2). The equations of motion in the systems S and S' are seen to be equivalent if the respective magnetic field strengths γ and γ' are related by

$$\gamma' = \gamma/\beta \quad . \tag{18}$$

Comparing with Eq. (6) shows that this corresponds to the behaviour of a homogeneous potential of degree two, and the square of the magnetic field strength plays the role of "field strength" F. For an atom ($d_1 = 1$, $F_1 = \text{const.}$) in a constant homogeneous magnetic field of strength γ, the scaled magnetic field strength $\tilde{\gamma}$ is thus defined via (16) with $F_j = \gamma^2$, $d_j = 2$, as

$$\tilde{\gamma}^2 = \gamma^2/|E|^3 , \quad \tilde{\gamma} = \gamma/|E|^{3/2} \quad . \tag{19}$$

The conditions for invariant classical dynamics of an atom in an external electric or magnetic field are conventionally stated as the condition of constant *scaled energy*, which is $E/\sqrt{f}$ for the electric field and $E/\gamma^{2/3}$ for the magnetic field.[2-9] The nomenclature evolved historically[10], and has probably been a mistake from the pedagogical point of view. This becomes clear when we consider an atom in a superposition of homogeneous electric and magnetic fields. We are then confronted with two different definitions of scaled energy, and usually the conditions of mechanical similarity are expressed as requiring one of these scaled energies and the ratio f^3/γ^4 to be constant.

The more natural statement of the conditions for mechanical similarity for an atom in a superposition of electric and magnetic fields is surely that both scaled field strengths, $\tilde{f}$ and $\tilde{\gamma}$, be constant. This of course implies the constance of the above mentioned scaled energies and of the ratio $\tilde{f}^3/\tilde{\gamma}^4$, which is equal to f^3/γ^4 and is independent of energy.

2.4 Time Dependent External Fields

In the presence of a time dependent external field the expression (9) for the potential energy of the system S must now be generalized, e.g. to

$$U(\vec{r},t) = \sum_{i=1}^{n} F_i V_i(\vec{r}) + F_0 V_0(\vec{r})\Phi(\omega t) \; ; \quad V_i(\alpha\vec{r}) = \alpha^{d_i} V_i(\vec{r}) \; , \quad i = 0,1,\ldots,n \quad , \tag{20}$$

where we have added a homogeneous potential V_0 with strength F_0 multiplied by a time dependent function $\Phi(\omega t)$, which is usually, but not necessarily, a harmonic function (sine, cosine or $\exp(\pm i\omega t)$). The time function Φ need not even be periodic, but the parameter ω (> 0) is included explicitly to give us a handle on the time scale. The corresponding potential energy in the system S' is

$$U'(\vec{r}',t') = \sum_{i=1}^{n} F_i' V_i(\vec{r}') + F_0' V_0(\vec{r}')\Phi(\omega' t') \quad . \tag{21}$$

Again we study the effect of the similarity transformation (2) on the kinetic and potential energy. The systems S and S' are mechanically similar, if kinetic and potential energies in S' differ from those in S by the same multiplicative factor. The time function Φ is generally assumed to be bounded, so it cannot be a homogeneous function. Hence we have no freedom to choose the parameter β connecting the times t and t'; if U' is to be proportional to U there is no choice but to set

$$\beta = \frac{\omega}{\omega'} \quad , \quad \text{so that} \quad \omega' t' = \omega t \quad . \tag{22}$$

The time scale parameter ω replaces the energy of the time independent case as additional parameter (beside the field strengths) determining the classical dynamics of the system. Whereas Eq. (7) fixes the ratio α/β in the time independent case, Eq. (22) fixes the time stretching parameter β in the time dependent case. This leaves one free parameter α and the $n+1$ conditions,

$$\alpha^{d_i} F_i' = \left(\alpha\frac{\omega'}{\omega}\right)^2 F_i \; , \quad i = 0,1,\ldots,n \quad . \tag{23}$$

Resolving for α now yields

$$\alpha = \left[\left(\frac{\omega'}{\omega}\right)^2 \frac{F_i}{F_i'}\right]^{\frac{1}{d_i-2}} \quad \text{for all} \quad i = 0,1,\ldots,n \quad . \tag{24}$$

For any pair (i,j) of labels this implies

$$\frac{|F_i/\omega^2|^{d_j-2}}{|F_j/\omega^2|^{d_i-2}} = \frac{|F_i'/\omega'^2|^{d_j-2}}{|F_j'/\omega'^2|^{d_i-2}} \quad , \tag{25}$$

in other words, mechanical similarity is given if

$$\frac{|F_i/\omega^2|^{d_j-2}}{|F_j/\omega^2|^{d_i-2}} = \text{const.} \tag{26}$$

The potential (20) may contain more than one time dependent contribution. As long as the dependence of each contribution on the coordinates is homogeneous, the results derived for the label $i = 0$ above are easily generalized to a finite number of time dependent terms. Note, however, that only one time scale parameter ω can be accommodated, because there can be only one time stretching factor β, see (22). If the potential contains e.g. a superposition of several harmonic terms with different frequencies, then the mutual ratios of these frequencies have to be the same in all mechanically similar systems, so that there is effectively only one parameter defining the time scale.

2.5 Examples with Time Dependence

For an atom (or ion) in a time dependent field and $n-1$ external static fields we again assume the label $i = 1$ to describe the constant (F_1 =const.) Coulomb field ($d_1 = -1$) of the atom, and this fixes the stretching parameter α via (24),

$$\alpha = \left(\frac{\omega}{\omega'}\right)^{2/3} \quad . \tag{27}$$

The conditions (26) now suggest the following definition for the scaled field strengths:

$$\tilde{F}_j \stackrel{\text{def}}{=} F_j/\omega^{\frac{2}{3}(d_j+1)} \quad . \tag{28}$$

With these definitions the n conditions for mechanical similarity can be expressed as the requirement

$$\tilde{F}_j = \text{const.} \quad j = 0, 2, 3 \ldots, n \quad . \tag{29}$$

For an atom described by a constant Coulomb field ($i = 1$) in a superposition of one ($j = 0$) time dependent and $n-1$ ($j = 2, \ldots, n$) static external fields, the classical dynamics is determined to within mechanical similarity by the values (29) of these n scaled field strengths.

The time dependent field is very often the oscillating electric field of microwave or laser radiation, so $F_0 = f_{\text{rad}}$ is the amplitude of the oscillating field of circular frequency ω, and $d_0 = 1$. The corresponding scaled field strength $\tilde{f}_{\text{rad}}$, which is constant under the conditions of mechanical similarity, is

$$\tilde{f}_{\text{rad}} = f_{\text{rad}}/\omega^{4/3} \tag{30}$$

according to (28). For an external static electric field of strength f, the scaled field strength $\tilde{f}$ is analogously given by $\tilde{f} = f/\omega^{4/3}$. For an additional magnetic field of strength γ (with F_j corresponding to γ^2), the scaled field strength $\tilde{\gamma}$ is given by

$$\tilde{\gamma}^2 = \gamma^2/\omega^2 \ , \quad \tilde{\gamma} = \gamma/\omega \quad , \tag{31}$$

according to (28). Under the conditions of mechanical similarity, $\tilde{f}_{\text{rad}}$, $\tilde{f}$ and $\tilde{\gamma}$ are constant, and so are $f_{\text{rad}}^3/\gamma^4$ and f^3/γ^4 as in the time independent case.

3. QUANTUM MECHANICS

The quantum mechanical system corresponding to the classical system S introduced in Sect. 2 is described by the Schrödinger equation,

$$-\frac{\hbar^2}{2m}\Delta\psi(\vec{r},t) + U(\vec{r},t)\psi(\vec{r},t) = i\hbar\frac{d}{dt}\psi(\vec{r},t) \quad , \tag{32}$$

and can be obtained by quantization via the *canonical commutation relations* between the coordinates r_i and the momenta $p_j = m\,dr_j/dt$,

$$[r_i, p_j] = i\hbar\delta_{i,j} \quad . \tag{33}$$

When the classical dynamics of the system S is related to the classical dynamics of the system S' via the *non-canonical* similarity transformation (2), the coordinates transform as $r_i' = \alpha r_i$ and the momenta as $p_j' = (\alpha/\beta)p_j$. [The latter also holds if the momentum p_j contains a term proportional to a vector potential describing a homogeneous magnetic field, because the vector potential must be proportional to a product of the magnetic field strength, transforming according to (18), and a linear function of the coordinates.] The same quantum mechanics is thus obtained by quantization of the system S' via the *non-canonical commutation relations*,

$$[r_i', p_j'] = i\hbar'\delta_{i,j} \quad , \tag{34}$$

where $\hbar'$ is an *effective Planck's constant*,

$$\hbar' = \frac{\alpha^2}{\beta}\hbar \quad . \tag{35}$$

If the field strengths F_i in S are varied under the conditions of mechanical similarity, then canonical quantization in the system S leads to the same quantum mechanics as non-canonical quantization in the mechanically similar "scaled system" system S' according to (34), with a variable effective Planck's constant (35).

3.1 Time Independent Homogeneous Potentials

For a system with one time independent homogeneous potential of degree d, $U(\vec{r}) = FV(\vec{r})$, the constants α and β are given by (8), and the effective Planck's constant in the scaled system S' is

$$\hbar' = \frac{\alpha^2}{\beta}\hbar = \left(\frac{F}{F'}\right)^{\frac{1}{d}}\left(\frac{E'}{E}\right)^{\frac{1}{d}+\frac{1}{2}}\hbar \quad . \tag{36}$$

It is interesting to study (for fixed energy E' and field strength F'), which combination of energy E and field strength F corresponds to the semiclassical limit, $\hbar' \to 0$. This obviously depends in the following way on the degree d of homogeneity of the potential:

$$0 < d: \quad F \to 0 \quad \text{or} \quad |E| \to \infty \tag{37}$$

$$-2 < d < 0: \quad |F| \to \infty \quad \text{or} \quad E \to 0 \tag{38}$$

$$d = -2: \quad |F| \to \infty \quad \text{and} \quad E \ \text{arbitrary} \tag{39}$$

$$d < -2: \quad |F| \to \infty \quad \text{or} \quad |E| \to \infty \quad . \tag{40}$$

When $d > 0$ or $d < -2$, the semiclassical limit for a given field strength F corresponds to the high energy limit $|E| \to \infty$. However, if the degree d of homogeneity of the potential lies between zero and -2, then the semiclassical limit of the Schrödinger equation for a given field strength corresponds to the limit of vanishing values of the energy E. This applies in particular to all Coulomb systems, where $d = -1$, and it is perhaps not surprising when remembering that the energies of the bound states of a one-electron atom vanish in the (semiclassical) limit of large quantum numbers.

The fact that the low energy limit, $E \to 0$, corresponds to the semiclassical limit in Coulomb systems also for scattering states, is not so widely appreciated, mainly because of the peculiar coincidence, that classical mechanics, quantum mechanics and the Born approximation all give the same answer for the Rutherford scattering cross section in three dimensions. This is not the case for the differential cross section $d\sigma/d\theta$ for scattering by a potential F/r in two spatial dimensions. As shown by Barton,[11] the exact quantum mechanical result is

$$\left(\frac{d\sigma}{d\theta}\right)_{\rm qm} = \frac{|\eta|}{2k\sin^2(\theta/2)}\tanh(\pi|\eta|) \quad , \tag{41}$$

where $\eta = Fm/(\hbar^2 k)$ is the Coulomb parameter ($k = \sqrt{2mE}/\hbar$) and θ is the scattering angle. In contrast, the classical scattering cross section in two dimensions is

$$\left(\frac{d\sigma}{d\theta}\right)_{\rm cl} = \frac{|\eta|}{2k\sin^2(\theta/2)} \quad , \tag{42}$$

which incidently is just the square root of the three-dimensional Rutherford cross section $d\sigma/d\Omega$. The result of the Born approximation for Rutherford scattering in two dimensions is[11]

$$\left(\frac{d\sigma}{d\theta}\right)_{\rm BA} = \frac{\pi\eta^2}{2k\sin^2(\theta/2)} \quad . \tag{43}$$

The quantum mechanical cross section (41) differs from the classical cross section (42) by the factor $\tanh(\pi|\eta|) = \tanh[\pi Fm/(\hbar^2 k)]$, which approaches unity in the low-energy limit $k \to 0$ and approaches $\pi|\eta|$, yielding the cross section obtained in the Born approximation, in the high-energy limit, $k \to \infty$, which is the anti-classical limit for the Coulomb potential. The classical and quantum cross sections are illustrated in Fig. 1 together with the result of the Born approximation. It is interesting to observe that both the classical approximation and the Born approximation always overestimate the cross section.

The fact that the semiclassical limit corresponds to vanishing total energy for any pure Coulomb system makes it seem natural, that classical and semiclassical methods are successful in describing electron impact ionization of one-electron atoms near threshold.[12] Note however, that substantial discrepancies between quantum and classical cross sections have been observed at small but finite total energies in model systems of the pure Coulomb type.[13]

3.2 Potential Consisting of Several Time Independent Homogeneous Contributions

Consider a potential U in (32) consisting of n contributions, $U(\vec{r}) = \sum_{i=1}^{n} F_i V_i(\vec{r})$, where V_i is a (time independent) homogeneous potential of degree d_i. The equivalence of the canonical Schrödinger equation for energy E and field strengths F_i with the non-canonical Schrödinger equation containing the modified Planck's constant (35) is maintained, as long as energy and field strengths are varied under the conditions of mechanical similarity described above. This implies

$$\hbar' = \frac{\alpha^2}{\beta}\hbar = \left(\frac{F_i}{F_i'}\right)^{\frac{1}{d_i}} \left(\frac{E'}{E}\right)^{\frac{1}{d_i}+\frac{1}{2}} \hbar \quad \text{for all} \quad i \quad . \tag{44}$$

The conditions of the semiclassical limit correspond in each contribution i to the limiting behaviour (37) – (40), depending on the degree d_i of homogeneity of the respective term. These conditions are compatible under the conditions of mechanical similarity (12). For

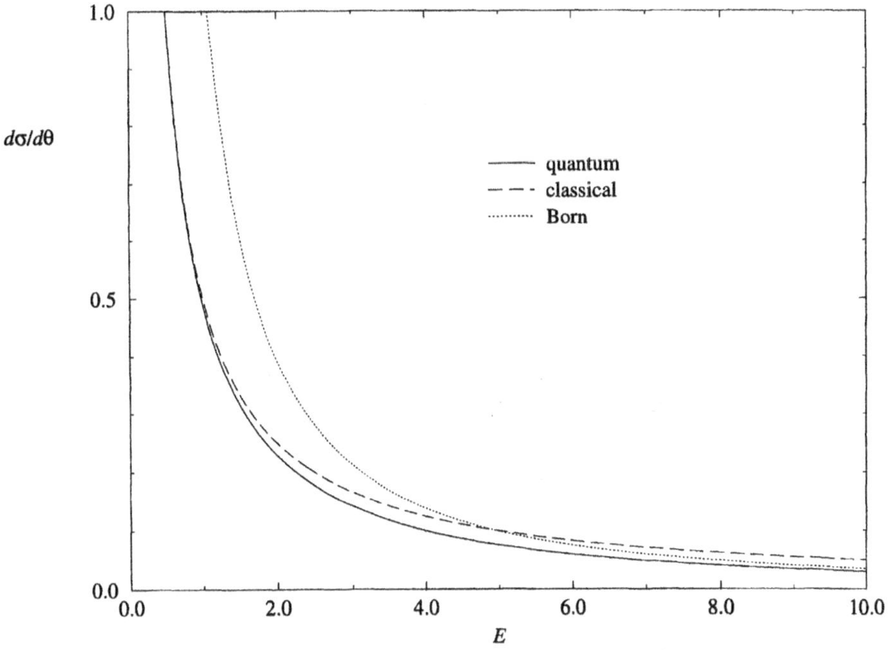

Figure 1. Energy dependence of the cross section for elastic scattering by the potential $U(r) = 1/r$ in two dimensions; $\hbar = m = 1$, $\theta = 90°$. Quantum (41), classical (42) and Born approximation (43) results are shown.

example, if the label $i = 1$ describes the fixed Coulomb potential in an atom, then the condition of constant scaled field strengths (16) implies

$$F_j \propto |E|^{1+d_j} \quad . \tag{45}$$

The semiclassical limit $\hbar' \to 0$ corresponds to $E \to 0$. For any further contributions with a positive degree of homogeneity, e.g. an external electric field with $d_j = 1$, or an external magnetic field with $d_j = 2$, the strengths F_j must tend to zero as prescribed by (45) in the semiclassical limit. Note in particular, that a fixed strength of the Coulomb potential and a non-vanishing external electric and/or magnetic field are incompatible with the conditions of the semiclassical limit.

3.3 Scaled Fields Spectroscopy

The energy and the $n-1$ strengths of the static external fields in which an atom is placed, have $n-1$ conditions to fulfill for mechanical similarity to hold, e.g. that the scaled field strengths (16) be constant. When the field strength of the Coulombic forces describing the atom is kept fixed, there remains one continuous parameter, which can be varied without changing the classical dynamics, except to within a similarity transformation (2). This makes it possible to study the variations of the quantum system corresponding to different values of the effective Planck's constant without changing the classical dynamics. Although the energy itself or any one of the external field strengths could be chosen as the variable parameter, a prudent choice is

$$\chi = \frac{\beta}{\alpha^2}\frac{1}{\hbar} = \frac{1}{\hbar'} \quad , \tag{46}$$

which has the dimensions of an inverse action and is just the inverse of the effective Planck's constant $\hbar'$.

A justification for this choice can be found in modern quantization techniques based on Gutzwiller's trace formula or variations thereof,[9,14] which generally express the energy spectrum or the energy level density or the semiclassical propagator in terms containing exponentials involving the actions $I_{\rm traj}$ along a multitude of trajectories,

$$\text{property of qm spectrum} = \text{function}\left[\exp\left(\frac{\rm i}{\hbar}I_{\rm traj}\right)\right], \quad I_{\rm traj} = \int_{\rm traj}\vec{p}\cdot d\vec{r} \quad . \tag{47}$$

Regarding both sides of Eq. (47) as functions of the variable χ defined by (46) leads to the following form of this general equation:

$$\text{property of qm spectrum}(\chi) = \text{function}\left(\exp\left({\rm i}\chi I'_{\rm traj}\right)\right) \quad , \tag{48}$$

where we have expressed the actions $I_{\rm traj}$ through the "scaled actions"

$$I'_{\rm traj} = \int_{\rm traj}\vec{p}'\cdot d\vec{r}' = \frac{\alpha^2}{\beta}I_{\rm traj} = \frac{\hbar'}{\hbar}I_{\rm traj} \quad . \tag{49}$$

The scaled classical actions (49) depend only on the fixed energy E', which defines the energy at which the effective Planck's constant $\hbar'$ assumes its physical value $\hbar$, and on the $n-1$ values of the scaled field strengths (16), which determine the classical dynamics. In the general formula (48) these scaled actions appear as Fourier conjugates to the variable χ. Applying a Fourier transform to Eq. (48) will thus reveal structures associated with classical trajectories at values of the conjugate variable corresponding to the scaled actions of the trajectories.

For an atom in external static fields the scaling parameters α and β are given by (14) and the natural variable (46) is,

$$\chi = \frac{1}{\hbar}\sqrt{\frac{E'}{E}} \propto \frac{1}{\hbar\sqrt{|E|}} \quad . \tag{50}$$

The definition of the natural variable χ depends on which field strength we are keeping constant, and not on which external fields (of variable strength) are present; the constance of the strength F_1 of the Coulombic ($d_1 = -1$) potential describing the atom leads to the simple result (50), $\chi \propto 1/\sqrt{|E|}$. For an external magnetic field of variable strength γ, this corresponds to $\chi \propto \gamma^{-1/3}$ when the scaled field strength $\tilde{\gamma}$ is kept constant, cf. (19). For an external static electric field of variable strength, $\chi \propto 1/\sqrt{|E|}$ corresponds to $\chi \propto f^{-1/4}$ when the scaled field strength $\tilde{f}$ is kept fixed, cf. (17). In a superposition of electric and magnetic fields both relations apply, which is consistent because f^3/γ^4 is constant under the conditions of mechanical similarity.

The technique of scaled fields spectroscopy is well established for the example of atoms in external electric and magnetic fields and has been been called "scaled energy spectroscopy"[15,16] and also "recurrence spectroscopy",[8,17] because of the dominating role which periodic and recurring classical orbits play in appropriately Fourier transformed spectra.

A scaled field spectrum is shown in Fig. 2 for the example of a lithium atom in a constant electric field (from Courtney *et al.*,[16] see also Delos and Schwieters[17]). In this figure, the Fourier transform of the measured photoabsorption spectrum is shown as a function of the "scaled action", which is the Fourier conjugate of the variable χ,

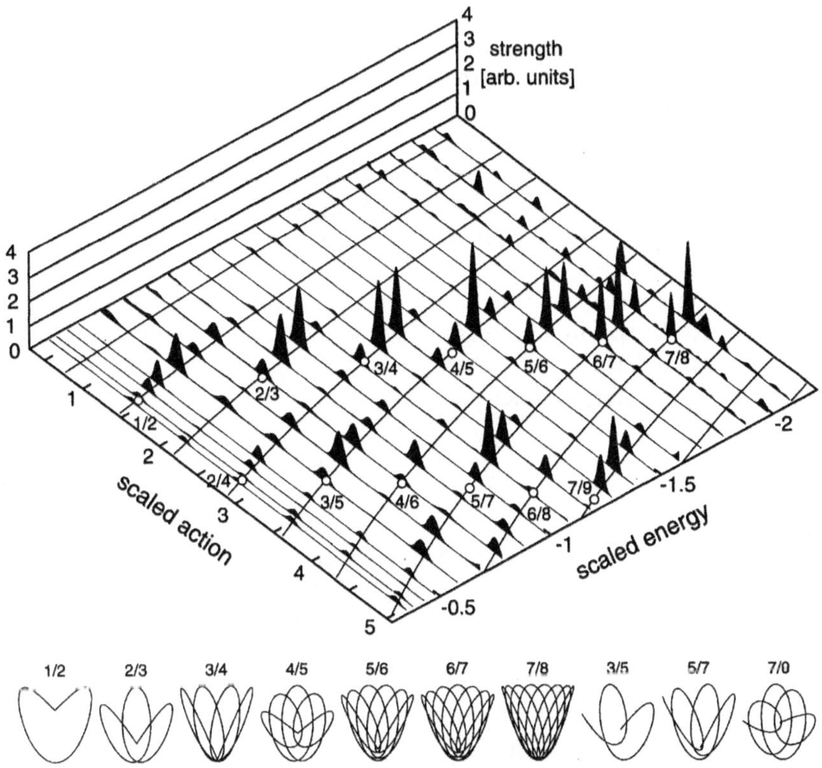

Figure 2. Fourier transformed photoabsorption cross section of lithium in an external electric field as function of the scaled action, which is the Fourier conjugate of the natural variable (50), for various values of the "scaled energy", which is minus the inverse square root of the scaled electric field strength (17) in this case. The smooth lines in the scaled action–scaled energy plane show the dependence on scaled energy of the scaled action of single, double, ..., eleven-fold traversals of the straight line orbit parallel to the direction of the electric field. The open circles show points of bifurcation where the the straight line "pregnant periodic orbit" gives birth to the new orbits, whose shape is illustrated at the bottom of the Figure. [From Courtney *et al.*[16]]

and the "scaled energy", which is plus or minus the inverse square root of the scaled electric field strength $\tilde{f}$; it is this quantity which determines the classical dynamics. The curves in the plane show the dependence on scaled energy of the scaled action of one or more traversals of the periodic straight line orbit parallel to the direction of the electric field; on this orbit the electron oscillates between the atomic nucleus and the classical turning point on the uphill side of the potential. The Fourier transformed spectra for fixed scaled energies clearly tend to be peaked at values of the scaled action variable corresponding to the scaled actions of one or more traversals of this periodic orbit. In the example, the peaks are especially strong at scaled energies beyond certain values (indicated as circles) at which the straight line orbit bifurcates and "gives birth" to further periodic orbits, which are illustrated at the bottom of the figure. "Ghost orbits" which are only realized in complex phase space prior to such points of birth, have been the focus of considerable attention in recent years.[5] The straight line orbit is not a ghost orbit, because it is present as a real orbit prior to the bifurcation points in Fig. 2, it is just that the "off-spring orbits" have not yet appeared. It would be consistent to introduce the term "pregnant periodic orbit" for the straight line orbit prior to the points of bifurcation in Fig. 2.

3.4 Time Dependent Potentials

The Schrödinger equation (32) with the time dependent potential (20) is equivalent to a non-canonical Schrödinger equation containing the effective Planck's constant (35) with the scaled potential (21) as long as the frequency parameter and the potential strengths obey the conditions (22) and (24) for mechanical similarity. If the label $i = 1$ describes the fixed ($F_1 = F_1'$) Coulomb potential ($d_1 = -1$) of an atom, the stretching parameter α is given by (27), and the conditions for mechanical similarity reduce to the requirement that the scaled field strengths (28) be constant.

For a concrete experiment with a one-electron atom in a time dependent field, the initial (unperturbed) state of the atom is described by a quantum number n_0, and $n_0\hbar$ is the classical action I of the electron on the corresponding orbit. The similarity transformation (2) transforms actions as

$$I' = \frac{\alpha^2}{\beta} I = \left(\frac{\omega}{\omega'}\right)^{1/3} I \quad , \tag{51}$$

according to (22) and (27), hence $n_0\omega^{1/3}$ is the corresponding scaled quantum number which remains constant under the conditions of mechanical similarity. Using the initial quantum number n_0 as reference rather than the frequency parameter ω leads to

$$\tilde{f} = f\, n_0^4 \tag{52}$$

as an alternative definition [instead of (30)] for the scaled strengths of the time dependent or static electric fields.[18,19] The corresponding alternative to (31) for the scaled strength of an external magnetic field is

$$\tilde{\gamma} = \gamma\, n_0^3 \quad . \tag{53}$$

With α and β given by (27) and (22), the effective Planck's constant is

$$\hbar' = \left(\frac{\omega}{\omega'}\right)^{1/3} \hbar \quad , \tag{54}$$

and the semiclassical limit $\hbar' \to 0$ corresponds to $\omega \to 0$. Note that a finite time scale for the time dependent part of the potential is incompatible with the semiclassical limit under the conditions of mechanical similarity. For fixed field strength of the Coulomb potential describing the atom, the semiclassical limit for an atom in external time dependent and/or time independent electric and magnetic fields corresponds to the static limit according to (54) and to vanishing field strengths according to (28), (45).

4. NON-COULOMBIC "ATOMS"

In the rather general treatment above, the Coulomb potential of the atom or ion on which the examples have been based, may readily be replaced by a different interaction potential describing an artificial atom. Let's assume our artificial atom is described by a homogeneous potential of degree d_1 with a constant field strength, and external time independent and time dependent fields may be present as in Sects. 2 and 3. Billiard systems in which a particle moves freely between specularly reflecting infinitely steep walls appear as the special case $d_1 \to +\infty$.

With the assumption that the field strength F_1 is constant, Eq. (11) for the stretching factor α for the time independent case becomes

$$\alpha = \left(\frac{E'}{E}\right)^{1/d_1} \quad , \tag{55}$$

corresponding to $\alpha = \sqrt{E'/E}$ for the harmonic potential and $\alpha = 1$ for billiards — the shape and size of trajectories does not depend on energy for a given billiard potential.

Adapting Eq. (13) for $i = 1$, $F_1 = \text{const.}$ leads to the following definition of scaled field strengths:

$$\tilde{F}_j = F_j/|E|^{1-d_j/d_1} \tag{56}$$

which replaces (16); the conditions for mechanical similarity are now, that these scaled field strengths be constant. Note that the scaled field strengths are equal to F_j/E regardless of d_j in the billiard limit $d_1 \to +\infty$.

For the special case of a harmonic potential, $d_1 = 2$, the scaled electric field strength $\tilde{f}$ is related to the real electric field strength f via ($d_j = 1$)

$$\tilde{f} = f/\sqrt{|E|} \quad , \tag{57}$$

while the scaled magnetic field strength ($d_j = 2$) is equal to the real magnetic field strength,

$$\tilde{\gamma} = \gamma \quad , \tag{58}$$

and does not depend on the energy.

The factor α^2/β, by which actions are transformed under the similarity transformation (2) is now

$$\frac{\alpha^2}{\beta} = \left(\frac{E'}{E}\right)^{\frac{1}{d_1}+\frac{1}{2}} = \begin{cases} E'/E & \text{for} \quad d_1 = 2 \quad , \\ \sqrt{E'/E} & \text{for} \quad d_1 \to \infty \quad , \end{cases} \tag{59}$$

and the natural variable for scaled field spectroscopy is

$$\chi \propto \frac{1}{\hbar}\,|E|^{\frac{1}{d_1}+\frac{1}{2}} \propto \begin{cases} |E| & \text{for} \quad d_1 = 2 \quad , \\ \sqrt{|E|} & \text{for} \quad d_1 \to \infty \quad . \end{cases} \tag{60}$$

These results are obvious, because the natural variable, being inversely proportional to the effective Planck's constant, is directly proportional to the integrated energy level density (i.e. the number of levels below a given energy) in the *one-dimensional* potential with the same degree of homogeneity as the "atomic potential" whose strength is kept constant.

For an artificial atom in a time dependent external field, the stretching factor α is given, instead of by (27), by

$$\alpha = \left(\frac{\omega'}{\omega}\right)^{\frac{2}{d_1-2}} \quad . \tag{61}$$

In the billiard limit $d_1 \to \infty$, α is unity as in the time independent case. For the "harmonic atom", $d_1 = 2$, Eq. (61) only makes sense if $\omega' = \omega$, and again $\alpha = 1$. The shape and size of trajectories does not depend on the time scale in billiard systems. In a harmonic potential of fixed strength, the energy independent period of oscillation defines a time scale, and hence no variation in the frequency of the external field is possible under the conditions of mechanical similarity.

Adapting Eq. (26) for $i = 1$, $F_1 = \text{const.}$ leads to the following definition of scaled field strengths:

$$\tilde{F}_j = F_j/\omega^{2\frac{d_1-d_j}{d_1-2}} \quad , \tag{62}$$

which replaces (28); the conditions for mechanical similarity are now, that these scaled field strengths be constant. Note that the scaled field strengths are F_j/ω^2 regardless of d_j in the billiard limit $d_1 \to +\infty$.

The scaling rules for a harmonic "atomic potential" would apply to a one-electron quantum dot, when the confining potential is quadratic potential of fixed strength. For several electrons confined in a quantum dot, the electron-electron interaction potential is Coulombic, i.e. homogeneous with degree -1, but the confining potential is in general something different. If we regard only the electron-electron part of the potential as "the atomic potential" whose strength is fixed and the confining potential as (one of the) external field(s), then all the results of Sects. 2 and 3 applying to ordinary Coulombic atoms hold without change for the quantum dot case, provided a way is found to vary the strength of the confining field. If the confining field is quadratic with oscillator parameter ω_0, then the corresponding strength parameter F_j is ω_0^2, and the scaled oscillator parameter $\tilde{\omega}_0$, which must be constant under the conditions of mechanical similarity, is given via (16),

$$\tilde{\omega}_0^2 = \omega_0^2/|E|^3 , \quad \tilde{\omega}_0 = \omega_0/|E|^{3/2} \quad . \tag{63}$$

In this way, the many-electron states in quantum dots, which were central topics in the talks by Vignale[20] and Maksym,[21] could, with or without additional external fields, be analyzed by the technique of scaled fields spectroscopy, if controlled variation of the strength of the confining potential can be realized.

5. SUMMARY

When the potential energy of a mechanical system can be written as a sum of n time independent homogeneous contributions, the classical dynamics of the system depends not on the energy and the strengths of all n contributions, but only on $n-1$ combinations of these $n+1$ numbers. If the strength of one of these homogeneous contributions, which we call the "atomic potential", is kept fixed, then this leaves a one-parameter family of systems whose classical dynamics is invariant to within a similarity transformation (2). The $n-1$ parameters determining the classical dynamics may e.g. be chosen to be the scaled field strengths (56), corresponding to (16) when the atomic potential is Coulombic. A homogeneous magnetic field of strength γ fits into this picture like a quadratic potential with strength parameter γ^2.

The quantum mechanics of mechanically similar systems with variable field strengths is identical to the quantum mechanics of a scaled system with fixed field strengths, derived via non-canonical quantization involving an effective Planck's constant (35). The semiclassical limit corresponds to the limit of vanishing effective Planck's constant. For time independent systems the inverse of the effective Planck's constant constitutes a convenient natural variable for recording spectra. When the strength of one "atomic potential" is kept fixed, the natural variable is uniquely defined to within multiplication by a constant, and it does not depend on which (variable) external fields are present. The semiclassical limit corresponds to vanishing total energy when the atomic potential is Coulombic. If quantum spectra are recorded under the conditions of mechanical similarity as functions of the natural variable, then the Fourier transformed spectra reveal structures related to prominent invariant quantities of the classical dynamics, such as the scaled actions of recurring orbits. This is the basis of "scaled fields spectroscopy".

For a time dependent potential the time scale of the time dependent contribution(s) replaces the energy as one of the parameters on which the classical dynamics may depend. If the "atomic potential" is homogeneous of degree d_1, then the parameters determining the classical dynamics may be chosen to be the scaled field strengths (62), which reduce to (28) when $d_1 = -1$.

The considerations above apply typically to atoms (or molecules or ions), described by a Coulombic potential ($d_1 = -1$) of fixed strength (F_1 = const.) in static or time dependent external electric ($d_j = 1$) or magnetic ($d_j = 2$) fields. They apply also to artificial atoms, such as quantum dots, when the "atomic potential" is homogeneous, but not necessarily Coulombic.

ACKNOWLEDGEMENT

I am grateful to Johannes Trost for critical reading of the manuscript.

REFERENCES

1. L.D. Landau and E.M. Lifschitz, *Lehrbuch der Theoretischen Physik*, Bd. I, *Mechanik* (Akademie-Verlag, Berlin, 1966).
2. H. Friedrich, *Theoretical Atomic Physics*, (Springer-Verlag, Heidelberg, 1991).
3. H. Ruder, G. Wunner, H. Herold and F. Geyer, *Atoms in Strong Magnetic Fields*, (Springer-Verlag, Heidelberg, 1994).
4. W. Hogervorst, K. Karremans, A. Kips and W. Vassen, contribution to this conference.
5. J. Main and G. Wunner, contribution to this conference.
6. J. Main, in: Classical, Semiclassical and Quantum Dynamics in Atoms, ed. H. Friedrich and B. Eckhardt, Lecture Notes in Physics, vol. **485**, 247 (Springer-Verlag, Heidelberg, 1997).
7. J. Main, contribution to this conference.
8. J. Main, V.A. Mandelshtam and H.S. Taylor, Phys. Rev. Lett. **78** (1997) 4351.
9. H. Friedrich and B. Eckhardt (eds.), Classical, Semiclassical and Quantum Dynamics in Atoms, Lecture Notes in Physics, vol. **485**, (Springer-Verlag, Heidelberg, 1997).
10. D. Wintgen and H. Friedrich, Phys. Rev. A **35**, 1464 (1987); H. Friedrich and D. Wintgen, Phys. Reports **183**, 37 (1989).
11. G. Barton, Am. J. Phys. **51**, 420 (1983).
12. J.-M. Rost and G. Tanner, in: Classical, Semiclassical and Quantum Dynamics in Atoms, ed. H. Friedrich and B. Eckhardt, Lecture Notes in Physics, vol. **485**, 273 (Springer-Verlag, Heidelberg, 1997).
13. J. Macek and W. Ihra, Phys. Rev. A **55**, 2024 (1997); W. Ihra, F. Mota-Furtado and P.F. O'Mahony, Phys. Rev. A **55**, 4263. (1997).
14. Chaos 2, No. 1 (1992), Special Focus Issue on Periodic Orbit Theory.
15. J. Main, G. Wiebusch and K.H. Welge, Comments on At. Mol. Phys. **XXV**, 233 (1991).
16. M. Courtney, H. Jiao, N. Spellmeyer, D. Kleppner, J. Gao and J.B. Delos, Phys. Rev. Lett. **74**, 1538 (1995).
17. J.B. Delos and C. Schwieters, in: Classical, Semiclassical and Quantum Dynamics in Atoms, ed. H. Friedrich and B. Eckhardt, Lecture Notes in Physics, vol. **485**, 232 (Springer-Verlag, Heidelberg, 1997).
18. P.M. Koch, Chaos **2**, 131 (1992).
19. D. Richards, in: Classical, Semiclassical and Quantum Dynamics in Atoms, ed. H. Friedrich and B. Eckhardt, Lecture Notes in Physics, vol. **485**, 172 (Springer-Verlag, Heidelberg, 1997).
20. M. Ferconi and G. Vignale, contribution to this conference.
21. P.A. Maksym, contribution to this conference.

TIME INDEPENDENT AND TIME DEPENDENT STATES OF ATOMS IN STATIC EXTERNAL FIELDS

P. F. O'Mahony,[1] I. Moser,[1] F. Mota-Furtado,[1] and J. P. dos Santos[2]

[1]Department of Mathematics,
Royal Holloway, University of London,
Egham, Surrey TW20 OEX, United Kingdom
[2]Centro de Física Atómica da Universidade de Lisboa
and Departamento de Física da Universidade Nova de Lisboa,
Av. Prof. Gama Pinto 2, P-1699 Lisboa Codex, Portugal

INTRODUCTION

The dynamics of a Rydberg atom in an applied static magnetic field is of interest because it provides an experimentally realizable prototype for studying classical and quantum chaos. The spectrum of such an atom is also of interest in other areas of physics for example in astrophysics when determining the magnetic field strengths present in white dwarfs and neutron stars.

The dynamics of Rydberg atoms in external fields has been extracted from traditional stationary state spectroscopy such as the measurement of the photoabsorbtion and the photoionization spectrum and more recently by explicitly time dependent techniques using pulsed lasers to excite non-stationary Rydberg wavepackets.

We present here (a) a method to calculate the photoionization spectrum of an atom in an arbitrary strength magnetic field but particularly for atoms in laboratory strength fields. This allows one, for the first time, to calculate the photoionization spectrum at arbitrarily high energies and over a very wide range of energies and field strengths. We do this by combining an adiabatic basis set approach with R matrix propagation. This method also allows a detailed analysis of complex resonances present in the spectrum via the use of multichannel quantum defect theory. We will present the method and results for photoionization of lithium in a field of 6.1 Tesla.

We also present (b) an analysis of recent Rydberg wavepacket experiments in parallel electric and magnetic fields. These experiments give direct information in the time domain on the most important periodic orbits in the classically regular and chaotic regimes. The magnitude of the time autocorrelation function between states excited by two Gaussian laser pulses is calculated for both hydrogen and rubidium atoms in parallel electric and magnetic fields and compared with experiment. Qualitative agreement is obtained with experiment.

Atoms and Molecules in Strong External Fields
Edited by Schmelcher and Schweizer, Plenum Press, New York, 1998

TIME INDEPENDENT STATES OF AN ATOM IN A MAGNETIC FIELD

Continuum states of atoms in a magnetic field have been of great interest since the observation of quasi Landau modulations above the ionization threshold in the now classic experiment of Garton and Tomkins[1]. These systems have an inherent non separability arising from the competing spherical symmetry of the atom and the cylindrical symmetry of the applied field. This competition leads to the corresponding classical system exhibiting chaotic behaviour. Since the experiments of Garton and Tomkins the experimental resolution has been increased by orders of magnitude using laser spectroscopy culminating in the recent experiments by Iu *et al* [2] on lithium in a field of about 6 Tesla.

Stationary states for the continuum spectrum of an atom in a magnetic field have been calculated using complex co-ordinate techniques[3], R-matrix propagation[4, 5, 6] and the diabatic-by-sector method[7]. Although all of these approaches have recreated the experimental spectrum of Iu *et al* [2] they are limited to studying energy regions close to the ionization threshold. We have designed a new method which combines the attractive features of both the diabatic-by-sector method and R-matrix propagation allowing us to extend the calculation of the photoionization cross section to energies far from the ionization threshold[8] . The multichannel quantum defect theory of Seaton[9] can also be incorporated allowing the resonances present in the spectrum to be analysed.

The Hamiltonian for hydrogen in a magnetic field (taken to be in the z direction) in the symmetric gauge, using atomic units ($\hbar = m = e = 1$) is

$$H = -\frac{\nabla^2}{2} - \frac{1}{r} + \beta L_z + \frac{1}{2}\beta^2 r^2 \sin^2\theta \tag{1}$$

where the magnetic field B given in atomic units is $\beta = B/B_0$ with $B_0 = 4.7 \times 10^5 T$. L_z is a conserved quantity.

At low r the Coulomb potential dominates while at large r the magnetic field potential is the dominant effect. Between these values a region of strong mixing occurs. For a general atom at a particular field strength there are four regimes of interaction for an excited electron. After excitation by a photon the electron moves initially through the core region of the atom and emerges into a region where the Coulomb potential dominates and the diamagentic potential is negligible. It then enters a region where both fields are of comparable strength and finally reaches the asymptotic region where the cylindrically symmetric magnetic field potential dominates.

The method presented here is based on the R matrix approach of O'Mahony and Mota-Furtado[4, 5, 6] in which these four regimes of interaction are treated in different ways. A major improvement on this method is presented in which adiabatic functions are used to represent the wavefunction in the strong mixing region. This allows for an order of magnitude saving in computer time and memory.

The Coulomb region

Outside the atomic core the Hamiltonian for the continuum electron is given by equation (1). At sufficiently low radii ($r < a$) this equation can be approximated by the field free Hamiltonian because the potential terms resulting from the influence of the magnetic field are negligible in comparison with the Coulomb potential. The wavefunction of a continuum electron at an energy ϵ and with an orbital angular momentum l can therefore be written in terms of the energy normalized Coulomb functions s and c as

$$\Psi_l(r) = \left(s_l^\epsilon(r) + c_l^\epsilon(r)\tan(\pi\mu_l)\right) Y_{lm}(\theta,\phi). \tag{2}$$

The quantum defects μ_l are used to represent the non-hydrogenic core[9] and $Y_{lm}(\theta, \phi)$ are spherical harmonics. The quantum defects can be calculated *ab initio* or taken from experiment. By calculating the wavefunction $\Psi_l(r)$ and its derivative at $r = a$, the R matrix or log-derivative matrix can be constructed on the outer boundary of this region.

$$\Psi_l(a) = \sum_{l'} R_{l,l'} \left. \frac{d\Psi_{l'}(r)}{dr} \right|_{r=a} \tag{3}$$

The effects of a non hydrogenic core are contained in μ_l.

The strong mixing region

In this region the Coulomb potential and the diamagnetic potential are of a comparable size. This region is defined by $a < r < b$ where the radius b is taken to be large enough such that the Hamiltonian is separable in cylindrical coordinates. The change in symmetry of the potential, from spherical to cylindrical, is therefore completely contained within this region. The method which is used aims to obtain the R matrix at the outer boundary of this region $r = b$ from the R matrix at the inner boundary $r = a$ (calculated in equation (3)).

In order to calculate the R matrix at $r = b$ from the R matrix at $r = a$ the range $r = a \rightarrow b$ is firstly split into N sectors with radii $a \rightarrow a_1, a_1 \rightarrow a_2, \cdots, a_{N-1} \rightarrow b$. Within each of these sectors the adiabatic Hamiltonian H_{ad}, i.e. the full Hamiltonian H (equation (1)) at a fixed radius r_n, is diagonalized in a basis of spherical harmonics with the radius r_n corresponding to a radius within that sector, i.e. for the nth sector $a_{n-1} < r_n < a_n$. In a particular sector the functions generated $\phi_\lambda(r_n; \theta, \phi)$ give an exact description of the angular solutions of the Schrödinger equation at $r = r_n$. The functions ϕ_λ are therefore a very good basis with which to represent the angular part of the wavefunction in the local region around r_n, namely within the sector n. These are produced by a diagonalization of the adiabatic Hamiltonian

$$H_{ad}(r_n; \theta, \phi) = \frac{\mathbf{L}^2}{2r_n^2} - \frac{1}{r_n} + \beta L_z + \frac{1}{2}\beta^2 r_n^2 \sin^2\theta \tag{4}$$

in a basis set of spherical harmonics such that

$$\phi_\lambda(r_n; \theta, \phi) = \sum_l c_l^\lambda Y_{lm}(\theta, \phi). \tag{5}$$

In each sector the full Hamiltonian (equation (1)) plus the Bloch operator[10] or surface term is diagonalized in a basis set consisting of a product of Legendre polynomials $f_n(r)$ and the adiabatic functions generated for that sector ϕ_λ. The eigenfunctions obtained from this diagonalization are therefore

$$\Psi^k = \sum_{n\lambda} c_{n\lambda}^k \frac{f_n(r)}{r} \phi_\lambda(r_n; \theta, \phi). \tag{6}$$

The eigenvalues E_k and the eigenvectors $c_{n\lambda}^k$ can be used to evaluate the quantity

$$U_{\lambda k}(r) = \sum_n c_{n\lambda}^k \frac{f_n(r)}{r} \tag{7}$$

on the inner and outer boundary of each sector ($r = a_{n-1}$ and $r = a_n$). Using these $U_{\lambda k}$'s we can construct four matrices, called sector R matrices, with i,jth elements given by

$$(r_1^n)_{ij} = \frac{1}{2}\sum_k \frac{U_{ik}(a_{n-1})\,U_{jk}(a_{n-1})}{E_k - \epsilon} \qquad (r_2^n)_{ij} = \frac{1}{2}\sum_k \frac{U_{ik}(a_{n-1})\,U_{jk}(a_n)}{E_k - \epsilon}$$

$$(r_3^n)_{ij} = \frac{1}{2}\sum_k \frac{U_{ik}(a_n)\,U_{jk}(a_{n-1})}{E_k - \epsilon} \qquad (r_4^n)_{ij} = \frac{1}{2}\sum_k \frac{U_{ik}(a_n)\,U_{jk}(a_n)}{E_k - \epsilon}. \tag{8}$$

It can be shown[11] that these matrices give a relationship between the R matrix on the inner boundary of the nth sector ($r = a_{n-1}$) and the R matrix on the outer boundary of the nth sector ($r = a_n$) such that

$$\mathbf{R}(a_n) = \mathbf{r}_4^n - \mathbf{r}_3^n\left(\mathbf{r}_1^n + \mathbf{R}(a_{n-1})\right)^{-1}\mathbf{r}_2^n \tag{9}$$

where $\mathbf{R}(a_n)$ and $\mathbf{R}(a_{n-1})$ are represented in the same adiabatic basis set as the sector R matrices.

To change the basis representation of the R matrix, the matrix with elements

$$T_{\lambda\lambda'} = \langle\phi_\lambda(r_{n-1};\theta,\phi)|\phi_\lambda(r_n;\theta,\phi)\rangle \tag{10}$$

is constructed. This relates the R matrix in the adiabatic basis evaluated at $r = a_{n-1}$ to the adiabatic basis evaluated at $r = a_n$ such that

$$\mathbf{R}^n = \overline{\mathbf{T}}\mathbf{R}^{n-1}\mathbf{T} \tag{11}$$

where $\overline{\mathbf{T}}$ is the transpose of $\mathbf{T}$.

The full R matrix can therefore be evaluated at the outer boundary of the first sector $r = a_1$ using the sector R matrices from the first sector ($\mathbf{r}_1^1, \mathbf{r}_2^1, \mathbf{r}_3^1$ and $\mathbf{r}_4^1$) and the R matrix at $r = a$ (evaluated from equation (3) in the Coulomb region). A frame transformation can then be performed to change the basis representation of the R matrix from the adiabatic basis used in the first sector to that which was used in the second sector. The propagation technique can then be carried out to evaluate the R matrix on the outer boundary of the second sector. By repeating this procedure through all of the sectors the R matrix can therefore be evaluated on the outer boundary of the region at $r = b$. (Although it is possible to evaluate the R matrix at each of the sector radii, it is more practical to derive global sector R matrices. These matrices allow the calculation of the R matrix at $r = b$ directly from the R matrix at $r = a$ and serve the same purpose as propagating the full R matrix through each sector individually[12].)

The asymptotic region

For large r the diamagnetic potential dominates and the Hamiltonian becomes separable in cylindrical coordinates. Since the motion in ρ is bounded, $-\frac{1}{r} \to -\frac{1}{z}$, and H becomes

$$H = -\frac{1}{2}\frac{d^2}{dz^2} - \frac{1}{z} + T_\rho + \beta L_z + \frac{1}{2}\beta^2\rho^2 + O\left(\frac{1}{z^3}\right) \tag{12}$$

where T_ρ is the kinetic energy in the ρ coordinate. The asymptotic region is therefore chosen to conform with the cylindrical symmetry present, $c \le z \le \infty, 0 \le \rho \le \infty$, where c is less than the radius $r = b$. For this region the wavefunction may be written in terms of a product of Landau states Φ_i for the two dimensional oscillator and Coulomb functions in z,

$$\Psi_{\epsilon j} = \sum_{ik} \Phi_i(\rho,\phi)\left(s_{ik}^{\epsilon_i}(z)\delta_{kj} + c_{ik}^{\epsilon_i}(z)K_{kj}\right), \tag{13}$$

where s and c are the energy normalized regular and irregular Coulomb functions evaluated at an energy $\epsilon_i = \epsilon - E_i$, E_i being the energies of the Landau states. The unknown matrix K_{kj} or K matrix is determined by matching the asymptotic solutions to the solutions obtained in the strong mixing region. The R matrix at $r = b$, found by propagation using the method described above, is matched through a two dimensional matching procedure[8] to these asymptotic solutions on an arc at $r = b$. Once the K matrix is known one has the solution over all space and it is therefore straightforward to calculate the photoionization cross section for any desired transition.

Multichannel quantum defect theory

We can treat weakly closed channels (i.e. channels with $\epsilon - E_i$ just below zero) using multichannel quantum defect theory[9] by using Coulomb functions s and c instead of Whittaker functions in (13). The matching procedure then produces a 'smooth' K matrix , $\mathcal{K}$, which is slowly dependent on energy as it is shorn of all resonance structure converging to the weakly closed Landau thresholds. The open part of the physical reactance matrix K can be recovered from the matrix $\mathcal{K}$ by the formula[9]

$$\mathbf{K}_{oo} = \mathcal{K}_{oo} - \mathcal{K}_{oc}\Big(\tan(\pi\nu) + \mathcal{K}_{cc}\Big)^{-1}\mathcal{K}_{co}. \tag{14}$$

The o and c subscripts refer to the open and closed channels of $\mathcal{K}$ and the $\tan \pi\nu's$ form a diagonal matrix where the ν's are related to the channel energies by $\nu_i = 1/\sqrt{2\,|\varepsilon_i|}$. Once the physical reactance matrix K has been obtained the photoionization cross section can be calculated as before.

This way of obtaining the cross section has two major advantages. Since $\mathcal{K}$ varies slowly with energy compared to K, this allows one to calculate $\mathcal{K}$'s on a coarse energy mesh with fairly large energy spacings. These can then be interpolated onto a finer energy mesh where the full K matrix can be obtained from equation (14). This procedure requires the propagation stage to be performed at fewer energy points ultimately speeding up the calculation. The second major advantage is that the resonance structure due to a single Landau threshold can easily be removed from the cross section. This is done by keeping 'open' the relevant Landau channel in the evaluation of equation (14). By comparing a spectrum with all resonances converging on a particular Landau threshold removed with that of the full spectrum it is possible to determine which resonances converge to which thresholds.

Results

The methods described in the previous sections are used to calculate the photoionization spectrum of lithium in a magnetic field of $6.1143T$. This corresponds to the field strength used in the experiment of Iu *et al.* For this field strength the radii $r = a$ and $r = b$ were taken to be 200 and 12000 a.u. and 64 sectors were used. The size of basis sets for the expansion of the wavefunction (equation (6)) was 10 Legendre polynomials and up to 40 adiabatic angular functions. The largest matrices to be diagonalized were therefore of order 400. The number of channels required to perform the propagation was quite small and in general varies with energy and radius. The number of channels retained in any one sector is the number of open channels plus a few of the closed ones. To produce the graphs shown in this section, the number of energy points at which the propagation was conducted was 400 per Landau threshold and the total number of energy points at which the physical reactance matrix K was constructed (equation (14)) was 10000 per Landau threshold. Note that this second process of evaluating

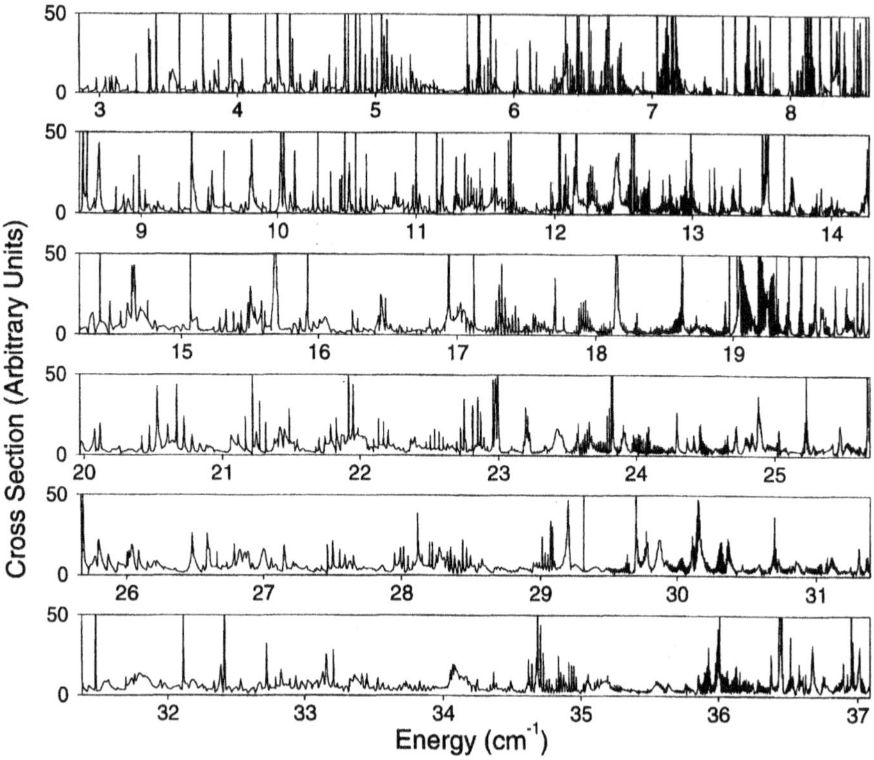

Figure 1. The photoionization spectrum of lithium in a magnetic field of $6.1143T$ extending over 6 Landau thresholds. The Landau thresholds are at the right hand side of each panel.

equation (14) only contributes a tiny amount to the CPU time of the calculation so an arbitrary resolution can be obtained by simply taking a greater density of energy points.

In figure 1 the black areas on the graph correspond to Rydberg series converging on the nearest Landau thresholds. These consist of a lot of very sharp resonances with extremely small decay widths. In the spectrum the background ionization cross section can be seen to rise as the energy increases. Also the resonances converging on the nearest threshold only slightly perturb the background of resonances converging on higher thresholds.

It is possible to conduct a detailed analysis of the resonances that are observed in the cross section by calculating $\mathcal{K}$ as described above[8]. This contains the information about the photoionization cross section with resonances converging on several thresholds removed. By only including some channels in the evaluation of equation (14) it is therefore possible to construct the cross section with the resonances converging on the nearest threshold, the second nearest threshold etc. removed.

These results are shown in figure 2 where the full cross section of the first Landau threshold is shown in the top panel. The second panel shows the cross section averaged over the resonances converging on the nearest threshold i.e. the threshold located at the right of the graph at 8.5 cm^{-1} (3.9×10^{-5} atomic units). The next panel shows the cross section with the resonances that are converging on the threshold at 14.2 cm^{-1} removed. The bottom panel shows the spectrum with the resonances converging on the

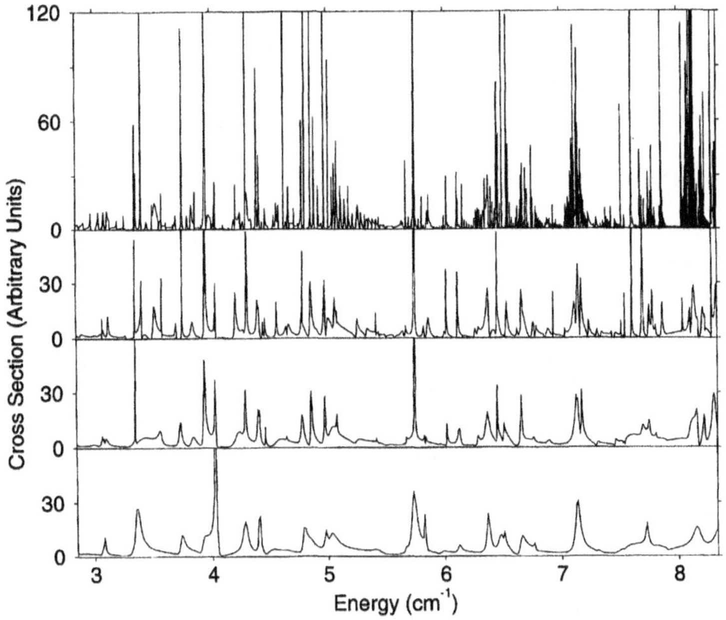

Figure 2. Photoionization spectrum of lithium in a field strength of $6.1143T$. The top panel shows the full spectrum and the second, third and fourth panels show the spectrum with all the resonances converging on the nearest, second nearest and third nearest threshold removed. The energy range is between the first and second Landau thresholds

threshold at 19.9 cm^{-1} removed. Comparison of various panels of the graphs therefore builds up an explanation of the complicated structure that is present in the full cross section. Removal of the Rydberg series of resonances shows a cross section dominated by relatively smooth modulations.

TIME DEPENDENT STATES IN EXTERNAL FIELDS

The frequencies or periods associated with classical periodic orbits play an important role in interpreting the spectra of atoms in external fields. These frequencies are in particular associated with unstable periodic orbits in the classically chaotic region of the spectrum. Until recently experiments have identified these periodicities indirectly by measuring the absorption spectrum over a wide energy range and then taking a Fourier transform of this frequency domain spectrum[13, 14]. Over the last few years, new experiments have attempted to identify these periodicities directly in the time domain[15-21]. This is done by using short pulsed lasers to create Rydberg wavepackets whose return times to the vicinity of the nucleus or core are measured giving directly the periods of the classical orbits. The initial Rydberg wavepacket experiments used a pump - probe technique which experimentally produces a poor signal to noise ratio.

This drawback prompted a new method based on a phase sensitive technique[22, 23]. This method uses two laser pulses separated by a short time delay t_d. The first pulse excites a wavepacket, and the second pulse creates an identical wavepacket which can interfere with the time evolved initial wavepacket. If the wavepacket excited by the first pulse is far from the core of the atom when the second wavepacket is excited there is no overlap between the two wavepackets and no interference will occur. If however the initial wavepacket has returned to near the core, the second wavepacket will interfere with the first in such a way that the Rydberg population is either enhanced or depleted. The final Rydberg population, after the pulses have interacted with the atom, is measured by field ionization. This is a very efficient measurement process giving a much better signal to noise ratio than the pump-probe technique.

Rydberg wavepackets in parallel electric and magnetic fields

Marmet *et al*[16] have used the above technique to measure the magnitude of the auto-correlation function, M, as a function of the time delay t_d for the rubidium atom in both magnetic and parallel electric and magnetic fields. They have also evaluated M semi-classically. The semi-classical theory, which was carried out for hydrogen, failed to agree with the experimental results for parallel fields and it was stated that this may be due to the neglect of the core potentials for rubidium.

Here we present a calculation (see also Moser *et al*[24]) of M as a function of t_d for an atom in parallel electric and magnetic fields subject to two Gaussian laser pulses. As in the experiment, the field strengths and energies are chosen in a regime where one observes a transition from regular to irregular behavior in the spectrum corresponding to a classical transition from regularity to chaos. M is evaluated via an *ab initio* quantum mechanical calculation for both hydrogen and rubidium in parallel fields allowing a direct examination of non-hydrogenic core effects. The peaks in M are compared with orbits whose periods have been determined from classical calculations.

The auto-correlation function

An atom is excited from its ground state by a pair of identical laser pulses separated by a time t_d. The first pulse, at time $t = 0$, creates a Rydberg wavepacket represented by $|\Psi(0)\rangle$ which can be expanded as a superposition of eigenstates of the atom. The coefficients of this superposition depend both on the form of the pulse and the dipole moments characterizing transitions from the ground state to the relevant Rydberg states. The state $|\Psi(0)\rangle$ evolves for a time t_d at which time a second laser pulse excites an identical Rydberg wavepacket. The total wavefunction is therefore $|\Psi_{Ryd}\rangle = |\Psi(0)\rangle + |\Psi(t_d)\rangle$. The total Rydberg population P_{tot} is proportional to $\langle\Psi_{Ryd}|\Psi_{Ryd}\rangle$ or

$$P_{tot} \propto \langle\Psi(0)|\Psi(0)\rangle + |\langle\Psi(0)|\Psi(t_d)\rangle| \cos\chi(t_d) \tag{15}$$

where χ denotes the phase difference between the wavepackets excited by the first and second pulses. The quantity which can be extracted from the experiment is the magnitude of the auto-correlation function $M = |\langle\Psi(0)|\Psi(t_d)\rangle|$ which measures the overlap between the initial wavepacket, $|\Psi(0)\rangle$, and the wavepacket at some time t_d later. This function plays an important role in the semi-classical theory of chaos[25]. Clearly M will be large whenever the overlap between the first and second wavepackets is large. In this case the total Rydberg population has in general a rapid sinusoidal time dependence around an average value, the amplitude of which is proportional to M. By using an additional phase delay, M can be extracted from the experiments and measured by field ionization of the Rydberg states[19, 16].

The auto-correlation function for an atom excited by two Gaussian pulses

When the intensity of the laser pulses is low, as in the experiment of Marmet *et al*[16], the interaction potential between the laser field and the atom, $V(t)$, can be treated by time dependent perturbation theory. The total Hamiltonian for the system is then written as

$$H = H_o + V(t), \tag{16}$$

where H_o is the time independent Hamiltonian for the atom in the presence of the static electric and magnetic fields. The ground state of H_o will be denoted by $|g\rangle$ and the excited eigenstates by $|j\rangle$. The potential $V(t) = -\underline{\mu}.\underline{E}(t)$ where $\underline{\mu}$ is the electric dipole vector and $\underline{E}(t)$ the classical electric field due to the laser.

We take the atom, in the presence of the static fields, to be initially in its ground state $|g\rangle$. Expanding the total wavefunction in terms of the unperturbed states, the time dependent wavefunction $|\Psi\rangle$is given by

$$|\Psi\rangle = |g\rangle a_g(t)e^{-i\omega_g t} + \sum_j |j\rangle a_j(t)e^{-i\omega_j t}, \tag{17}$$

where the sum is taken over all excited states and $a_g(t), a_j(t)$ represent the time dependent amplitudes of the ground and excited states with energies ω_g, ω_j. Note that the sum over $|j\rangle$ must contain a sum over all the relevant quantum numbers.

The time dependent electric field $\underline{E}(t)$ for the excitation scheme of two Gaussian pulses of pulse width τ and angular frequency ω_l is given by

$$\underline{E}(t) = \underline{E_0}e^{i\omega_l t}e^{-2ln2\left(\frac{t}{\tau}\right)^2} + \underline{E_0}e^{i\omega_l(t-t_d)}e^{-2ln2\left(\frac{t-t_d}{\tau}\right)^2}, \tag{18}$$

hence the time dependent potential $V(t)$ is known. By substituting $V(t)$ and the expansion in equation (17) into the time dependent Schrödinger equation, a standard set of equations for the time dependent Rydberg amplitudes is obtained. Using the rotating wave approximation and neglecting depletion of the ground state[26], the excited state amplitudes a_j as $t \to \infty$ can be calculated in first order in perturbation theory. Hence the wavefunction of the excited or Rydberg wavepacket can be represented by

$$|\Psi_{Ryd}\rangle = \sqrt{\frac{\pi}{2ln2}}\tau \sum_j e^{-i\omega_j t}|j\rangle\langle j|\ \underline{\mu}.\underline{E_0}\ |g\rangle e^{-\frac{(\omega_j-\omega_g-\omega_l)^2}{8ln2}\tau^2}\left(1+e^{i(\omega_j-\omega_g)t_d}\right). \tag{19}$$

The total Rydberg wavefunction can be written as the sum of the first and second excited wavepackets

$$|\Psi_{Ryd}\rangle = |\Psi(0)\rangle + |\Psi(t_d)\rangle \tag{20}$$

where

$$|\Psi(0)\rangle = \frac{1}{\sqrt{2}}\sum_j c_j|j\rangle e^{-\left(\frac{\Delta_j}{\Delta}\right)^2} e^{\ i\omega_j t}$$

$$|\Psi(t_d)\rangle = \frac{1}{\sqrt{2}}\sum_j c_j|j\rangle e^{-\left(\frac{\Delta_j}{\Delta}\right)^2} e^{-i\omega_j t}e^{-i(\omega_j-\omega_g)t_d} \tag{21}$$

and the following abbreviations have been used

$$c_j = \sqrt{\frac{\pi}{ln2}}\langle j|\ \underline{\mu}.\underline{E_0}\ |g\rangle, \qquad \Delta = \frac{\sqrt{8ln2}}{\tau}, \qquad \Delta_j = \omega_l - (\omega_j - \omega_g). \tag{22}$$

c_j is proportional to the dipole integral from the ground state $|g\rangle$ to the state $|j\rangle$, Δ represents an energy window of states which can be excited and Δ_j is the difference between the laser frequency and the frequency associated with the atomic transition from $|g\rangle$ to $|j\rangle$.

Substituting the above expressions for the wavefunctions into the expression M ($M =|\ \langle\Psi(0)|\Psi(t_d)\rangle\ |$) yields

$$M = \left|\sum_j |\ c_j\ |^2\ e^{-2(\frac{\Delta_j}{\Delta})^2}e^{i\Delta_j t_d}\right|. \tag{23}$$

The function can be calculated both semi-classically and quantum mechanically. We have used quantum theory to evaluate it.

Evaluation of the auto-correlation function for an atom in parallel electric and magnetic fields

A Gaussian pulse only excites, with any significant probability, a group of quantum states within an energy range of the order $1/\tau$ about the central excitation energy which is given by the laser frequency ω_l. To evaluate equation (23) for a given pulse width τ, only the energies and the dipole integrals very near to the frequency of the laser will therefore need to be evaluated.

The Hamiltonian for hydrogen in parallel electric and magnetic fields directed along the z-axis is

$$H = -\frac{1}{2}\nabla^2 - \frac{1}{r} + \frac{1}{2}\beta^2\rho^2 + fz, \tag{24}$$

where β is the magnetic field strength and f is the electric field strength in atomic units. The linear Zeeman term has been omitted because L_z is a conserved quantity for this Hamiltonian and hence m the corresponding quantum number is conserved. For hydrogen, the eigenvalues ω_j and eigenfunctions $|j\rangle$ of this Hamiltonian are evaluated for given fields by expanding the wavefunction in a basis set

$$\Psi = \sum_{nl} c_{nl}\frac{S^{\zeta}_{nl}(r)}{r}Y_{lm}(\theta,\phi), \tag{25}$$

i.e. in products of Sturmian functions and spherical harmonics. The fields have a negligible effect on the ground state $|g\rangle$ which is taken to be the hydrogenic ground state. Once the states $|j\rangle$ have been found the dipole integrals $\langle j|\,\underline{\mu}.\underline{E_0}\,|g\rangle$ are readily calculated, hence all of the quantities entering equation (23) are known.

For rubidium, the method developed by O'Mahony and Taylor[27] for non-hydrogenic atoms in fields is used. The Hamiltonian in (24) is only valid at a radius, $r = b$, outside of the atomic core and hence the expansion in equation (25) is now employed over a semi-infinite region $r = b$ to ∞. The eigenvalues and eigenfunctions of H plus a surface term or Bloch operator at $r = b$, are used to construct an R-matrix or log-derivative matrix at $r = b$. This R-matrix is matched to phase shifted Coulomb functions at $r = b$ and an energy search is conducted to find eigenenergies w_j such that the total wavefunction and its derivative are continuous over the boundary. The corresponding eigenvectors $|j\rangle$ can then be constructed. Note that it is at this stage that the fingerprint of the non-hydrogenic core is introduced via the quantum defects used to calculate the phase shifted Coulomb functions.

RESULTS

In the experiment a rubidium atom is excited from its ground state by a pulse sequence using linearly polarized light and in the presence of a fixed magnetic field of $B = 1.5$ Tesla and a static electric field ranging from $F = 0$ to $28V/cm$. (Note $\beta = B/B_c$ and $f = F/F_c$ where $B_c = 4.7\times10^5T$ and $F_c = 5.14\times10^9V/cm$.) The frequency of the laser is such that final Rydberg states centered around the binding energy $-36.1cm^{-1}$ are excited by the pulses. For $F = 0$, this energy range corresponds classically to regular motion and quantum mechanically to the region where the external fields are but a perturbation. However as F increases one goes over to the strong field mixing region where the quantum spectrum becomes irregular and the classical mechanics becomes chaotic. Equation (23) has been evaluated theoretically for hydrogen and rubidium, for the same set of field strengths and conditions as in the experiment. The pulse width τ is taken to be $4ps$. The polarization of the laser

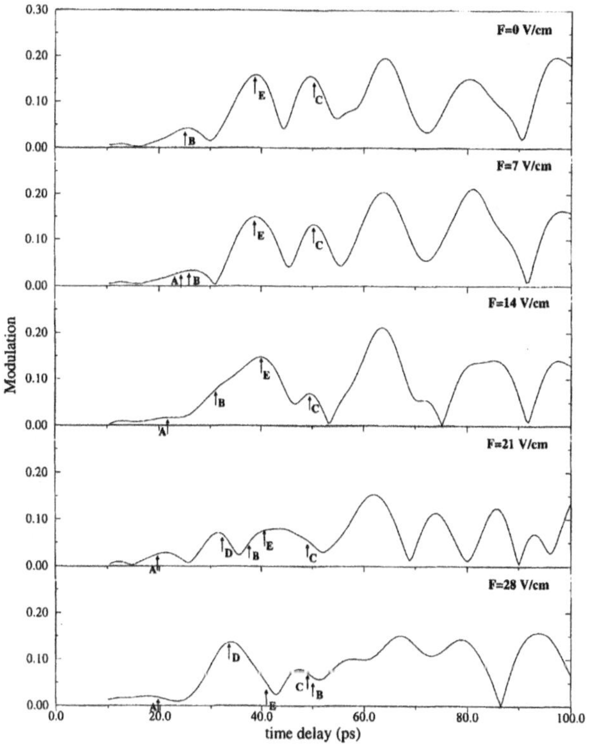

Figure 3. Theoretical results showing the magnitude of the auto-correlation function, M, as a function of the time delay in picoseconds between pulses for a rubidium atom excited from the ground state in parallel electric and magnetic fields. The magnetic field strength is 1.5 Tesla in every panel and the electric field strength, F, varies from 0 to $28V/cm$. The periods of the classical orbits A, B, C, D and E are indicated. A and B are orbits along the $+z$ and $-z$ directions respectively.

light and the fact that L_z is conserved means that only $m = 0$ final states need to be considered in the basis set expansion in (25). The condition that the Gaussian energy window $\exp(-2\left(\frac{\Delta_i}{\Delta}\right)^2)$ decays off to a negligible value, requires that all Rydberg states within about $7cm^{-1}$ of the above total energy need to be calculated. The relevant energy levels and dipole integrals are obtained by the basis set methods described. These quantities are used to calculate the corresponding amplitude from (23), for time delays from 0 to 100 picoseconds. The results for rubidium are given in Figure 3 which shows the appearance of broad peaks at certain time delays. These times correspond to the shortest periodic orbits in the system. For specific values of the static fields, the periods of these orbits can be determined from the classical equations of motion. Four of these periods are indicated on the diagram by the letters A to D. The periodic orbit designated by A is the orbit parallel to the z-axis while B is the orbit anti-parallel to the z axis. The orbit D results from a bifurcation of the orbit B at about $20V/cm$. This feature is also seen in the experiment of Marmet *et al*[16]. However an additional peak, labeled by E, is also seen in the theory at about $36ps$. This peak, which is not present in the experiment, can be identified with a particular classical periodic orbit[24]. In addition the relative peak intensities that we find are quite different to those found in the experiment.

The main differences between hydrogen and rubidium, and hence the effect of the core, is for long time delays $t_d > 60ps$ as has been seen previously in Fourier transform of the frequency domain spectra[28]. These differences can be viewed as due to scattering of the wavepacket from the core for longer time delays. The discrepancies between theory and experiment cannot be explained by the rubidium core as this has been fully included via the R-matrix method. There are several possible explanations for the differences. In the experimental paper[16] it is mentioned that a poor signal to noise ratio is obtained when the laser field is parallel to the magnetic field as in the case studied here. Also any errors in the pulse shape may alter the spectrum obtained because of the different distribution of excited Rydberg levels. It is however clear that this phase-sensitive technique provides a very promising method to observe directly in the time domain the frequencies of the shortest periodic orbits for classically chaotic systems in general.

Acknowledgments

IM was supported by a postgraduate studentship from EPSRC, UK. Financial support from the Junta Nacional de Investigacao Cientifica e Tecnologica (JNICT), Portugal through the Praxis XXI programme and the EU Human Capital and Mobility programme is gratefully acknowledged. PFOM would like to thank Prof. F. Parente for his hospitality at the Centro de Física Atómica da Universidade de Lisboa where this work was completed.

REFERENCES

1. W.R.S. Garton and F.S. Tomkins, *Astrophys. J.* **159**, 839 (1969).
2. C. Iu, G.R. Welch, M.M. Kash, D. Kleppner, D. Delande, and J.C. Gay, *Phys. Rev. Lett.* **66**, 145 (1991).
3. D. Delande, A. Bommier, and J.C. Gay, *Phys. Rev. Lett.* **66**, 141 (1991).
4. P.F. O'Mahony and F. Mota-Furtado, *Phys. Rev. Lett.* **67**, 2283 (1991).
5. P.F. O'Mahony and F. Mota-Furtado, *Comm. At. Mol. Phys.* nos. 4-6, 309 (1991).
6. F. Mota-Furtado, P.F. O'Mahony, and H. Marxer, *Atomic Physics 13* **275**, 449 (1993).
7. S. Watanabe and H. Komine, *Phys. Rev. Lett.* **67**, 3227 (1991).
8. I. Moser and P.F. O'Mahony, submitted to *Phys. Rev. A* (1997).
9. M.J. Seaton, *Rep. Prog. Phys.* **46**, 167 (1983).
10. C. Bloch, *Nucl. Phys.* **4**, 503 (1957).
11. K.L. Baluja, P.G. Burke, and L.A. Morgan, *Computer Phys. Com.* **27**, 299 (1982).
12. E.B. Stechel, R.B. Walker, and J.C. Light, *J. Chem. Phys.* **69**, 3518 (1978).
13. J. Main, G. Weibusch, A. Holle, and K.H. Welge, *Phys. Rev. Lett.* **57**, 2789 (1986).
14. A.Holle, J. Main, G. Weibusch, H. Rottke, and K.H. Welge, *Phys. Rev. Lett.* **61**, 161 (1988).
15. J.A.Yeazell, G. Raithel, L. Marmet, H. Held, and H. Walther, *Phys. Rev. Lett.* **70**, 2884 (1993) and references therein.
16. L. Marmet, H. Held, G. Raithel, J.A.Yeazell, and H. Walther, *Phys. Rev. Lett.* **72**, 3779 (1994).
17. G. Raithel, H. Held, L. Marmet, and H. Walther, *J. Phys. B* **27**, 2849 (1994).
18. J. Wals, H.H. Fielding, J.F. Christian, L.C. Snoek, W.J. van der Zande, and H.B. van Linden van den Heuvell, *Phys. Rev. Lett.* **72**, 3783 (1994).
19. B.Broers, J.F. Christian, and H.B. van Linden van den Heuvell, *Phys. Rev. A* **49**, 2498 (1994).
20. H.H. Fielding, J. Wals, W.J. van der Zande, and H.B. van Linden van den Heuvell, *Phys. Rev. A* **51**, 611 (1995).
21. G.M. Lankhuijzen and L.D. Noordam, *Phys. Rev. A* **52**, 2016 (1995).
22. L.D. Noordam, D.I. Duncan and, T.F. Gallagher, *Phys. Rev. A* **45**, 4734 (1992).
23. J.F. Christian, B. Broers, J.H. Hoogenraad, W.J. Van der Zande, and L.D. Noordam, *Opt. Comm.* **103**, 79 (1993).
24. I. Moser, P.F. O'Mahony, F. Mota-Furtado, and J.P. dos Santos, *Phys. Rev. A* **55**, 3724 (1997).
25. S. Tomsovic and E.J. Heller, *Phys. Rev. E* **47**, 282 (1993) and references therein.
26. G. Alber and P. Zoller, *Phys. Rep.* **199**, 231 (1991).
27. P.F. O'Mahony and K.T. Taylor, *Phys. Rev. Lett.* **57**, 2931 (1986).
28. P.F. O'Mahony, *Phys. Rev. Lett.* **63**, 2653 (1989).

SECULAR MOTION OF 3-D RYDBERG STATES IN A MICROWAVE FIELD

Andreas Buchleitner

Max-Planck-Institut für Quantenoptik
Hans-Kopfermann-Straße 1
D-85478 Garching

INTRODUCTION

The ionization of atomic Rydberg states by microwave fields allows for the experimental and theoretical study of quantum transport under the condition of a highly nonlinear interaction between the atomic degrees of freedom and the driving field. In a classical picture, the ionization of the Rydberg electron is due to chaotic transport in classical phase space,[1] within a wide range of microwave frequencies. In a quantum description, classical chaos manifests itself in a strong mixing of the unperturbed atomic eigenstates and the partial loss of good quantum numbers.[2] Due to the large number of atomic states involved in the chaotic ionization of highly excited Rydberg states (principal quantum number between $n_0 \simeq 20$ and $n_0 \simeq 130$), the treatment of the full quantum problem has been prohibitive until very recently.[2] However, one-dimensional quantum models[2, 3] of the atom turned out to recover the experimentally observed ionization yields[4, 5] surprisingly well. As chaos is "born" for classical initial conditions *along* the z-axis (parallel to the microwave field axis), it has been argued that chaotic ionization will merely reflect the dynamics along this one-dimensional subspace of the 3-D configuration space.[1, 3] In a more quantal language, initial atomic states elongated along the polarization axis exhibit the largest oscillating dipole and do therefore most effectively interact with the driving field.

Hence, our understanding of state of the art *ionization* experiments[4, 5] seems rather complete. Conversely, up to now we do not have much insight in the history of the electron on its way through classical phase space to the continuum, or on the rôle of the different degrees of freedom in the quantum mechanical excitation process.

It is the purpose of the present contribution to reveal the "backbone" of the quantum mechanical ionization process as a novel structure in the exact quantum spectrum, which will be accessible in a new generation of experiments on Rydberg states in microwave fields. Our analysis is built on the dynamical symmetry of the hydrogen atom[6] and therefore a priori not applicable for non-hydrogenic Rydberg states, e. g., of rubidium.[7] For the sake of simplicity we focus on microwave frequencies which are *nonresonant* with the unperturbed Kepler motion of the Rydberg electron. We shall

Atoms and Molecules in Strong External Fields
Edited by Schmelcher and Schweizer, Plenum Press, New York, 1998

see that the individual eigenstates of the atom in the field reflect the structure of classical phase space in their localization properties, their energies, and their ionization rates. We also show that the restricted 1-D dynamics are naturally embedded in the full dynamics of the 3-D atom and finally sketch the experimental strategy to probe this structure. The extension to resonant driving, which directly relates to the creation of nondispersive wave packets,[8, 9, 10] is straightforward but technically a little tedious, and will be presented elsewhere.

QUANTUM AND CLASSICAL DYNAMICS

The Hamiltonian describing the hydrogen atom in a monochromatic, linearly polarized microwave field of constant amplitude F and frequency ω reads, in the length gauge and in atomic units:

$$H = \frac{\mathbf{p}^2}{2} - \frac{1}{r} + Fz\cos\omega t. \tag{1}$$

As described in earlier work, a complex dilation of the corresponding Floquet-Hamiltonian allows us to obtain the energies, the ionization rates, and the associated Floquet eigenstates of the atom dressed by the microwave photons.[2] Due to the azimuthal symmetry around the z-axis, the problem is effectively 2-D with a 5-D phase space (because of the explicit time dependence of H). For vanishing microwave amplitude, the dressed eigenstates are products of spherical (n, ℓ, m) states with Fock states of photon number K, where we choose $m = 0$ for the conserved magnetic quantum number. For nonresonant microwave frequency, each energy level (n, K) has a degeneracy equal to the principal quantum number n in the angular momentum $0 \le \ell \le n-1$. For non-vanishing field amplitude, the term $Fz\cos\omega t$ in the Hamiltonian couples different states with the selection rules $\Delta\ell = \pm 1, \Delta K = \pm 1$. At first order in F, there is no coupling inside the (n, K) manifold and the degeneracy is not lifted. At second order, there are non-zero terms which couple the state (n, ℓ, K) to the states $(n, \ell \pm 2, K)$, through the intermediate states $(n', \ell \pm 1, K \pm 1)$; these terms lift the degeneracy. A crucial point is that the second-order term is *not diagonal* in ℓ. The eigenstates of the 3-D atoms dressed by the microwave field are consequently not (n, ℓ, K) states, even *at lowest non-vanishing order.*

¿From a classical point of view, the electronic trajectories are simple Kepler ellipses, for $F = 0$. For nonvanishing field amplitude, the structure of phase space is affected by the resonances between the microwave and the Kepler frequency. At small field, the various resonance zones are separated by regular non-resonant tori. The non-resonant situation studied here corresponds to an electron on its Kepler orbit subject to a microwave field oscillating at a non-commensurate frequency. Hence, the electron cannot efficiently exchange energy with the field. At first order, it exhibits oscillations (driven by the microwave field) around its unperturbed Kepler ellipse. At second order, the resulting trajectory is an instantaneous elliptical trajectory the parameters (eccentricity, spatial orientation) of which slowly evolve in time. This is nothing but the familiar secular approximation which separates the fast motion at the Kepler frequency from the slow secular motion of the ellipse.[11] During the slow motion, the unperturbed energy is conserved, which is the classical analog of n remaining an approximate good quantum number.

The classical secular motion can be described by an effective Hamiltonian, after appropriate averaging over the oscillation of the microwave field.[12] We start from the expansion of (1) in the canonical action-angle coordinates n (total action), θ (conjugate angle), ℓ (angular momentum) and ψ (conjugate angle measuring the angle between the z axis and the major axis of the ellipse) of the hydrogen atom: [1]

$$H = -\frac{1}{2n^2} + Fn^2 \cos\omega t \times \sum_{M\geq 0} \{a_M(L_0)\cos M\theta \cos\psi + b_M(L_0)\sin M\theta \sin\psi\}. \quad (2)$$

$L_0 = \ell/n$ is the scaled angular momentum,

$$a_M(x) = -\frac{2J'_M(M\sqrt{1-x^2})}{M} \quad (M \neq 0);\ a_0 = \frac{3}{2}\sqrt{1-x^2},$$
$$b_M(x) = \frac{2xJ_M(M\sqrt{1-x^2})}{M\sqrt{1-x^2}} \quad (M \neq 0);\ b_0 = 0, \quad (3)$$

J_M represents the Bessel function and the prime the derivative. At lowest nonvanishing order in F we obtain:

$$\bar{H}_2 = \frac{F^2n^6}{8}\left\{h_1(L_0,\omega_0)\cos^2\psi + h_2(L_0,\omega_0)\sin^2\psi\right\}, \quad (4)$$

where $\omega_0 = \omega n_0^3$ is the scaled frequency (ratio of the microwave to the Kepler frequency), and:

$$h_1 = \sum_{M\geq 0}\left(\frac{M^2(a_M^2 - 2L_0a_Ma'_M) - 2Ma_Mb'_M}{\omega_0^2 - M^2} - \frac{6M^4a_M^2}{(\omega_0^2 - M^2)^2}\right). \quad (5)$$

Here, a_M and b_M are evaluated for the argument $L_0 = \ell/n$. h_2 is obtained by exchanging a_M and b_M in (5). The lower order terms of the effective Hamiltonian in F read $\bar{H}_0 = -1/2n^2$ and $\bar{H}_1 = 0$.

CLASSICAL PHASE SPACE STRUCTURE

The secular electronic motion as the temporal evolution of the conjugate variables ℓ and ψ, generated by $\bar{H}_2$, can be represented by the motion of the vector (L_0, A_ρ, A_z) on the unit sphere.[13] A_z and A_ρ design the components of the Runge-Lenz vector (directed along the major axis of the ellipse with a modulus equal to its eccentricity) along and perpendicular to the field axis, respectively, defined by

$$A_z^2 = (1 - L_0^2)\cos^2\psi \ ; \ A_\rho^2 = (1 - L_0^2)\sin^2\psi. \quad (6)$$

Three fixed points, two stable and one unstable, characterize the structure of of classical phase space within the ℓ-ψ dimension. Each of them corresponds to either one of three distinct types of classical periodic motion. The first two ones defined by $L_0 = 1$ and $A_\rho = 1$ are stable. They correspond to the circular orbit in the plane containing the field polarization vector, and to the straight line orbit normal to that plane. The electronic motion is trapped within elliptic islands around these stable fixed points. The fixed point defined by $A_z = 1$ corresponds to the straight line orbit along the field direction – the precise analog to the restricted 1-D dynamics of the atom – and is *unstable*: an initial trajectory close to the field axis will diverge from it.

EXACT VS. SEMICLASSICAL ENERGY LEVELS

We can now predict the quantum spectrum of the atom dressed by the microwave field, by WKB quantization of $\bar{H}_2$. The semiclassical energies are defined by $\oint \ell \, d\psi =$

$p+\frac{1}{2}$, the integral being evaluated along a classical secular trajectory (fixed value of $\bar{H}_2$) and p an integer ranging from 0 to $n-1$. The validity of this approach can be tested by a direct comparison to the "exact" numerical quantum spectrum. Indeed, after diagonalization of (1), one finds "multiplets" which originate from the unperturbed energy levels. The energies of these states are plotted in Fig. 1(a), for $n_0 = 23$, $F_0 = Fn_0^4 = 0.03$, $\omega_0 = 1.304$,[4] and compared to the semiclassical prediction. The agreement is quite good, with some quantitative disagreement for the lower part of the manifold. For both, the semiclassical and the exact result, there are 10 eigenstates (labelled from $p = 0$ to 9 by increasing energy) corresponding to secular motion around the $L_0 = 1$ fixed point, and 13 eigenstates (from $p = 10$ to 22) corresponding to motion around the $A_\rho = 1$ fixed point. The two series are separated by a minimum in the energy spacing, the quantum consequence of the slowing down of the classical motion[1] near the unstable fixed point. As indicated by an arrow in Fig. 1(a), it is *exactly* at this energy which separates the L_0- and the $A_\rho-$motion that we find the Floquet eigenstate originating from $n_0 = 23$, in the restricted 1-D dynamics.

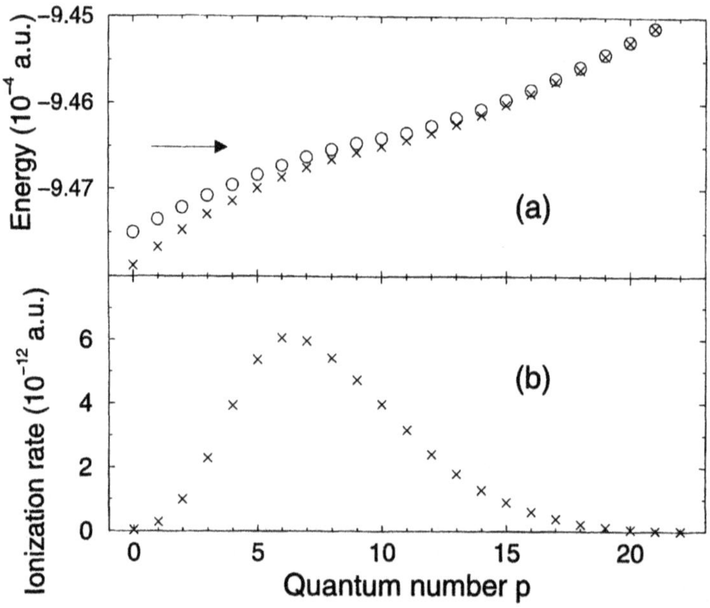

Figure 1. Energy levels (originating from the $n = 23$ manifold) of a hydrogen atom exposed to a microwave field with scaled frequency $\omega_0 = 1.304$ and scaled field $F_0 = 0.03$. (a) Comparison between the exact energy levels obtained from numerical diagonalization (crosses) and the semiclassical prediction using the effective Hamiltonian $\bar{H}_2$ (circles); the lowest (resp. highest) energy levels $p = 0$ to 9 (resp. 10 to 22) correspond to states localized around the fixed point $L_0 = 1$ (resp. $A_\rho = 1$) on the unit sphere (see text); the arrow indicates the position of the corresponding energy level for the 1-D model of the atom. (b) Ionization rates obtained from numerical diagonalization. There is a sharp maximum for states localized in the vicinity the microwave field axis.

LIFE TIMES OF INDIVIDUAL FLOQUET STATES

The semiclassical structure of the manifold is also reflected by the ionization rate of the individual Floquet states, as shown in Fig. 1(b). The states with pronounced L_0- or A_ρ-character have extremely small ionization rates, the A_ρ-states being actually more stable than the L_0-states. On the other hand, the states which correspond to motion close to the unstable fixed point do exhibit the largest ionization rates. In our semiclassical picture, we interpret this p-dependence of the widths as due to tunneling ionization out of the regime of regular motion in the action variable n. Since $F_0 = 0.03$ corresponds to classically near-integrable motion in the restricted 1-D dynamics, and as chaotic ionization sets in for $F_0 > 0.03$,[2] the ionization rates of the A_z-states are enhanced by tunneling through a relatively narrow potential barrier separating regular from irregular motion (at higher classical action n), as compared to the L_0- and A_ρ-states. Interestingly, the maximum ionization rate of the 3-D atom is not observed for the eigenstate which comes closest to A_z-motion, i.e. to the restricted 1-D dynamics, but slightly displaced to p-values below the separatrix. According to our presently accumulated numerical data (for different values of F_0 and ω_0), this displacement towards L_0-motion seems systematic but is not yet understood.

¿From Fig. 1 (b) it is now clear why experiments[4] which initially prepare the atoms in superpositions of substates of the hydrogenic multiplet of Fig. 1(a) can be well approximated by 1-D model calculations: the onset of ionization is due to the first eigenstates to ionize, the ones dominated by the A_z-motion of the 3-D atom. Note, however, that at $F_0 = 0.03$ the ionization rate of the corresponding eigenstate of the 1-D model is found to be two orders of magnitude smaller than the maximum ionization rate observed in Fig. 1(b). 1-D and 3-D widths acquire comparable values only in the vicinity of the 10% ionization threshold, i. e. at $F_0 \simeq 0.05$,[2, 4] when strong n-mixing prevails.

LOCALIZATION PROPERTIES OF FLOQUET STATES

A further test of our semiclassical picture is to investigate the localization properties of the quantum eigenstates in configuration (and in phase[13]) space. This is done in Fig. 2 for the 3 states with $p = 2, 8, 20$. The plots are contours of the electronic densities averaged over one field period. As expected, the state $\mid p = 2 >$ is localized close to the circular orbit, $\mid p = 8 >$ near the field axis, and $\mid p = 20 >$ close to the plane perpendicular to the field. As a matter of fact, the states $\mid p = 2 >$ and $\mid p = 20 >$ are almost pure eigenstates of the unperturbed hydrogen atom.[6] Furthermore, "flat" electronic densities like $\mid p = 20 >$ which *do* probe the Coulomb singularity are systematically more stable against ionization than states like $\mid p = 2 >$ which make no contact with the nucleus. Since flat eigenstates are compatible with the cylindrical symmetry imposed by a superintense field,[14] they are good candidates to exhibit similar behaviour in the microwave as well as in the optical regime. This is consistent with classical studies[15] which suggest flat electron trajectories to be quite stable against ionization at very high field frequencies ($\omega_0 \simeq 100$) and amplitudes ($F_0 \simeq 1000$).

OUTLOOK AND CONCLUSION

We have seen that the quasi one dimensional ionization of 3-D Rydberg states of atomic hydrogen is embedded in a transverse structure of 5-D phase space which can

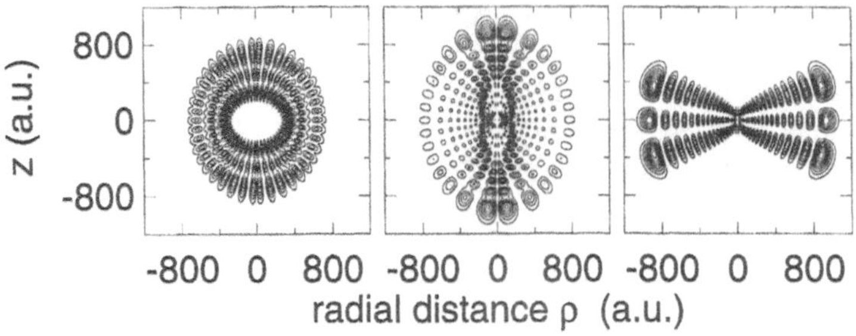

Figure 2. Isovalue curves for the electronic densities (averaged over one field period) of three eigenstates $p = 2, 8$ and 20 (from left to right) of Fig. 1. As predicted by the secular calculation (see text), they are localized near a circular orbit, the field axis, and the plane perpendicular to the field, respectively; only the $p = 8$ state significantly contributes to the ionization yield.

be described by a secular approximation. In future experiments, the details of this structure could be revealed by direct photospectroscopy[16] of the Floquet spectrum[9] from low lying atomic states. In such kind of experiments, the underlying phase space geometry should be manifest in the positions and widths of the atomic resonances. When Rydberg states of a different atomic species are used,[7] the interaction of the electron with the ionic core should destroy the multiplets and produce Floquet eigenstates with completely different localization properties, since the described organization of the resonances strongly relies on the degeneracy of the energy levels of the hydrogen atom.

REFERENCES

1. G. Casati, I. Guarneri, and D. L. Shepelyansky, IEEE J. Quant. Electron. **24**, 1420 (1988).
2. A. Buchleitner, D. Delande, and J.-C. Gay, J. Opt. Soc. Am. B **12**, 505 (1995).
3. R. Blümel and U. Smilansky, Z. Phys. **D6**, 83 (1987); R. V. Jensen, S. M. Susskind, and M. M. Sanders, Phys. Rev. Lett. **62**, 1476 (1989).
4. P. M. Koch and K. A. H. van Leeuwen, Phys. Rep. **255**, 289 (1995) and references therein.
5. J. E. Bayfield, Chaos **1**, 110 (1991).
6. D. Delande and J.C. Gay, J. Phys. B **17**, 335 (1984)
7. M. Arndt, A. Buchleitner, R. N. Mantegna, and H. Walther, Phys. Rev. Lett. **67**, 2435 (1991).
8. A. Buchleitner, thèse de doctorat, Université Pierre et Marie Curie, Paris (1993).
9. A. Buchleitner, D. Delande, and J. Zakrzewski, *in*: "Multiphoton Processes 1996", P. Lambropoulos and H. Walther, eds., IOP, Bristol (1997).
10. K. Hornberger, T. Uzer, both this volume.
11. B.I. Meerson, E.A. Oks and P.V. Sasorov, J. Phys. B **15**, 3599 (1982).
12. A.J. Lichtenberg and M.A. Lieberman, "Regular and Stochastic motion", Springer-Verlag, New York (1983).
13. A. Buchleitner and D.Delande, Phys. Rev. **A55**, R1585 (1997).
14. R. Shakeshaft, Comm. At. Mol. Phys. **28**, 179 (1992).
15. D. Shepelyansky, "AIP conference proceedings 275", H. Walther et al., eds., New York (1992).
16. W. Hogervorst, this volume.

SPONTANEOUS DECAY OF NONDISPERSIVE WAVE PACKETS

Klaus Hornberger and Andreas Buchleitner

Max-Planck-Institut für Quantenoptik
Hans-Kopfermann-Straße 1
D-85748 Garching

For a Rydberg atom conventional microwave amplitudes of several V/cm are strong in the sense that the exerted force is comparable to the Coulomb attraction. In this regime of highly nonlinear coupling between the external driving and the atomic motion nondispersive electronic wave packets can exist.

NONDISPERSIVE WAVE PACKETS

Usually the term wave packet refers to the coherent superposition of the stationary eigenstates of a time-independent Hamiltonian forming an initially localized wave function. The generically unharmonic energy spacing of the stationary eigenstates leads to the apparently unavoidable *dispersion* of these objects.[1]

However, it has been shown recently that in non-linear time periodic systems wave packets can exist that *do not disperse*, because they are periodic eigenstates of the time-dependent problem.[2–6] They faithfully trace a classical trajectory and maintain their localization properties. In a more general perspective, these nondispersive wave packets are the quantum analog of a classical nonlinear resonance in a periodically driven Hamiltonian system.[2, 3] They can be considered as direct generalizations of (stationary) eigenstates associated with elliptic islands in the classical phase space of nonlinear, autonomous Hamiltonian systems.[7]

A simple and experimentally accessible system where this classical stabilization phenomenon can be exploited is a highly excited hydrogen atom subject to a monochromatic microwave field.[8, 9] For driving field amplitudes comparable to the Coulomb attraction, only the elliptic island associated with the principal nonlinear resonance survives, embedded in the chaotic domain of classical phase space. It is due to the resonant coupling of the microwave field to the fundamental harmonic of the atomic Kepler motion. In a quantum calculation one finds[3, 5, 6] eigenstates of the atom dressed by the microwave field, which are perfectly localized on the elliptic island.

As a consequence of the field-induced coherent coupling of Rydberg states to the atomic continuum, these dressed states acquire a *finite lifetime*.[3, 5, 6] However, the width (the inverse of the life time) of the wave packet eigenstates is extremely small (allowing

Atoms and Molecules in Strong External Fields
Edited by Schmelcher and Schweizer, Plenum Press, New York, 1998

for millions of Kepler orbits) and decreases on the average[5] exponentially with increasing principal quantum number n. The latter can be understood semi-classically if one views ionization as due to the tunneling of the wave function through the classical invariant tori isolating the elliptic island from the surrounding "chaotic sea".

The apparently almost eternal life time[3, 5] of wave packet eigenstates makes them very interesting objects in the diagnostics of complicated atomic and molecular dynamics.[10, 11] It also encourages speculations about the observability of extremely slow classical processes like Arnold diffusion in real quantum systems.[12] However, the discussion in the literature neglected any *decoherence effects* so far, which originate from the unavoidable coupling to the environment. Though, the life times predicted for the wave packet eigenstates are comparable to the radiative life times of Rydberg states. Hence, it is still an open question to what extent they will be observable in a laboratory experiment.

In the following, we want to elucidate the rôle of spontaneous emission. For this purpose, we introduce a novel implementation of complex scaling[13] which allows for the simultaneous description of the exact coherent coupling between the atom and the dressing field, together with a perturbative description of the coupling to the electromagnetic vacuum. The resulting master equation for coherences and populations of the dressed states provides a general framework for the incoherent coupling of decaying states to a Markovian environment.

RADIATIVE COUPLING TO THE ENVIRONMENT

For the sake of simplicity, we restrict ourselves to Rydberg atoms confined to one spatial dimension, driven by a linearly polarized microwave field. If we set aside for a moment the coupling to the vacuum field, the periodically driven Rydberg electron is described by the Hamiltonian

$$\mathrm{H_{sys}}(t) = \frac{\mathrm{p}^2}{2} - \frac{1}{\mathrm{x}} + F\,\mathrm{x}\cos(\omega t), \qquad (x > 0) \tag{1}$$

i.e. we treat the driving microwave as a classical field in dipole approximation, neglect relativistic effects, assume an infinite nuclear mass, and employ atomic units. These simplifications are perfectly justified in the Rydberg regime.

Complex scaling of (1), by the positive angle θ, together with the Floquet theorem, allows us to determine the exact quasi-energies E_α, widths Γ_α, and wave functions $|\chi_\alpha\rangle$ of the dressed states of the atom in the coherent field.[14, 15] These can be expressed in terms of the discrete $2\pi/\omega$-periodic eigenstates $|\chi_{\nu\theta}\rangle$ (with Fourier components $|\chi^k_{\nu\theta}\rangle$) of the dilated Floquet Hamiltonian $\mathcal{H}^\theta_{\mathrm{sys}}$. The latter is obtained from the Floquet Hamiltonian $\mathcal{H}_{\mathrm{sys}} = \mathrm{H_{sys}}(t) - i\partial_t$ by a non-unitary, complex scaling transformation $\mathrm{R}(\theta)$. The associate left eigenvectors $\overline{\langle\chi_{\nu\theta}|}$ are the complex conjugate of $\langle\chi_{\nu\theta}|$, i. e. the transpose of the right eigenvectors. Together with the complex eigenvalues $\mathrm{E}_\nu - i\Gamma_\nu/2$ they provide, for $t > 0$, the time evolution operator[14] generated by $\mathrm{H_{sys}}(t)$:

$$\mathrm{U_{sys}}(t) = \sum_{\nu k_1 k_2} e^{-\Gamma_\nu t/2} e^{-i(\mathrm{E}_\nu + k_2\omega)t}\, \mathrm{R}(-\theta)|\chi^{k_2}_{\nu\theta}\rangle\overline{\langle\chi^{k_1}_{\nu\theta}|}\mathrm{R}(\theta). \tag{2}$$

The adjoint $\mathrm{U}^\dagger_{\mathrm{sys}}$ is given by the complex conjugate expression.[14] These are needed in our treatment of the coupling of the dressed states to a reservoir,

$$\mathrm{H_{res}} = \sum_j \omega_j (b^\dagger_j b_j + \frac{1}{2}), \tag{3}$$

by a dipole interaction,

$$\mathrm{H_{int}} = \sum_j \mathrm{x}(g_j b_j + g_j^* b_j^\dagger), \tag{4}$$

with a symbolic sum over the continuum of reservoir modes j, the associated creation and annihilation operators $b_j^\dagger$ and b_j, and the coupling constants g_j.

In the product basis spanned by the eigenstates of $\mathcal{H}_{\mathrm{sys}}^\theta$ and the reservoir Fock states, the evolution $\mathrm{U_0} := \mathrm{U_{sys}} \otimes \exp(-i\mathrm{H_{res}}t)$ is explicitly known. This suggests to proceed in the interaction picture, to formulate the temporal evolution of the density operator $\hat{\rho}(t)$ generated by the total Hamiltonian $\mathrm{H}(t) = \mathrm{H_{sys}}(t) + \mathrm{H_{res}} + \mathrm{H_{int}}$.

At this point some care has to be taken due to the finite widths Γ_ν of the eigenstates of $\mathcal{H}_{\mathrm{sys}}^\theta$: as a consequence, any square-integrable initial atomic state will leak out to infinity on a finite time scale. Hence, its norm within any finite volume around the nucleus is *not* conserved,[16] what explains that $\mathrm{U}_{\mathrm{sys}}^\dagger(t)\mathrm{U_{sys}}(t) \neq \mathrm{id}$ for $t > 0$, since the expressions for $\mathrm{U_{sys}}$ and $\mathrm{U}_{\mathrm{sys}}^\dagger$ only operate on square-integrable states.[13, 14] Consequently, since we want to formulate time dependent perturbation theory in the basis represented by the $|\chi_{\mu\theta}\rangle$, the reduced density operator in the interaction picture, $\tilde{\sigma}$, has to be defined by explicitly employing the inverse operators of $\mathrm{U_{sys}}$ and $\mathrm{U}_{\mathrm{sys}}^\dagger$, $\mathrm{U}_{\mathrm{sys}}^{-1}$ and $[\mathrm{U}_{\mathrm{sys}}^\dagger]^{-1}$, respectively. Those are obtained, for $t > 0$, through the identification $\mathrm{U}_{\mathrm{sys}}^{-1}(t) := \mathrm{U_{sys}}(-t)$ and $[\mathrm{U}_{\mathrm{sys}}^\dagger]^{-1}(t) := \mathrm{U}_{\mathrm{sys}}^\dagger(-t)$. The appropriate definition

$$\tilde{\rho} := \mathrm{U}_0^{-1}\, \hat{\rho}\, [\mathrm{U}_0^\dagger]^{-1} \tag{5}$$

then leads to the equation of motion for the coherent evolution of the total density operator

$$\frac{\mathrm{d}}{\mathrm{dt}}\tilde{\rho} = -i\,(\mathrm{U}_0^{-1}\mathrm{H_{int}}\mathrm{U}_0\,\tilde{\rho} - \tilde{\rho}\,\mathrm{U}_0^\dagger\mathrm{H_{int}}[\mathrm{U}_0^\dagger]^{-1}). \tag{6}$$

Note that (5) and (6) coincide with the usual equations for vanishing widths Γ_ν.

Now a perturbative treatment of the reservoir action on the electron, up to second order in the coupling constants can be performed.[17] We trace over the reservoir variables, $\tilde{\sigma} := \mathrm{tr_{res}}\{\tilde{\rho}\}$, since we are only interested in the atomic degrees of freedom. Rotating wave and Markov approximation, together with the neglect of any systematic degeneracy in the Floquet spectrum, yield a differential equation for $\tilde{\sigma}$ which we shall derive and discuss in detail in a separate publication. Back in the Schrödinger picture, $\hat{\sigma} := \mathrm{U}_0\tilde{\sigma}\mathrm{U}_0^\dagger$, the matrix elements of the system's density operator $\hat{\sigma}_{\alpha\beta} = \langle\overline{\chi_\alpha(t)}|\hat{\sigma}(t)|\chi_\beta(t)\rangle$ obey the final master equation

$$\frac{\mathrm{d}}{\mathrm{dt}}\hat{\sigma}_{\alpha\alpha} = -\,\hat{\sigma}_{\alpha\alpha}\Gamma_\alpha - \hat{\sigma}_{\alpha\alpha}\sum_\mu \mathrm{M}_{\alpha\mu} + \sum_\mu \hat{\sigma}_{\mu\mu}\mathrm{M}_{\mu\alpha}, \tag{7}$$

$$\frac{\mathrm{d}}{\mathrm{dt}}\hat{\sigma}_{\alpha\beta} = -i\,(\mathrm{E}_\alpha - \mathrm{E}_\beta)\,\hat{\sigma}_{\alpha\beta} - \frac{1}{2}\{\Gamma_\alpha + \Gamma_\beta + \sum_\eta (\mathrm{M}_{\alpha\eta} + \mathrm{M}_{\beta\eta})\}\,\hat{\sigma}_{\alpha\beta}, \qquad (\alpha \neq \beta), \tag{8}$$

with transition rates $\mathrm{M}_{\mu\nu}$ from $|\chi_\mu\rangle$ to $|\chi_\nu\rangle$. Hence, populations and coherences separate and the latter decay exponentially. These equations are valid for any initial condition $\hat{\sigma}(t_0) := \sum_l p_l|\psi_l\rangle\langle\psi_l|$, $\sum_l p_l = 1$, $|\psi_l\rangle$ square-integrable. They completely describe the temporal evolution of the system in the *discrete* set of dressed states $|\chi_\alpha\rangle$ and *exactly* incorporate the coupling to the atomic continuum.

Focusing on the special case of spontaneous emission as the only source of decoherence, the transition rates from $|\chi_\mu\rangle$ to $|\chi_\nu\rangle$ read

$$\mathrm{M}_{\mu\nu} = \sum_k \mathrm{Re}(X_{\mu\nu}^2(k))\; K(|\Delta_{\mu\nu}^k|)\, \frac{1}{2}(1 + \mathrm{sgn}\Delta_{\mu\nu}^k). \tag{9}$$

They involve the (θ-independent) Fourier components of the dipole matrix element,

$$X_{\mu\nu}(k) := \sum_{\ell} \overline{\langle \chi^{\ell}_{\mu\theta} | \mathrm{R}(\theta) \mathrm{x} \mathrm{R}(-\theta) | \chi^{\ell-k}_{\nu\theta} \rangle}, \tag{10}$$

the corresponding quasi-energy spacings,

$$\Delta^{k}_{\mu\nu} := \mathrm{E}_{\mu} - \mathrm{E}_{\nu} + k\omega, \tag{11}$$

and the response function of the reservoir at resonant frequency:

$$K(\Omega) := 2\pi \sum_{m,n,\lambda} \rho^{(\lambda)}_{mn}(\Omega) |g^{(\lambda)}_{mn}(\Omega)|^2. \tag{12}$$

In our numerical calculations the mode densities $\rho^{(\lambda)}_{mn}$ and coupling constants $g^{(\lambda)}_{mn}$ in (12) are chosen as for a typical waveguide experiment.[17] Note, however, that $K(\Omega)$ is virtually identical to the response function in free space for large values of Ω which dominate the spontaneous decay of the wave packet eigenstates.

NUMERICAL IMPLEMENTATION

In order to access the time evolution of the atomic population initially prepared in the wave packet eigenstate, we first have to extract the eigenvalues and matrix elements which enter in (7). This is done by the numerical diagonalization of $\mathcal{H}^{\theta}_{\mathrm{sys}}$ in a real Sturmian basis.[14] For comparison to previously published results[5] we define the amplitude and the frequency of the driving field by $Fn^4 = 0.04442$ and $\omega n^3 = 1$, yielding a fixed phase space structure. Here n indicates the principal quantum number of the Rydberg manifold the wave packet originates from. In our 1D model we typically obtain 200 Floquet eigenstates per Floquet zone, with about 100 Fourier components each. With these ingredients (7) has been solved for n between 5 and 130. Initially, at $t = 0$, the wave packet eigenstate was assumed to carry all available atomic population.

It turns out that transitions to low lying atomic states, which remain unaffected by the driving field, dominate the spontaneous decay of the wave packet. Asymptotically, all population which has not been lost through direct ionization to the atomic continuum is accumulated in the atomic ground state. In particular, there is no diffusion of population in the Rydberg regime. Therefore, the radiative decay of the wave packet can be described effectively by a rate $\overline{\mathrm{M}}$ defined as the sum over all radiative decay rates depopulating the wave packet state.

RADIATIVE DECAY VS. COHERENT CONTINUUM COUPLING

As already mentioned above, the wave packet's width Γ_{wp} is known to decrease *exponentially* (on the average) with n, since it essentially represents a tunneling probability. In contrast, the spontaneous decay rates of unperturbed Rydberg states are known to decrease *algebraically* like n^{-a}, with a between 3 and 5, depending on the quantum numbers of the Rydberg state.[18, 19] Hence, for sufficiently large values of n, spontaneous emission must dominate the population loss out of the wave packet eigenstate. Fig. 1 summarizes the comparison of spontaneous rates $\overline{\mathrm{M}}$, resonance widths Γ_{wp}, and total decay rates $\overline{\mathrm{R}} = \overline{\mathrm{M}} + \Gamma_{\mathrm{wp}}$ extracted from the solution of (7) for the wave packet states over a large range of principal quantum numbers. Indeed, for n between 5 and 93 the coherent coupling to the atomic continuum is generally dominant over the spontaneous decay, except for the occasional appearance of extremely long-lived

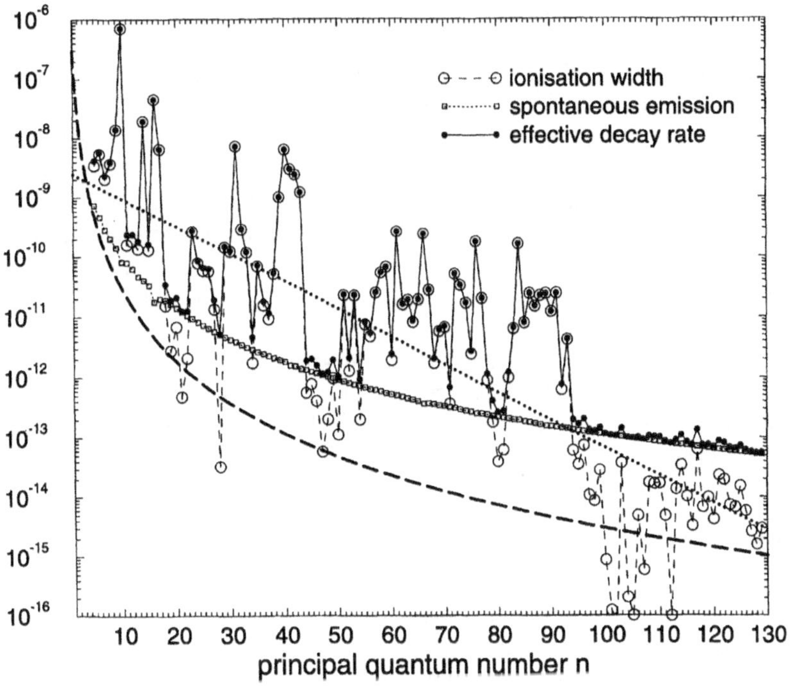

Figure 1. Comparison of the decay rates to the atomic continuum Γ_{wp}, the radiative decay rates $\overline{M}$, and the total decay rates $\overline{R}$ of the nondispersive wave packets, as a function of the principal quantum number n. One observes a cross-over from coherent to radiative decay. The dotted line indicates the overall *exponential* decay of Γ_{wp} as opposed to the *algebraic* decay of $\overline{M}$ proportional to n^{-3} (1D, squares) or n^{-4} (3D, dashed line). The chaos-induced[5] fluctuations of Γ_{wp} are partially amended for intermediate and virtually suppressed for large values of n.

wave packet states.[5] In contrast, for principal quantum numbers $n \geq 94$ spontaneous emission dominates over the coherent continuum coupling and therefore makes the non-dispersive wave packet a rigorously classical object at *finite* values of n. The wave packet dynamics is now strictly confined to the atomic bound space with its analogy in the classical phase space of bounded trajectories. It is a remarkable result that this behavior, which one would expect in the quasi-classical limit, can already be expected for quantum numbers accessible by experiment.

The effective spontaneous emission rate in Fig. 1 scales as n^{-3}, like 3D Rydberg states of low angular momentum. However, our numerical results are obtained for 1D wave packet eigenstates which have their direct 3D analog in wave packets tightly confined to the vicinity of the polarization axis of the driving field and are essentially composed of extremal parabolic states.[6] The latter exhibit spontaneous decay rates proportional[18] to n^{-4}. Whereas our 1D description of the atom is perfectly justified in the Rydberg regime,[20] it implies an unrealistic representation of the atomic ground state, which acquires a permanent dipole moment. As a consequence, transitions from an extremal parabolic state to the ground state are overestimated. To compensate for this artifact of the 1D picture we rescale the radiative transitions to low lying states by replacing the dominant dipole matrix element by its actual 3D value.[18] The dashed line in Fig. 1 shows the resulting estimate for the quasi 1D wave packet of the real atom.[6] Since this curve only accounts for transitions within the ladder of extremal parabolic states, it underestimates the actual spontaneous decay rates. Therefore, we predict the crossover from dominant continuum coupling to dominant spontaneous decay to occur in the interval $n = 94 \ldots 145$. The upper limit has been obtained from an extrapolation of the dotted and the dashed lines in Fig 1.

Note that this transition will occur at even higher values of n for wave packets which trace classical trajectories with larger angular momentum. The low angular momentum wave packets we are considering are most sensitive to radiative decay. Hence, the fluctuations of $\overline{\mathrm{R}}$, which are chaos-induced in the sense that they are due to the interference of transition amplitudes associated with ionizing irregular paths originating from the principal resonance,[5] remain an observable phenomenon even in the least favorable case.

Finally, since spontaneous emission should affect any "exotic" state in the Rydberg domain, our findings should be relevant also for nondispersive wave packets in combined microwave and static magnetic fields.[10] They also imply limits for the actual observability of extremely slow classical processes such as Arnold diffusion.[12]

REFERENCES

1. G. Alber and P. Zoller, Phys. Rep. **199**, 231 (1991).
2. H.-P. Breuer and M. Holthaus, Ann. Phys. (NY) **211**, 249 (1991).
3. A. Buchleitner, thèse de doctorat, Université Pierre et Marie Curie, Paris 1993; D. Delande and A. Buchleitner, Adv. At. Mol. Opt. Phys. **35**, 85 (1994).
4. I. Bialynicki-Birula et al., Phys. Rev. Lett. **73**, 1777 (1994).
5. J. Zakrzewski et al., Phys. Rev. Lett. **75**, 4015, (1995).
6. A. Buchleitner and D. Delande, Phys. Rev. Lett. **75**, 1487 (1995); A. Buchleitner et al., in *Proceedings of the 7th International Conference on Multiphoton Processes*, edited by P. Lambropoulos and H. Walther, (Institute of Physics Publishing, Bristol, 1997).
7. K. Richter and D. Wintgen, Phys. Rev. Lett. **65**, 1965 (1990).
8. J. E. Bayfield et al., Phys. Rev. A **53**, R12 (1996).
9. P. M. Koch, Physica D **83**, 178 (1995).
10. A. F. Brunello et al., Phys. Rev. Lett. **76**, 2874 (1996); C. Cerjan et al., Phys. Rev. **A55** 2222 (1997); T. Uzer, this volume.
11. I. Bialynicki-Birula and Z. Bialynicka-Birula, Phys. Rev. Lett. **77**, 4298 (1996).
12. J. v. Milczewski et al., Phys. Rev. Lett. **76**, 2890 (1996).
13. E. Balslev and J. M. Combes, Comm. Math. Phys. **22**, 280 (1971); K. Yajima, Comm. Math. Phys. **87**, 331 (1982).
14. A. Buchleitner et al., J. Opt. Soc. Am. B **12**, 505 (1995).
15. A. Buchleitner et al., J. Phys. B **27**, 2663 (1994).
16. Y. B. Zeldovich, Sov. Phys. JETP **12**, 542 (1961); N. Hokkyo, Prog. Theo. Phys. **33**, 1116 (1965); J. W. Pang, D. Neuhauser, and N. Moiseyev, to be published in J. Chem. Phys. (1997).
17. R. Blümel et al., Phys. Rev. A **44**, 4521 (1991).
18. H. A. Bethe and E. E. Salpeter, *Quantum Mechanics of One- and Two-Electron Atoms* (Plenum Publishing Corp., New York, 1977), p. 276.
19. H. Marxer and L. Spruch, Phys. Rev. A **43**, 1268 (1991).
20. R. Blümel and U. Smilansky, Z. Phys. D **6**, 83 (1987).

IONIZATION OF HELIUM BY STATIC ELECTRIC FIELDS AND SHORT PULSES

Armin Scrinzi *

Institut für theoretische Physik, Universität Innsbruck,
Technikerstrasse 25, A-6020 Innsbruck, Austria

INTRODUCTION

In this paper we present calculations of the ionization probability of Helium in external electric fields that are comparable in strength to the Coulomb fields between the constituents of *He*. We consider static fields as well as short DC– and laser pulses.

The problem is difficult, because one deals with five–dimensional partial differential equations, when rotational symmetry is broken and only cylindrical symmetry is preserved. In practice, the difficulty strongly depends on the states of *He* that are investigated. For example, in higher singly excited states the correlation between the two electrons can be neglected or corrected for, in which case the problem is effectively reduced to two dimensions. However, the energies of the ground state and the low lying doubly excited states contain significant contributions from correlation. Our purpose is to fully include correlation effects in order to detect their role in ionization and to determine the limits of validity of widely used approximations like the single active electron model[1].

Due to the high dimensionality, a purely numerical representation of the wave function leads to very large systems of equations. When one expands the wave function with respect to total angular momentum, the resulting system of three–dimensional differential equations grows quadratically with the maximum angular momentum included. In an effort to obtain the most compact representation of the three–dimensional wave function components, we use a basis set expansion in terms of functions of the distances between the three particles. We find that a few hundred functions per angular momentum are sufficient to accurately describe bound and doubly excited states. Thus the numerical solution of the Schrödinger equation with strong external fields becomes feasible.

After briefly introducing our basis set expansion we give the *non–perturbative Stark shifts and widths* of the ground state and the lowest doubly excited state of Helium for fields up to and beyond the critical field strength. As a second example we present the results of the *integration of the time dependent Schrödinger equation* with a very short

*present address: Einsiedlerplatz 2/10, A–1050 Vienna, Austria
e–mail: scrinzi@amuon.imep.univie.ac.at

DC pulse for a range of field strengths. Finally, we show the ionization by *femtosecond medium intensity laser pulses.*

REPRESENTATION OF THE HELIUM WAVE FUNCTION

We discretize the Schrödinger equation by expanding the wave function in a Hylleraas–like explicitly correlated basis

$$\Psi(\vec{r}_1,\vec{r}_2;t) = P_s \sum_{L=0}^{L_{\max}} \sum_{l_1=0}^{L} G_{Ll_1}(\vec{r}_1,\vec{r}_2) \sum_i c_i^{Ll_1}(t) r_1^{k_i} r_2^{m_i} |\vec{r}_1 - \vec{r}_2|^{n_i} \exp(-\alpha_i r_1 - \beta_i r_2) \quad (1)$$

The vectors $\vec{r}_1$ and $\vec{r}_2$ denote the electron coordinates relative to the nucleus and the operator P_s projects on the singlet states. The $c_i^{Ll_1}(t)$ are the linear expansion coefficients. The two–electron angular factors G_{Ll_1} for total angular momentum L and z–component $L_z = 0$ are given by

$$G_{Ll_1} = r_1^{l_1} r_2^{L-l_1} \sum_m C^{L,0}_{l_1,m;L-l_1,-m} Y^{l_1}_m(\hat{r}_1) Y^{L-l_1}_{-m}(\hat{r}_2). \quad (2)$$

The $C^{L,0}_{l_1,m;L-l_1,-m}$ are Clebsch–Gordan coefficients and Y^l_m are spherical harmonics. For each (L,l_1) several sets of exponents (α_i,β_i) were used and the combination of powers (k_i,m_i,n_i) was constrained by[2]

$$k_i + m_i + n_i + |k_i - m_i|(1 - \delta_{0n_i}) \leq p. \quad (3)$$

Note that for each L there are only $L+1$ angular functions G_{L,l_1}. Angular correlation, which in the usual atomic physics basis requires a large number of combinations of single–electron angular momenta l_1 and l_2, is here contained in the inter–electron coordinate $|\vec{r}_1 - \vec{r}_2|$. To determine ionization rates and to avoid reflections in the finite range of space that can be numerically represented we use the method of complex scaling[3].

It has been demonstrated that such a basis can efficiently and accurately describe He bound states[2, 4]. For our calculations we used an average of 300 expansion functions for each L up to $L_{\max} = 7$. The discretized atomic Hamiltonian matrix contained 6 to 8 bound states for each L, out of which the lowest 3 to 4 had an accuracy in energy of $\lesssim 10^{-8}\,au$. The energies and widths of the lowest bound and doubly excited states as obtained with that basis are compared with literature values in Table 1.

IONIZATION BY A STATIC ELECTRIC FIELD

The complex scaled Schrödinger equation of the Helium atom in an external electric field is (in atomic units)

$$\left[-\frac{e^{-2i\theta}}{2}(\Delta_1 + \Delta_2) - \frac{2e^{-i\theta}}{r_1} - \frac{2e^{-i\theta}}{r_2} + \frac{e^{-i\theta}}{|\vec{r}_1 - \vec{r}_2|} + e^{i\theta} F_0 (z_1 + z_2)\right] \Psi = (E - i\frac{\Gamma}{2})\Psi. \quad (4)$$

Fig. 1 shows the energies E and ionization widths Γ as a function of the electrical field strength F_0 as obtained with the expansion (1). The basis set truncation errors are $< 5\%$, which includes dependence on the scaling angle θ as well as the limitation of the angular momenta to $L_{\max} = 7$. Somewhat counter–intuitively, the highest angular momenta are required to obtain good relative accuracy for the ground state ionization rate at the lowest field strengths $F_0 < 0.1$. For the rest of the data $L_{\max} = 3$ suffices. Interestingly, the autoionization rate of the doubly excited state is slightly suppressed by small external fields $F_0 < 0.02$.

Table 1. Energies and autoionization widths as obtained in our calculations compared to literature values. The states given are the two lowest bound states and the lowest doubly excited states below the $He^+(n=2)$ and $He^+(n=3)$ ionization thresholds, respectively, for angular momenta $L=0,\ldots,4$. All states are singlet with parity $(-1)^L$. The basis size is $\sim$ 300 functions for each L.

	present		literature	
L	E	Γ	E	Γ
0	-2.903724372	—	-2.903724375[a]	—
	-2.145974045	—	-2.14597403[a]	—
	-0.77780	4.45×10^{-3}	-0.777868	4.53×10^{-3} [c]
	-0.35356	3.004×10^{-3}	-0.353537	3.004×10^{-3} [c]
1	-2.123843086	—	-2.12384308[a]	—
	-2.055146361	—	-2.0551463[a]	—
	-0.69312	1.345×10^{-3}	-0.6931349	1.3733×10^{-3} [c]
	-0.33553	6.7×10^{-3}	-0.3356259	7.023×10^{-3} [c]
2	-2.055620726	—	-2.05562073307[b]	—
	-2.031279842	—	-2.03127984619[b]	—
	-0.70192	2.44×10^{-3}	-0.7019457	2.3622×10^{-3} [c]
	-0.34302	5.30×10^{-3}	-0.343173	5.155×10^{-3} [c]
3	-2.031255143	—	-2.031255144[b]	—
	-2.020002936	—	-2.020002937[b]	—
	-0.55828	1.17×10^{-5}	-0.55828	1.28×10^{-5} [d]
	-0.30419	3.30×10^{-3}	-0.304215	3.25×10^{-3} [c]
4	-2.020000710	—		
	-2.013889344	—		
	-0.53236	9.64×10^{-6}		
	-0.31042	6.84×10^{-3}	-0.30705	6.70×10^{-3} [c]

[a]Accad, Pekeris and Schiff, Ref. 5, [b]Sims and Martin, Ref. 4, [c]Ho et al., as tabulated in Ref. 6, [d]Lindroth, Ref. 6

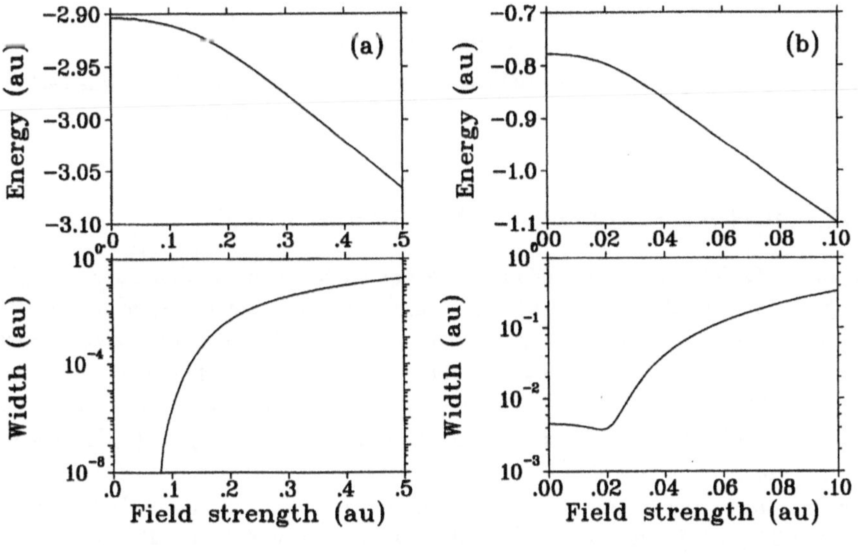

Figure 1. Energies and ionization widths in static electric field: (a) ground state of Helium, (b) lowest doubly excited S–state.

IONIZATION BY A DC–PULSE

We solved the Schrödinger equation for an external field with a $\sin^2$ time dependence

$$F(t) = F_0 \left[\sin(\pi t/T)\right]^2 . \tag{5}$$

The pulse duration was chosen to be $T = 10\,au$. This time is comparable to the orbit period of the electrons and therefore the adiabatic approximation for the ionization yield, defined as the integral over the static rates

$$\text{Yield}_{\text{adiabatic}} = \int_0^T \Gamma_{static}(F(t))dt, \tag{6}$$

must be expected to fail. Fig. 2 compares the ionization yield as obtained by time–integrating the Schrödinger equation with the adiabatic approximation. The adiabatic approximation underestimates the actual yield by more than one order of magnitude.

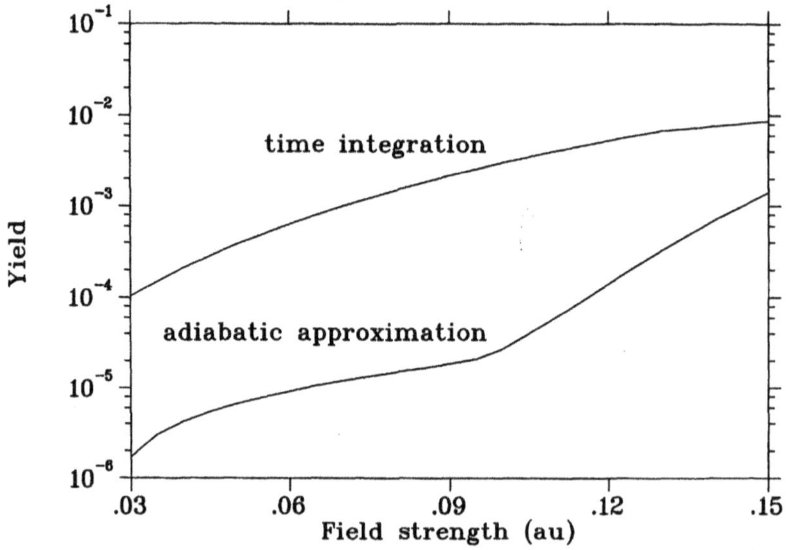

Figure 2. Ionization by a $\sin^2$–shaped DC–pulse of duration 10 *au* as a function of peak field strength.

IONIZATION BY A FEMTOSECOND LASER PULSE

While DC–fields as considered above are not available under laboratory conditions, laser pulses of such and higher intensities are routinely used in experiments. Pulse durations can be as short as a few femtoseconds. Reliable *ab initio* calculations for comparison with experiments need to include both, the time dependence of the pulse envelope and the correlation between the electrons. Our basis allows such calculations at moderate laser intensities and frequencies $\gtrsim 0.3\,au$, which correspond to wave lengths below $152\,nm$[7].

Fig. 3 shows the bound state excitation and ionization from the Helium ground state by a $3.8\,fs$ (157 *au*) laser pulse with peak intensity of $1.49 \times 10^{14} W/cm^2$ (peak field strength 0.092 *au*) at frequencies between 0.3 and 1.15 *au* (wave lengths from 152 *nm* to 40 *nm*). The two photon and single photon ionization thresholds are visible as a rapid increase of first excitation and then ionization. Beyond the threshold

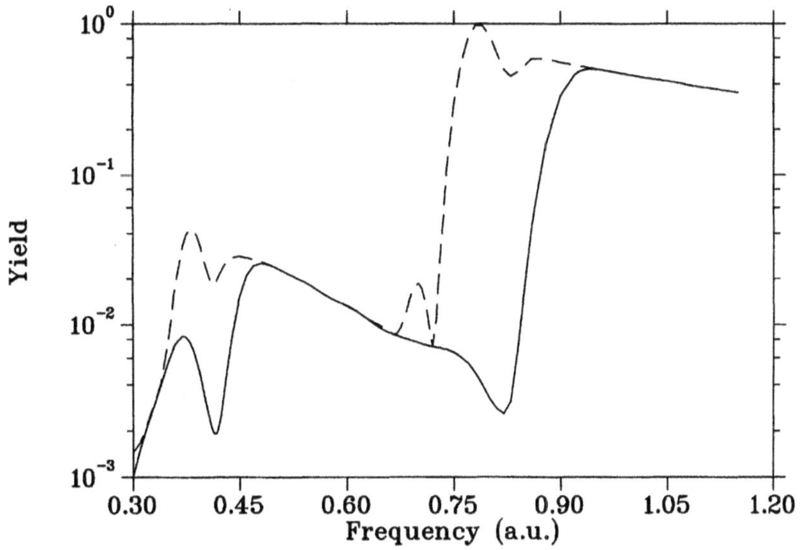

Figure 3. Single excitation and ionization from the ground state of He by $\sin^2$–shaped pulses of duration $3.8\,fs$ and peak intensity $1.49 \times 10^{14} W/cm^2$. Lower curve: ionization yield, upper curve: sum of ionization and excitation.

ionization decreases slowly with frequency. Bound state excitation occurs dominantly through resonances. The large excitation peak below the two–photon threshold is due to resonance from the ground state to the first excited S–state, the peak below the single–photon threshold is a resonance with the lowest P–state. The dips in excitation originate in Rabi–like oscillations. The basis set truncation error remains below 5% for the whole range of frequencies.

CONCLUSION

The three examples presented here provide benchmark data for the ionization of Helium by a strong electric field. All calculations could be performed on a work station. For example, the time–integration of one point in Fig. 3 required about 40 CPU minutes on a 500 MHz DEC–Alpha work station.

The application of the same method for magnetic fields is straight forward. A study of the Helium bound state spectrum in moderate (by astronomical standards) magnetic fields is currently being made.

REFERENCES

1. K.J.Schafer et al., Phys. Rev. Lett. **70**, 1599 (1993).
2. A. Kono and Sh. Hattori, Phys. Rev. A **29**, 2981 (1984).
3. W. Reinhardt, Ann. Rev. Phys. Chemistry **33**, 223 (1978).
4. J. S. Sims and W. C. Martin, Phys. Rev. A **37**, 2259 (1988).
5. Y. Accad, C. L. Pekeris, and B. Schiff, Phys. Rev. A **4**, 516 (1971).
6. E. Lindroth, Phys. Rev. A **49**, 4473 (1994).
7. A. Scrinzi and B. Piraux, Phys. Rev. A **56**, R13 (1997).

ADIABATIC INVARIANTS OF RYDBERG ELECTRONS IN CROSSED FIELDS

Jan von Milczewski and T. Uzer
School of Physics, Georgia Institute of Technology.
Atlanta, Georgia 30332-0430, USA

INTRODUCTION

Rydberg atoms in external fields constitute atomic-scale laboratories where the quantum mechanics of strongly nonlinear systems can be tested [1,2]. While the problem of a Rydberg atom interacting with a strong magnetic field has been fairly well understood as a result of sustained research in the past two decades [3,4], the superficially similar scenario resulting from the addition of a perpendicular electric field – the so-called crossed fields arrangement [5,6,7,8,9,10,11] – remains the least understood of all Rydberg problems. This is all the more regrettable in view of the prominence of crossed fields in diverse areas of physics ranging from excitonic systems to plasmas and neutron stars. This problem is so complex because no continuous symmetry survives the extensive symmetry breaking [12] induced by the two fields. The result is a wealth of new physics which is only possible beyond two degrees of freedom, such as Arnol'd diffusion [1,13]. This absence of symmetry also allows localizing electronic wavepackets in all spatial dimensions, and the observation of these wavepackets [14] has led to new insights into the dynamics of the electron in the Correspondence Principle regime. It has also been found that a velocity-dependent, Coriolis-like force in Newton's equations causes the ionization of the electron to exhibit chaotic scattering [15,16]. All these phenomena, as well as renewed interest in the motional Stark effect [17,18], make the crossed fields problem an experimentally accessible paradigm for a wide variety of outstanding issues in atomic and molecular physics, solid state physics [19,20], nuclear physics [21], astrophysics [22], and celestial mechanics [23].

Raithel, Fauth and Walther [8,24,25], in a recent landmark of experiments, have identified a class of quasi-Landau (QL) resonances in the spectra of rubidium Rydberg atoms in crossed electric and magnetic fields. Similar to the original QL resonances observed by Garton and Tomkins [26], this set of resonances is associated with a rather small number of planar orbits of the crossed fields Hamiltonian which is known to support an enormous number of mostly non-planar periodic motions [8]. In view of the large phase and parameter space of the problem (which is further enlarged by the choice of polarization of the stimulating laser), how does the experiment achieve such remarkable selectivity? What is special about these orbits and which systematics do they obey? In this Letter we will answer these puzzles by uncovering some remarkable

Atoms and Molecules in Strong External Fields
Edited by Schmelcher and Schweizer, Plenum Press, New York, 1998

adiabatic structures that govern the motion of the electron under the influence of three interactions (Coulomb, electric and magnetic fields) of comparable strength. Solving this problem quantum mechanically is not possible with current computational facilities — the number of states in the experimental energy range is very large and these are very extended states. Instead, we will use classical mechanics which does give reliable results because the experiments are performed in a regime where the Correspondence Principle is valid.

PERTURBATION THEORY

The Hamiltonian (in atomic units) of the crossed fields problem is [9, 10]

$$\mathrm{H} = \frac{1}{2}p^2 + \frac{B}{2}m + \frac{B^2}{8}(x^2+y^2) - \frac{1}{r} + Fx. \tag{1}$$

We use atomic units throughout, *i.e.*, the magnetic field B is given in units of $B_0 = (e/\hbar)^3 m_e^2 c = 2.35 \times 10^5$ T and points in the z direction whereas the electric field F is in units of $F_0 = e^5 m_e^2/\hbar^4 = 5.14 \times 10^{11}$ V/m and points in the x direction. The quantity m is the z-component of the angular momentum $\mathbf{L} = \mathbf{r} \times \mathbf{p}$. We set out by obtaining the Normal Form of this Hamiltonian which reduces the problem from three to two degrees of freedom without restricting the approximate motion to the xy-plane where most orbits accessed by the Garching experiment [8] reside. The technically involved derivation [27] is as follows. The regularization of the Coulomb singularity in (1) using the four Kustaanheimo-Stiefel coordinates $\mathbf{u}$ and their conjugate momenta $\mathbf{P}_u$ transforms that Hamiltonian into

$$\begin{aligned}\mathrm{K} = \frac{4}{\omega} &= \frac{1}{2}(\mathbf{P}^2 + \mathbf{u}^2) + \frac{2}{\omega^2}BL_z\mathbf{u}^2 \\ &+ 2\frac{B^2}{\omega^4}\mathbf{u}^2(u_1^2+u_4^2)(u_2^2+u_3^2) + \frac{8F}{\omega^3}\mathbf{u}^2(u_1u_3+u_2u_4)\end{aligned} \tag{2}$$

where

$$\omega = \sqrt{-8E} \tag{3}$$

which is then analyzed using the succession of canonical transformations of the F_2 type [28, 29]. When the enormous number of terms resulting from this treatment are expressed in terms of the two angular momenta $\mathbf{J}$ and $\mathbf{K}$, the Lie-algebraic generators of the group SU(2)⊗SU(2) (locally isomorphic to the symmetry group SO(4) of the Coulomb problem [30, 12]),

$$\mathbf{J} = (\mathbf{L}+\mathbf{A})/2; \quad \mathbf{K} = (\mathbf{L}-\mathbf{A})/2, \tag{4}$$

(where $\mathbf{A}$ is the Runge-Lenz-Laplace vector) a compact Normal Form is obtained [27]. It contains a pair of degenerate, strongly coupled asymmetric tops in these two generators:

$$
\begin{aligned}
\mathrm{E}^{(2)} = & -\frac{1}{2n^2} + \frac{B}{2}(J_z + K_z) - \frac{3Fn}{2}(J_x - K_x) \\
& + \frac{B^2 n^2}{16}[3n^2 - 4(J_z^2 + K_z^2 - J_z K_z) - 8(J_x K_x + J_y K_y)] \\
& + \frac{F^2 n^4}{16}[-17n^2 + 12(J_x^2 + K_x^2 + J_x K_x)] \\
& - \frac{FBn^3}{2}(J_z K_x - J_x K_z)
\end{aligned} \tag{5}
$$

Key features of the dynamics emerge when Poincare Surfaces of Section (PSOS) are displayed in terms of the canonical variables

$$
q_J = \frac{2J_x}{\sqrt{n - 2J_z}}; \quad p_J = \frac{-2J_y}{\sqrt{n - 2J_z}} \tag{6}
$$

where $n = 2|\mathbf{J}| = 2|\mathbf{K}|$ is the classical analog of the principal quantum number. The PSOS's (Fig. 4) are dominated by two fixed points, which we will call P_s and P_u (Fig. 5). P_s is stable and robust to changes of external parameters, whereas P_u may become unstable. This fixed point spawns two others (denoted by X_1 and X_2 in Fig. 4) signaling the onset of chaos when the fields are increased. Translating to cartesian coordinates, one finds that P_s represents the motion in the xy-plane which is invariant due to parity symmetry, i. e. an electron started with $z = p_z = 0$ remains forever in this plane. The Garching experiments [8] showed the importance of these motions as the most important QL oscillations could be explained by planar orbits. In contrast, P_u translates to non-planar orbits. The immediate contact we establish with experiment is an indication that our choice above of the "Extended Lissajous" variables [31] (invented and perfected by astronomers [32]) are best able to bring out relevant features of this problem.

The agreement in key aspects such as individual quantized energy levels and dynamical behavior (even in chaotic regions) between the predictions of the perturbation Hamiltonian, Eq. (9), and the exact Hamiltonian is excellent in the regime of small Bn^3 and Fn^4 and we confine our following remarks to this regime. The most pronounced difference is that the exact Hamiltonian does allow for ionization and mixing between n-states whereas the approximate one does not. The dimensionality of the problem can be further reduced by averaging over the first order motion, thus smoothing out the chaos.

ADIABATIC INVARIANTS AND THEIR ANALYSIS

A pair of rotations, one for $\mathbf{J}$,

$$
\begin{pmatrix} \tilde{J}_x \\ \tilde{J}_y \\ \tilde{J}_z \end{pmatrix} = \begin{pmatrix} \cos\phi & 0 & \sin\phi \\ 0 & 1 & 0 \\ -\sin\phi & 0 & \cos\phi \end{pmatrix} \begin{pmatrix} J_x \\ J_y \\ J_z \end{pmatrix} \tag{7}
$$

by an angle

$$
\tan\phi = a = \frac{3nF}{B}. \tag{8}
$$

and a corresponding one of $\mathbf{K} \to \tilde{\mathbf{K}}$ through an angle $-\phi$ simplifies the first order terms in Hamiltonian (5) and turns the Normal Form into

$$
\tilde{\mathrm{E}}^{(2)} = C_{nBF} + \omega_L(\tilde{J}_z + \tilde{K}_z) + \omega_L \epsilon\, e^{(2)}(\tilde{\mathbf{J}}, \tilde{\mathbf{K}}) \tag{9}
$$

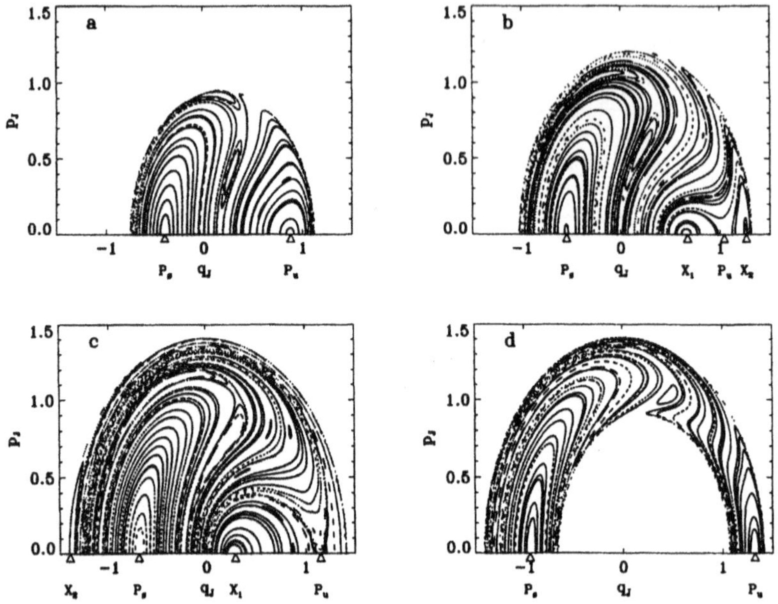

Figure 1. Poincare Surfaces of Section plots for Hamiltonian (9). The fields are fixed at $B = 0.40$, $F = 0.05$. The energy is $E =$ -0.60, -0.55, -0.50, -0.40 (panels a–d). The classical dynamics may be scaled from $n = 1$ in this figure to any excitation. The positions of the most important periodic orbits which perform one loop in **JK**-space are marked by the triangles.

where

$$e^{(2)}(\tilde{\mathbf{J}},\tilde{\mathbf{K}}) = -2a^2(\tilde{J}_x^2+\tilde{K}_x^2)+(3-a^2)(1+a^2)\tilde{J}_z\tilde{K}_z - 6(1+a^2)(\tilde{J}_x\tilde{K}_x+\tilde{J}_y\tilde{K}_y)+a(1+a^2)(\tilde{J}_z\tilde{K}_x-\tilde{J}_x\tilde{K}_z)+ 2(3+a^2)(\tilde{K}_x\tilde{K}_z-\tilde{J}_x\tilde{J}_z)+(a^4-3)(\tilde{J}_z^2+\tilde{K}_z^2) \quad (10)$$

and

$$C_{nBF} = -\frac{1}{2n^2}+\frac{3B^2n^4}{16}-\frac{17F^2n^6}{16}. \quad (11)$$

The motion of **J** and **K**, which is thus decoupled from the fast motion along Kepler ellipses, is given to first order in the fields by the frequency

$$\omega_L = \sqrt{\left(\frac{B}{2}\right)^2+\left(\frac{3Fn}{2}\right)^2} = \frac{B}{2}\sqrt{1+a^2}, \quad (12)$$

separated from the second order motion by a factor of

$$\epsilon = \frac{Bn^2}{6(1+a^2)^{3/2}}. \quad (13)$$

Here, the fast motion is the first order motion, and we average over a period $2\pi/\omega_L$ of this motion, assuming ϵ to be small. The averaged motion is governed by the adiabatic Hamiltonian

$$E_{ad} = C_{nBF}+\omega_L(\tilde{J}_z+\tilde{K}_z)+ \epsilon\,\omega_L\left[\frac{3(1+a^2)}{2}\Omega-\frac{3-4a^2-a^4}{4}(\tilde{J}_z+\tilde{K}_z)^2-\frac{n^2a^2}{2}\right] \quad (14)$$

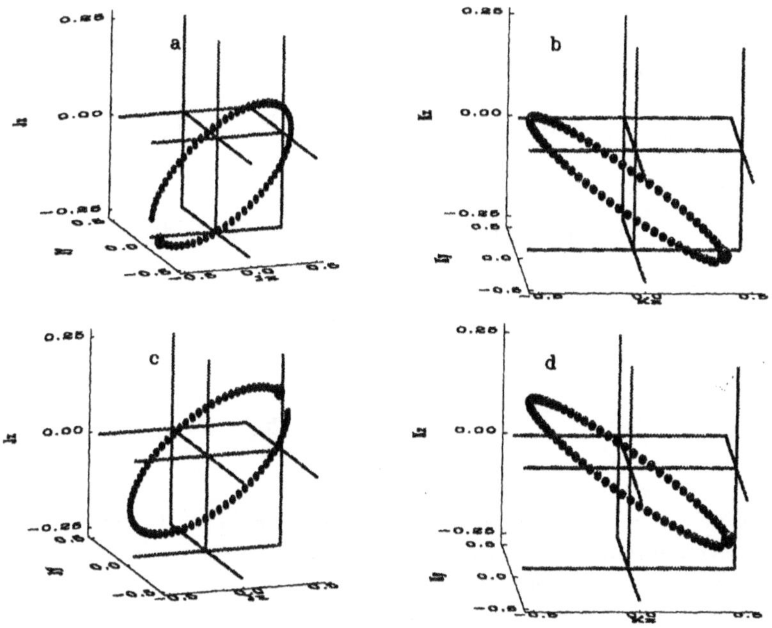

Figure 2. The two orbits P_s (top) and P_u (bottom) in the approximate motion (5). The parameters are $E = -\frac{1}{2}$, $B = 0.4$, $F = 0.05$, and $n = 1$. In order to show the sense of rotation, the starting point of the orbits is marked by a big ball, and the orbits are not completely closed. Panels (a) and (c) respectively (b) and (d) show the motion of **J** respectively **K**. Both orbits are very similar except for the phase difference between the **J** rotor and the **K** rotor, which is π for P_s and 0 for P_u.

where

$$\Omega = \gamma(\tilde{J}_z - \tilde{K}_z)^2 - 4(\tilde{J}_x\tilde{K}_x + \tilde{J}_y\tilde{K}_y), \tag{15}$$

$$\gamma = \frac{a^4 - 3}{2(1+a^2)}, \quad -\frac{3}{2} \leq \gamma < \infty. \tag{16}$$

Hamiltonian (14) is equivalent to the quantal perturbative treatment [33]. Using the canonical variables $(\phi_{\tilde{L}_z}, \tilde{L}_z)$, $(\phi_{\tilde{A}_z}, \tilde{A}_z)$, E_{ad} can be rewritten as

$$\begin{aligned} E_{ad} &= C_{nBF} + \omega_L \tilde{L}_z + \\ &\epsilon\,\omega_L \left[\frac{3(1+a^2)}{2}\Omega - \frac{3 - 4a^2 - a^4}{4}\tilde{L}_z^2 - \frac{n^2 a^2}{2}\right] \end{aligned} \tag{17}$$

where

$$\Omega = \gamma \tilde{A}_z^2 - \sqrt{\left[n^2 - (\tilde{A}_z + \tilde{L}_z)^2\right]\left[n^2 - (\tilde{A}_z - \tilde{L}_z)^2\right]}\,\cos(2\phi_{\tilde{A}_z}). \tag{18}$$

It is now evident that $\tilde{L}_z$ is an adiabatic constant of the motion and so is Ω — in other words, E_{ad} is an integrable approximation and incidentally similar to a pendulum [31].

The motion of $(\phi_{\tilde{A}_z}, \tilde{A}_z)$ occurs along the contour lines of Ω. This means that the system evolves according to three timescales of different orders in the fields:

1. The fastest, "zeroth order motion" is the motion conjugate to n along instantaneously fixed Kepler ellipses. Due to the conservation of n in the perturbative Hamiltonians, this motion is uncoupled from

2. the intermediate evolution, linear in the fields, of the elements of the ellipses, represented by **J** and **K**, or more specifically, by the motion conjugate to the action $\tilde{L}_z$.

3. The slowest motion, quadratic in the fields, is that of the conjugate pair $(\phi_{\tilde{A}_z}, \tilde{A}_z)$. The critical points of the invariant Ω are given by

$$\tilde{A}_z = 0, \quad \cos(2\phi_{\tilde{A}_z}) = -1 \quad \Rightarrow \Omega = (n^2 - \tilde{L}_z^2) \tag{19}$$

which gives P_s and

$$\tilde{A}_z = 0, \quad \cos(2\phi_{\tilde{A}_z}) = 1 \quad \Rightarrow \Omega = -(n^2 - \tilde{L}_z^2). \tag{20}$$

which gives the non-planar orbit P_u. This orbit becomes unstable if and only if

$$\gamma < -1 \Leftrightarrow a < 0.644 \Leftrightarrow Fn < 0.215B \tag{21}$$

and

$$|\tilde{L}_z| < n\sqrt{\frac{|\gamma| - 1}{|\gamma| + 1}}. \tag{22}$$

The condition $\gamma = -1$ corresponds to the subgroup chain $O(4) \supset O(3)_\lambda \supset O(2)_q$ for which the second-order perturbation expression for the energy is separable [34]. In a symmetric manner, P_s becomes unstable when

$$\gamma > 1 \Leftrightarrow a > 1.857 \Leftrightarrow Fn > 0.619B, \tag{23}$$

and the corresponding condition $\gamma = 1$ corresponds to the subgroup chain $O(4) \supset O(3)_Q \supset O(2)_q$ which also renders the second-order expansion separable [34]. Both these periodic orbits are stable when $|\gamma| < 1$.

CORE-VISITING ORBITS

The key to understanding the selectivity of the photoabsorption experiment is to realize that in the semiclassical regime, photon absorption or emission by the electron can only take place close to the nucleus [35]. Therefore, one needs to investigate which set(s) of orbits are accessible to an electron when it approaches the nucleus closely[11]. When $r = 0$, $E = -\frac{1}{2n^2}$, and since the angular momentum vanishes at the nucleus,

$$\tilde{L}_z = \frac{-2a}{\sqrt{1+a^2}} J_x, \quad \tilde{A}_z = \frac{2}{\sqrt{1+a^2}} J_z. \tag{24}$$

The invariant Ω becomes

$$\Omega = n^2 - \tilde{L}_z^2 - \frac{3a^4 + 6a^2 + 5}{2(1+a^2)} \tilde{A}_z^2, \tag{25}$$

and we will determine which values of the invariant Ω can be reached at the nucleus and relate these to P_s and P_u.

The values of Ω are bounded by the curves

$$\Omega_{up} = n^2 - \tilde{L}_z^2 \tag{26}$$

and

$$\Omega_{dn} = n^2 - \tilde{L}_z^2 - \frac{3a^4 + 6a^2 + 5}{2(1+a^2)} \left(\frac{n^2}{1+a^2} - \frac{\tilde{L}_z^2}{a^2} \right) \tag{27}$$

and they are shown in Fig. 6. The position of P_s is given by the upper parabola Ω_{up} and P_u lies on the dashed line given by Eq. (20). This curve is reached at $r = 0$ only

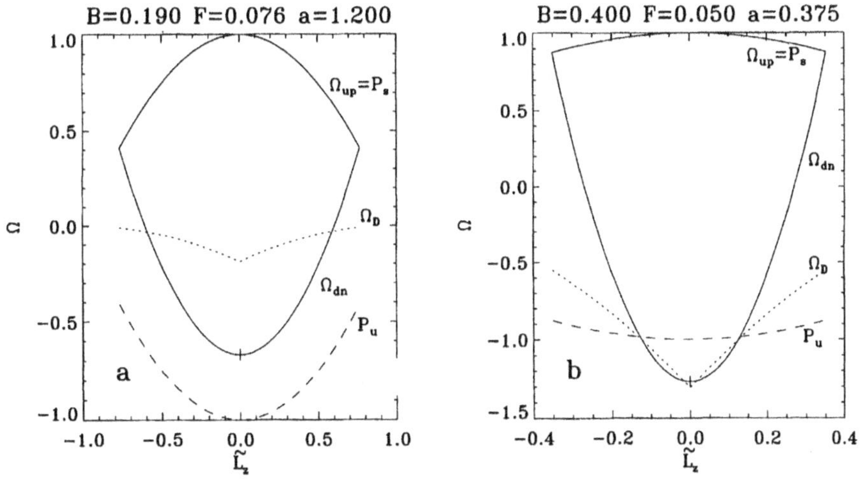

Figure 3. The accessible values of Ω are bounded by the two solid parabolas Ω_{up} and Ω_{dn}. The upper parabola coincides with the position of P_s, while the dashed line denotes the position of P_u. Orbits moving around P_s are separated from orbits moving around P_u by the dotted line Ω_D. Left panel: The Garching experiment. Right panel: for the parameters in Fig. 4. In this case P_u can be accessed, but it is unstable.

if $\gamma < -1$ and then P_u is unstable. The demarcation between orbits based on P_s and those based on P_u is given by

$$\Omega_D = \Omega(\tilde{A}_z = n - |\tilde{L}_z|) = \gamma(n - |\tilde{L}_z|)^2. \tag{28}$$

The left panel of Fig. 6 shows this for the parameters of the Garching experiment [8]. The periodic orbit P_u is never accessed whereas P_s is, and this constitutes the reason for the preferential excitation of planar orbits based on P_s. The right panel of Fig. 6 shows the same demarcation for the parameters of Fig. 4. This time, the chaotic region around P_u can be accessed in addition to P_s orbits which are, however, still accessed preferentially. P_u is unstable in this regime.

CONCLUSIONS

In summary, we have shown that the Hamiltonian of an electron in combined Coulomb and crossed external electric and magnetic fields can be usefully approximated by combining Normal Form theory and Lie algebra. Despite the enormous size of the parameter and phase space of the original problem, the special status assigned to some intricate periodic motions of the electron by recent experiments is connected to adiabatic invariants through our treatment. While deviations from the Coulomb potential require slight changes in our arguments due to finite core size, the success of the adiabatic invariants in explaining key features of the experiment shows their potential value for simplifying other aspects of this complex problem in the future.

ACKNOWLEDGMENTS

JvM is grateful to the Alexander von Humboldt Foundation for a Feodor-Lynen Research Fellowship. We thank the National Science Foundation for its support of this research.

REFERENCES

10

1. M. C. Gutzwiller, *Chaos in Classical and Quantum Mechanics* (Springer–Verlag, New York, 1990).
2. P. M. Koch and K. A. H. van Leeuwen, Phys. Rep. **255**, 289 (1995).
3. H. Hasegawa, M. Robnik, and G. Wunner, Prog. Theor. Phys. **98**, 198 (1989).
4. H. Friedrich and D. Wintgen, Phys. Rep. **183**, 37 (1989).
5. E. A. Solov'ev, Sov. Phys. JETP **58**, 63 (1983).
6. P. A. Braun and E. A. Solov'ev, Sov. Phys. JETP **59**, 38 (1984).
7. G. Wiebusch *et al.*, Phys. Rev. Lett. **62**, 2821 (1989).
8. G. Raithel, M. Fauth, and H. Walther, Phys. Rev. A **44**, 1898 (1991).
9. J. von Milczewski, G. H. F. Diercksen, and T. Uzer, Phys. Rev. Lett. **73**, 2428 (1994).
10. J. von Milczewski, G. H. F. Diercksen, and T. Uzer, Int. J. Bifurcation and Chaos **4**, 905 (1994).
11. J. von Milczewski, D. Farrelly, and T. Uzer, Phys. Rev. Lett. **78**, 2349 (1997).
12. D. Delande and J.-C. Gay, in *The Hydrogen Atom*, edited by G. Bassani, M. Inguscio, and T. Haensch (Springer–Verlag, Berlin, 1989), pp. 323–334.
13. J. von Milczewski, G. H. F. Diercksen, and T. Uzer, Phys. Rev. Lett. **76**, 2890 (1996).
14. J. A. Yeazell *et al.*, Phys. Rev. Lett. **70**, 2884 (1993).
15. J. Main and G. Wunner, Phys. Rev. Lett. **69**, 586 (1992).
16. T. Uzer and D. Farrelly, Phys. Rev. A **52**, R2501 (1995).
17. B. R. Johnson, J. D. Hirschfelder, and K. H. Yang, Rev. Mod. Phys. **55**, 109 (1983).
18. D. Farrelly, Phys. Lett. A **191**, 265 (1994).
19. M. M. Dignam and J. E. Sipe, Phys. Rev. B **45**, 6819 (1992).
20. P. Schmelcher, Phys. Rev. B **48**, 14642 (1993).
21. A. Bohr and B. R. Mottelson, *Nuclear Structure, Vol II* (Benjamin Reading, MA, 1975).
22. G. Mathys, Fundamentals of Cosmic Physics **13**, 143 (1989).
23. F. Mignard, Icarus **49**, 347 (1982).
24. G. Raithel, M. Fauth, and H. Walther, Phys. Rev. A **47**, 419 (1993).
25. G. Raithel and H. Walther, Phys. Rev. A **49**, 1646 (1994).
26. W. R. S. Garton and F. S. Tomkins, Ap. J. **158**, 839 (1969).
27. M. J. Gourlay, T. Uzer, and D. Farrelly, Phys. Rev. A **47**, 3113 (1993), erratum Phys. Rev. A **48**, 2508 (1993).
28. F. G. Gustavson, Astron. J. **71**, 670 (1966).
29. W. Dittrich and M. Reuter, *Classical and Quantum Dynamics* (Springer, Berlin, 1992).
30. M. J. Englefield, *Group Theory and the Coulomb Problem* (Wiley-Interscience, New York, 1972).
31. D. Farrelly *et al.*, Phys. Rev. A **45**, 4738 (1992).
32. A. Deprit, Cel. Mech. Dynam. Astr. **51**, 201 (1991).
33. E. A. Solov'ev, Sov. Phys. JETP **55**, 1017 (1982).
34. J. Main, Ph.D. thesis, Universität Bielefeld, 1991.
35. J. H. Eberly, J. Javanainen, and K. Rzazewski, Phys. Rep. **204**, 331 (1991).

HIGHLY EXCITED CHARGED TWO-BODY SYSTEMS IN A MAGNETIC FIELD: A PERTURBATION THEORETICAL APPROACH TO THE CLASSICAL DYNAMICS

W. Becken and P. Schmelcher

Theoretische Chemie
Physikalisch-Chemisches Institut
INF 253
69120 Heidelberg, FRG

INTRODUCTION AND PHEMOMENA

In the presence of an external magnetic field neutral as well as charged two-body systems possess an inherent two-body character, i.e. their center of mass (CM) and relative motion cannot be separated but are intimately coupled [1, 2].

Very recently interesting effects due to the coupling of the CM and electronic motions in one-electron atomic ions have been studied [3, 4, 5, 6, 7] by solving the classical equations of motion (eqms) for electronically highly excited atomic ions. Two major effects have been reported: the self-stabilization and self-ionization effects. For *regular motion* (Coulomb interaction dominates over magnetic forces) and *vanishing initial CM velocity the highly excited ion stabilizes on a cyclotron orbit.* Due to the above-mentioned coupling the CM motion gains a finite mean kinetic energy.

Observed phenomena for He^+ are :

- oscillating flow of energy between the CM and electronic degrees of freedom on three, by orders of magnitude different time scales T_1, T_2, T_3
- repetition of similar patterns of the CM trajectory on the time scales T_1, T_2, cyclotron orbit on an even larger time scale T_4

The Hamiltonian can be decomposed into the sum of an electronic part, a CM part describing a charged pseudoparticle in the external field and a coupling part. The effect of the self-stabilization is shown to occur for any initial conditions in the regular regime.

CONSTANTS OF MOTION AND CANONICAL TRANSFORMATION OF THE HAMILTONIAN

Our goal is to derive low-dimensional effective Hamiltonian equations of motion which describe the averaged classical motion on the different time scales. Furthermore, we will reveal approximate constants of motion.

Our starting point is the Hamiltonian for a charged two-body system of two interacting particles in a homogeneous external magnetic field ($\mathbf{A}(\mathbf{r}) = \frac{1}{2}\mathbf{B} \times \mathbf{r}$ with $\mathbf{B} = B(0,0,1)$).

$$H(\{\mathbf{p}^{(i)}\},\{\mathbf{r}^{(i)}\}) = \sum_{i=1}^{2} \frac{1}{2m_i}\left(\mathbf{p}^{(i)} - q_i\mathbf{A}(\mathbf{r}^{(i)})\right)^2 + V\left(|\mathbf{r}^{(1)} - \mathbf{r}^{(2)}|\right) \tag{1}$$

Due to the nonvanishing magnetic field the only remaining conserved quantities of this nonintegrable Hamiltonian are the energy H, the z-component of the total angular momentum[12] $L_z = \sum_{i=1}^{2} x^{(i)}p_y^{(i)} - y^{(i)}p_x^{(i)}$ and the pseudomomentum $\mathbf{K} = \sum_{i=1}^{2}\left(\mathbf{p}^{(i)} + \frac{q_i}{2}\mathbf{B} \times \mathbf{r}^{(i)}\right)$. A maximal set of commuting constants of motion is (H, $\mathbf{K}_\perp^2$, K_z, L_z). Our aim is to introduce these quantities by canonical transformations as canonical momenta and to transform the remaining degrees of freedom to variables which are well-suited for the perturbation theoretical approach with respect to the magnetic field.

We therefore perform the following steps:

Step 1: Transformation from laboratory system to CM ($\mathbf{R}_{cm}, \mathbf{P}_{cm}$) and relative ($\mathbf{r}, \mathbf{p}$) variables

Step 2: Power-Zienau-Wooley transformation [2, 9, 15, 16], leading to $H = H^{[1]} + H^{[2]} + H^{[3]}$,

$$H^{[1]} = \frac{1}{2M}\left(\mathbf{P}_{cm} - \frac{Q}{2}\mathbf{B} \times \mathbf{R}_{cm}\right)^2 \; ; \quad H^{[2]} = -\frac{\beta}{M}\left(\mathbf{P}_{cm} - \frac{Q}{2}\mathbf{B} \times \mathbf{R}_{cm}\right)\mathbf{B} \times \mathbf{r}$$
$$H^{[3]} = \frac{\mathbf{p}^2}{2\mu} + \gamma_L B l_z + \lambda B^2(x^2 + y^2) + V(\mathbf{r}) \tag{2}$$

where l_z is the z-component of the relative angular momentum ($\gamma_L = -\frac{q_1}{2m_1} - \frac{q_2}{2m_2} + \frac{Q}{2M}$, $\beta = \frac{q_1 m_2 - q_2 m_1}{M}$, $\lambda = \frac{1}{8}\left\{\frac{q_1^2}{m_1} + \frac{q_2^2}{m_2} - \frac{2Q}{M^2}(q_1 m_1 + q_2 m_2) + Q^2\frac{m_1^3 + m_2^3}{M^4}\right\}$). $H^{[1]}$ is the CM Hamiltonian for a free pseudoparticle with charge Q and mass M in a magnetic field. $H^{[2]}$ contains the coupling of the CM and relative degrees of freedom. $H^{[3]}$ is the purely electronic part.

Step 3: Introduction of suitable variables for the CM and the internal motion, separately

CM motion :

$$p_1 = \frac{\mathbf{K}_{cm\perp}^2}{2B} \; ; \; p_2 = L_{cm\,z} \; ; \; p_3 = K_{cm\,z} \tag{3}$$

relative motion: action-angle variables

$$I_1 = l_z = xp_y - yp_x \; ; \; I_2 = |\mathbf{l}| = \sqrt{(\mathbf{r} \times \mathbf{p})^2} \; ; \; I_3 = \sqrt{-\frac{\mu k^2}{2H^{[3]}(\mathbf{p},\mathbf{r})}} \tag{4}$$

where $\mu = \frac{m_1 m_2}{M}$ and $V(r) = -\frac{k}{r}$. The conjugated angles belonging to p_1, p_2, p_3 are q_1, q_2, q_3, and the conjugated angles belonging to I_1, I_2, I_3 are ϕ_1, ϕ_2, ϕ_3.

Step 4: Effects of the coupling

The relative and CM motion exchange angular momentum due to the coupling $\Rightarrow$ only the sum $L_{cm\,z} + l_z$ is conserved. Thus we introduce $\mathcal{L} = \frac{1}{2}(p_2 + I_1)$ and $\bar{\mathcal{L}} = \frac{1}{2}(p_2 - I_1)$ as canonical momenta with the corresponding coordinates $\xi = q_2 + \phi_1$ and $\bar{\xi} = q_2 - \phi_1$.

The final Hamiltonian reads ($\gamma_Z := -\frac{Q}{M}$):

$$H^{[1]}(\mathcal{L}, p_1, p_3\,;\ \bar{\mathcal{L}}) = \frac{p_3^2}{2M} + B\left(\gamma_Z(\mathcal{L}+\bar{\mathcal{L}}) + \frac{p_1}{M}\right)$$

$$H^{[2]}(p_1, \mathcal{L}; \bar{\mathcal{L}}, I_2, I_3, \bar{\xi}, \phi_2, \phi_3) = -\frac{\sqrt{2}\beta}{M} B^{3/2}\sqrt{p_1 - Q(\mathcal{L}+\bar{\mathcal{L}})}\ \times\ r\left(\sin\bar{\xi}\cos\varphi - \cos\bar{\xi}\sin\varphi\frac{I_1}{I_2}\right)$$

$$H^{[3]}(\mathcal{L}; \bar{\mathcal{L}}, I_2, I_3, \phi_2, \phi_3) = -\frac{\mu k^2}{2I_3^2} + \gamma_L B I_1 + \lambda B^2 r^2\left(1 - \sin^2\varphi\left(1 - \frac{I_1^2}{I_2^2}\right)\right) \quad (5)$$

with $I_1 = \mathcal{L} - \bar{\mathcal{L}}$ and $\varphi = \phi_2 + \chi$. The arguments of the Hamiltonian are the conserved momenta separated by a semicolon from the dynamical variables. The coordinates q_1, q_3, ξ are cyclic (The special case of infinite nuclear mass which corresponds to the Hamiltonian $H^{[3]}$ for $m_2 \to \infty$ has in detail been investigated in ref.[13]).

PERTURBATION THEORETICAL CONCEPTS

We rearrange the Hamiltonian H in powers of the perturbation parameter $\sigma := B^{1/2}$:

$$H(\mathcal{L}, p_1, p_3\,;\ \bar{\mathcal{L}}, I_2, I_3, \bar{\xi}, \phi_2, \phi_3) = H_0(p_3\,;\ I_3) + \frac{\sigma^2}{2!}H_2(\mathcal{L}, p_1\,;\ \bar{\mathcal{L}})$$
$$+\frac{\sigma^3}{3!}H_3(\mathcal{L}, p_1\,;\ \bar{\mathcal{L}}, I_2, I_3, \bar{\xi}, \phi_2, \phi_3) + \frac{\sigma^4}{4!}H_4(\mathcal{L}\,;\ \bar{\mathcal{L}}, I_2, I_3, \phi_2, \phi_3) \quad (6)$$

with
$$H_0(p_3\,; I_3) = \frac{p_3^2}{2M} - \frac{\mu k^2}{2I_3^2}\,; \quad H_2(\mathcal{L}, p_1\,;\ \bar{\mathcal{L}}) = 2\left(\gamma\mathcal{L} + \frac{p_1}{M}\right) + 2\bar{\gamma}\bar{\mathcal{L}} \quad (7)$$

$$H_3 = -6\frac{\sqrt{2}\beta}{M}\sqrt{p_1 - Q(\mathcal{L}+\bar{\mathcal{L}})} \times \frac{I_2^2}{\mu k}\,\frac{1}{1+\epsilon\cos\chi}\left(\sin\bar{\xi}\cos\varphi - \cos\bar{\xi}\sin\varphi\frac{I_1}{I_2}\right) \quad (8)$$

$$H_4(\mathcal{L}\,;\ \bar{\mathcal{L}}, I_2, I_3, \phi_2, \phi_3) = 24\lambda\,\frac{I_2^4}{\mu^2 k^2}\,\frac{1}{\left(1+\epsilon\cos\chi\right)^2} \times \left(1 - \sin^2\varphi\left(1 - \frac{I_1^2}{I_2^2}\right)\right) \quad (9)$$

with $I_1 = \mathcal{L} - \bar{\mathcal{L}}, \varphi = \phi_2 + \chi$. Furthermore $\gamma := \gamma_Z + \gamma_L$ and $\bar{\gamma} := \gamma_Z - \gamma_L$, $\omega := \sigma^2\gamma, \bar{\omega} := \sigma^2\bar{\gamma}$

Physical meaning of the H_i :

H_0: - CM energy parallel to the magnetic field
- pure Kepler Hamiltonian, fast Kepler-motion of the angle ϕ_3, time scale T_1

H_2: - cyclotron motion of the ion with charge Q and mass M
- Zeeman term of $H^{[3]}$, slower motion of the angle $\bar{\xi}$, time scale T_2

H_3: - coupling between the CM and electronic motion

H_4: - diamagnetic electronic part of $H^{[3]}$

Properties:
- H is integrable up to second order (σ^2)
- I_3 is the only dynamical variable in H_0
- H depends 2π-periodically on the dynamical angles $\phi_3, \bar{\xi}$

Perturbation theoretical approach to H:

1.) Series expansion due to the **explicit** and **implicit** dependence for all relevant quantities (Hamiltonian and variables) on σ.

2.): Oscillatory motions on different time scales T_i are superimposed on each other $\Rightarrow$ successive averaging of the eqms over the time scales T_i

Problem: To reveal the effects of H_4 which are masked by those of H_3 more profound mathematical methods have to be elaborated [15, 16].

THE PERTURBATION THEORETICAL METHOD

Characteristics of the starting situation:

1. The underlying exact Hamiltonian can be represented in a series (in our case up to fourth order) in a small parameter σ.
2. In the lowest order σ^0 only a single variable (ϕ_3) is not constant.
3. The total Hamiltonian is a periodic function of this variable.

These features are not as special as they might seem from a first glance. Indeed, if possible one would always choose as a zeroth order Hamiltonian an integrable one which meets in suitable action angle variables the first two of the above conditions.

A Expansion

Series for H (**explicit** dependence on σ) and formal solution:

$$H(\sigma;\mathbf{p},\mathbf{q}) = \sum_{l=0}^{\infty} \sigma^l/l! H_l(\mathbf{p},\mathbf{q}) \; ; \quad p_i(\sigma;t) = \sum_{r=0}^{\infty} \sigma^r/r!\, p_{ir}(t) \; ; \quad q_i \textit{ analogous} \tag{10}$$

The motion of the r-th order is determined by

$$\dot{p}_{ir} = -\frac{\partial}{\partial q_{i0}} H^{(r)}(\mathbf{p}_0,\mathbf{q}_0|\mathbf{p}_1,\ldots,\mathbf{p}_r,\mathbf{q}_1,\ldots,\mathbf{q}_r) \,, \tag{11}$$

and $\dot{q}_{ir} = \frac{\partial}{\partial p_{i0}} H^{(r)}$, where $H^{(r)}$ is defined as

$$H^{(r)}(\mathbf{p}_0,\mathbf{q}_0|\mathbf{p}_1,\ldots,\ldots,\mathbf{q}_r) = \sum_{l+s=r} \binom{r}{s} \sum\nolimits^{(s)}(\mathbf{p}_1,\ldots,\mathbf{p}_s,\mathbf{q}_1,\ldots,\mathbf{q}_s)\; H_l(\mathbf{p}_0,\mathbf{q}_0) \;, \tag{12}$$

$$\sum\nolimits^{(s)}(\mathbf{p}_1,\ldots,\ldots,\mathbf{q}_s) = \begin{cases} 1 & ; \quad s=0 \\ \sum_{\nu=1}^{s} 1/\nu! \sum_{\substack{r_\alpha\geq 1 \\ r_1+\ldots+r_\nu=s}} \left\{\frac{s!}{r_1!\ldots r_\nu!} D_{r_1}(\mathbf{p}_{r_1},\mathbf{q}_{r_1})\ldots \right. & \\ \qquad \left. \ldots D_{r_\nu}(\mathbf{p}_{r_\nu},\mathbf{q}_{r_\nu})\right\} & ; \quad s\neq 0 \end{cases} \,, \tag{13}$$

where $D_r(\mathbf{p}_r,\mathbf{q}_r) = \left(p_{ir}\cdot\frac{\partial}{\partial p_{i0}} + q_{ir}\cdot\frac{\partial}{\partial q_{i0}}\right)$ for $r \geq 1$.

Problem: Coupling of r-th order of $\dot{\mathbf{p}}$, $\dot{\mathbf{q}}$ to the same order in the argument of $H^{(r)}$.

Simplifications: Due to the property 2 of the Hamiltonian H its components $H^{(r)}$ depend only on the first $(r-1)$ orders for all degrees of freedom except for say (p_0,q_0) (in our case (I_3,ϕ_3)). Due to the property 3 the behaviour of q_0 can be averaged out.

B Averaging

Represent H (and also $H^{(r)}$) as a Fourier series in q_0 and perform the time integration of the eqms over one time cycle of q_0. The result is a system of difference equations which can, under appropriate conditions, be replaced by a system of differential equations. These equations can, up to a certain order $\tilde{r}$ depending on the special structure of the given Hamiltonian H, be considered as the eqms of one **effective Hamiltonian** H_{eff}.

Important property: H_{eff} does not depend on the averaged quantity $q_0 \Rightarrow p_0$ is an **approximate** (up to $\tilde{r}$) **constant of motion** $\Rightarrow$ **dimensional reduction introduced by averaging**

RESULTS AND DISCUSSION

We apply our method up to the order $r = 4$: The eqms are generated by:

$$\begin{aligned} H^{(0)}(p_3\,;\,I_{30}) &= H_0(p_3\,;\,I_{30})\,; \quad H^{(1)} = 0\,; \quad H^{(2)}(\mathcal{L},p_1\,;\,\bar{\mathcal{L}}_0) = H_2(\mathcal{L},p_1\,;\,\bar{\mathcal{L}}_0) \\ H^{(3)}(\mathcal{L},p_1\,;\,...|I_{33}) &= H_3(\mathcal{L},p_1\,;\,...\bar{\mathcal{L}}_0,I_{20},I_{30},\bar{\xi}_0,\phi_{20},\phi_{30}) + I_{33}\frac{\partial}{\partial I_{30}}H_0(p_3\,;\,I_{30}) \\ H^{(4)}(\mathcal{L};...|I_{33}) &= H_4(\mathcal{L};\bar{\mathcal{L}}_0,I_{20},I_{30},\phi_{20},\phi_{30}) + I_{34}\frac{\partial}{I_{30}}H_0(p_3;I_{30}) \end{aligned} \tag{14}$$

The eqms (14) for $r = 0,1,2$ repeat the fact that the time scales T_1,T_2 belong to the motion of ϕ_{30} and $\bar{\xi}_2$, respectively. But now we can obtain more information about the higher orders by averaging. As described above, we obtain an effective Hamiltonian

$$\begin{aligned} H_{eff}(\sigma;\mathcal{L},p_1,I_3\,;\,\bar{\mathcal{L}},I_2,\bar{\xi},\phi_2) &:= \frac{\sigma^2}{2!}H_{20}(\mathcal{L},p_1\,;\,\bar{\mathcal{L}}) \\ +\frac{\sigma^3}{3!}H_{30}(\mathcal{L},p_1,I_3\,;\,\bar{\mathcal{L}},I_2,\bar{\xi},\phi_2) &+ \frac{\sigma^4}{4!}H_{40}(\mathcal{L},I_3\,;\,\bar{\mathcal{L}},I_2,\phi_2) \end{aligned} \tag{15}$$

(H_{20}, H_{30}, H_{40} are the zeroth Fourier components of H_2, H_3, H_4 with respect to ϕ_3);

Property:

- I_3 is up to order $r = 4$ an exact constant of motion and in general an **approximate constant of motion** $\Rightarrow$ Only $(\bar{\mathcal{L}}, I_2, \bar{\xi}, \phi_2)$ are dynamical variables $\Rightarrow$ **reduction of the number of degrees of freedom from 3 to 2**

Though H_{eff} is simpler than H, it does not yet show directly the system's behaviour in the orders $r > 2$.

- Idea: Repeat the averaging procedure but now with respect to the oscillations of $\bar{\xi}$ on the second larger time scale T_2.
- Preparation: Rescaling of time $\vartheta := \sigma^2 \cdot T$ in order to meet the features 1,2,3 of the starting situation.
- Result of the second averaging process: a second effective Hamiltonian :

$$\begin{aligned} H_{effeff}(\sigma;I_2,\phi_2) \;=\; & \frac{\sigma^2}{2!}\frac{2!}{4}\frac{\lambda}{\mu^2k^2}I_3^4\frac{1}{(I_2/I_3)^2} \times \\ \Bigg\{ & \left\{-3\eta_2(I_2/I_3)^4 + \left(5\eta_1 - 3\eta_2\frac{(\mathcal{L}-\bar{\mathcal{L}})^2}{I_3^2}\right)(I_2/I_3)^2 + 5\eta_1\frac{(\mathcal{L}-\bar{\mathcal{L}})^2}{I_3^2}\right\} \\ & + \left\{-5\eta_1(I_2/I_3)^4 + 5\eta_1\left(1+\frac{(\mathcal{L}-\bar{\mathcal{L}})^2}{I_3^2}\right)(I_2/I_3)^2 - 5\eta_1\frac{(\mathcal{L}-\bar{\mathcal{L}})^2}{I_3^2}\right\}\cos 2\phi_2 \\ & - 20\,\delta\frac{(\mathcal{L}-\bar{\mathcal{L}})}{I_3^2}(I_2/I_3)^2\left(\frac{p_1}{e} - \frac{Q}{e}(\mathcal{L}+\bar{\mathcal{L}})\right)\Bigg\} \end{aligned} \tag{16}$$

with $\delta = \frac{72}{5}\frac{\mu^2}{M^2}$, $\eta_1 = (1 + \delta\frac{Q}{e})$ and $3\eta_2 = (3 + 5\,\delta\frac{Q}{e})$.

Properties:

- $\bar{\mathcal{L}}$ is up to order $r = 4$ an exact constant of motion and in general an **approximate constant of motion** $\Rightarrow$ Only (I_2, ϕ_2) are dynamical variables $\Rightarrow$ **reduction of the number of degrees of freedom from 2 to 1** $\Rightarrow H_{eff_{eff}}$ is **integrable**
- For $\delta \to 0$ $(M \to \infty)$ $H_{eff_{eff}}$ is identical with the one-particle Hamiltonian in ref.(13). $H_{eff_{eff}}$ has, apart from the term containing -20δ, a very similar structure to that one-particle Hamiltonian also for finite δ

Analogously to the case $\delta = 0$ the solutions of our eqms are librators or rotators. Their time scales depend strongly on the initial conditions but possess a strict lower bound $T_3 = \frac{\pi\kappa}{2\sqrt{5}}\frac{\mu^2 k^2}{\lambda} \cdot \frac{1}{B^2 I_3^3}$ (κ is a factor of order 1 measuring the failure of δ to be zero) which occurs as a new third time scale additionally to T_1 and T_2.

Considering the obtained results in Cartesian coordinates by using our canonical transformation equations and also the relation $\dot{\mathbf{R}}_{cm} = \frac{1}{M}\mathbf{\Pi}_{cm} - \frac{\beta}{M}\mathbf{B} \times \mathbf{r}$ we find a fourth time scale T_4 corresponding to a CM motion on a circular orbit with radius

$$R_Z = \left| \frac{M}{QB^2}\mathbf{B} \times \dot{\mathbf{R}}_{cm}(0) - \frac{\beta}{Q}\mathbf{r}_\perp(0) \right| \quad , \tag{17}$$

and cyclotron frequency $\omega_Z := \gamma_Z \sigma^2$ which now confirms the self-stabilization effect rigorously. (Remark: R_Z depends on the initial relative distance due to the coupling!) The order of magnitude of those four time scales is shown in Fig.1.

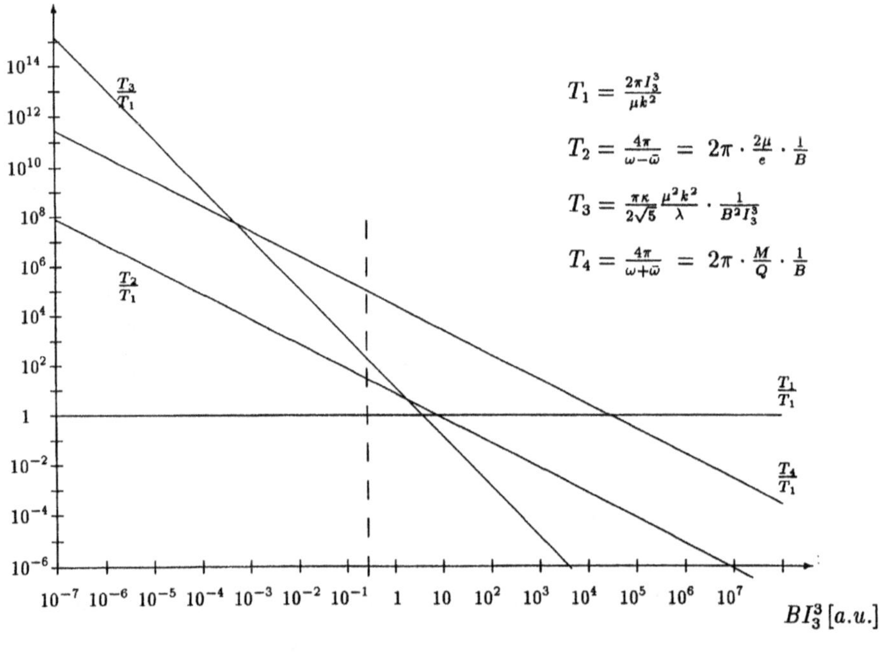

Figure 1. Representation of the four time scales $T_1, ..., T_4$ in units of T_1, depending on the critical parameter BI_3^3; its value $BI_3^3 = 0,278\,a.u.$ for the calculations of Schmelcher and Cederbaum [11] is marked with a dashed line and is located in the regular regime $BI_3^3 \lesssim 1$

SUMMARY AND CONCLUSIONS

- The classical behaviour of charged two-particle systems, in particular He^+, in an external field B is for small B regular and can be treated by perturbation theory.
- All the constants of motion except H (i.e.$\mathbf{K}_\perp^2, K_z, L_z$) can be introduced as canonical momenta.
- H can be decomposed into a sum of an electronic part, a CM part for a charged pseudoparticle in an external field and a coupling Hamiltonian.
- The phenomena observed in the regular regime were fast Kepler-oscillations, Larmor precession, modulation of the CM-energy and a cyclotron rotation of the CM on four different time scales T_1, T_2, T_3, T_4, respectively.
- By averaging the eqms succesively over T_1 and T_2 we obtain two approximative constants of motion (I_3,$\mathcal{L}$) and arrive at a one-dimensional integrable effective Hamiltonian.
- The oscillations on the time scale T_1 are described by the unaveraged Hamiltonian eqms, the modulations on the time scale T_2 by the once averaged ones. The behaviour on the time scale T_3 is best described by the twice averaged and integrable effective Hamiltonian. The motion on the time scale T_4 does not have to do anything with the averaging procedure but enters when transforming back to Cartesian coordinates.
- All the observed phenomena and in particular the self-stabilization effect can be reproduced by our perturbation theory.
- Up to order B^2 no further phenomena are to be expected

REFERENCES

1. J.E.Avron, I.W.Herbst and B.Simon, *Ann.Phys.(NY)* **114**, 431 (1978)
2. B.R.Johnson, J.O.Hirschfelder and K.H.Yang, *Rev.Mod.Phys.* **55**, 109 (1983)
3. P.Schmelcher and L.S.Cederbaum, *Phys.Lett.* **A164**, 305 (1992); *Zeitschr.f.Phys.***D24**, 311 (1992) and P.Schmelcher, *J.Phys.* **B25**, 2697 (1992).
4. D.Baye, N.Clerbaux and M.Vincke, *Phys.Lett.* **A166**, 135 (1992)
5. I.Dzyaloshinskii, *Phys.Lett.***A165**, 69 (1992)
6. P.Schmelcher and L.S.Cederbaum, *Chem.Phys.Lett.***208**, 548 (1993); Phys.Rev.**A47**,2634 (1993) and P.Schmelcher, *Phys.Rev.***B48**, 14642 (1993)
7. O.Dippel, P.Schmelcher and L.S.Cederbaum, *Phys.Rev.* **A49**, 4415 (1994)
8. D.Baye and M.Vincke, *Phys.Rev.***A42**, 31 (1990)
9. P.Schmelcher and L.S.Cederbaum, *Phys.Rev.* **A43**, 287 (1991)
10. P.Schmelcher, L.S.Cederbaum and U.Kappes, Kluwer Academic Publishers, Conceptual Trends in Quantum Chemistry, 1-51 (1994)
11. P.Schmelcher and L.S.Cederbaum, *Phys.Rev.Lett.* **74**, 662 (1995) and P.Schmelcher, *Phys.Rev.***A52**, 130 (1995).
12. T.P.Mitchell, *J.Math.Phys.***22**, 1948 (1981)
13. J.B.Delos, S.K.Knudson, D.W.Noid, *Phys.Rev.* **A28**, 7 (1983)
14. A.S.Dickinson and J.M.Patterson, *J.Phys.***A19**, 1811 (1985)
15. W.Becken, Diplomarbeit 1995, Theoretische Chemie, Institut für Physikalische Chemie der Universität Heidelberg.
16. W.Becken, P.Schmelcher, *Phys. Rev. A*, **54**, 4868 (1996)

ANALYSIS OF QUANTUM SPECTRA BY HARMONIC INVERSION

J. Main,[1] G. Wunner,[1] V. A. Mandelshtam,[2] and H. S. Taylor[2]

[1]Institut für Theoretische Physik I, Ruhr-Universität Bochum, D-44780 Bochum, Germany

[2]Department of Chemistry, University of Southern California, Los Angeles, California 90089

INTRODUCTION

Semiclassical *periodic orbit* (Gutzwiller, 1990) and *closed orbit* (Du and Delos, 1988; Bogomolny, 1989) theory have become the key for the interpretation of quantum spectra of classically chaotic systems. The semiclassical spectra at least in low resolution are given as the sum of a smooth background and a superposition of modulations whose amplitudes, frequencies, and phases are solely determined by the closed or periodic orbits of the classical system. For the interpretation of quantum spectra in terms of classical orbits it is therefore most natural to obtain the recurrence spectrum by Fourier transforming the energy spectrum into the time domain. Each closed or periodic orbit should show up as a sharp δ-peak at the recurrence time (period), provided, first, the classical recurrence times do not change within the whole range of the spectrum and, second, the Fourier integral is calculated along an infinite energy range. The first condition can be fulfilled in systems possessing a classical scaling property in that the classical dynamics does not change if the energy and an external scaling parameter are simultaneously varied. An example is the hydrogen atom in crossed magnetic and electric fields given by the Hamiltonian [in atomic units, $\gamma = B/(2.35 \times 10^5\,\mathrm{T})$, $f = F/(5.14 \times 10^9\,\mathrm{V/cm})$]

$$H = \frac{1}{2}\mathbf{p}^2 - \frac{1}{r} + \frac{1}{2}\gamma L_z + \frac{1}{8}\gamma^2(x^2+y^2) + fx\ . \tag{1}$$

The classical dynamics does not depend on three parameters, the energy E and the two field strengths γ and f but solely on the scaled energy $\tilde{E} = E\gamma^{-2/3}$ and the scaled field strength $\tilde{F} = f\gamma^{-4/3}$. Applying the method of *scaled energy spectroscopy* (Main et al., 1994) the quantum spectra can be created at a constant scaled energy as a function of the scaling parameter $w \equiv \gamma^{-1/3}$. The scaled spectra can be Fourier transformed along arbitrarily long ranges of w to generate Fourier transform recurrence spectra of in principle arbitrarily high resolution. By varying the scaled energy a direct comparison of the quantum recurrence spectra with the bifurcation diagram of the underlying classical system is possible (Main et al., 1994). However, the second condition is never fulfilled in

Atoms and Molecules in Strong External Fields
Edited by Schmelcher and Schweizer, Plenum Press, New York, 1998

practice, i.e., the length of quantum spectra is always restricted either by experimental limitations or, in theoretical calculations, by the growing dimension of the Hamilton matrix which has to be diagonalized numerically. The given length of a quantum spectrum determines the resolution of the quantum recurrence spectrum due to the uncertainty principle, $\Delta E \cdot \Delta T \sim \hbar$, when the conventional Fourier transform is used. Only those closed or periodic orbits can be clearly identified quantum mechanically which appear as isolated non-overlapping peaks in the quantum recurrence spectra. This is especially not the case for orbits which undergo bifurcations at energies close to the bifurcation point.

Here we present a method to calculate *high resolution recurrence spectra* from experimental or theoretical quantum spectra of *finite length* thereby circumventing the restrictions of the uncertainty principle to the resolution of finite range Fourier transforms (Main et al., 1997a). It allows, e.g., to identify real orbits with nearly degenerate periods and to detect complex "ghost" orbits which can be of particular importance in the vicinity of bifurcations (Kuś et al., 1993; Main and Wunner, 1997).

HARMONIC INVERSION OF QUANTUM SPECTRA

According to *periodic orbit* theory (Gutzwiller, 1990) the semiclassical density of states can be written as the sum of a smooth background $\varrho_0(w)$ and oscillatory modulations induced by the periodic orbits,

$$\varrho(w) = \varrho_0(w) + \mathrm{Im} \sum_k A_k e^{-2\pi i \tilde{S}_k w} \, . \tag{2}$$

The amplitude A_k and scaled action $\tilde{S}_k = (2\pi w)^{-1} \oint_k \mathbf{p} d\mathbf{r}$ of periodic orbit k are obtained from classical calculations and are in general complex quantities. The amplitude A_k contains the phase information determined by the Maslov index of the orbit. A non-real action S_k indicates a "ghost" orbit (Kuś et al., 1993; Main and Wunner, 1997) which exists in the complex continuation of the classical phase space. A similar formula as Eq. (2) also holds in *closed orbit* theory (Du and Delos, 1988; Bogomolny, 1989) to calculate semiclassically the photoabsorption cross section for dipole transitions from an energetically low lying initial state. Oscillatory modulations of the cross section are induced by closed orbits of the electron starting at and returning back to the nucleus. (In closed orbit theory the oscillatory terms of the photoabsorption cross section are proportional to w^α with $\alpha = -1/2$ for the hydrogen atom in a magnetic field and $\alpha = -1$ for the atom in crossed magnetic and electric fields. To obtain the functional form of Eq. 2 we multiply the cross section by $w^{-\alpha}$.)

As mentioned above the low resolution nature of the Fourier transform for a short signal does not permit one to resolve closely spaced peaks, neither does it allow to distinguish between real and ghost orbits. Instead of using the standard Fourier analysis, to extract the amplitudes and actions we essentially fit a finite range of the quantum spectrum by the semiclassical expression (2) with unknown and in general complex parameters A_k and $\tilde{S}_k$. The problem of fitting a "signal" $\varrho(w)$ to the functional form (2) is known as *harmonic inversion* with a large variety of applications in various fields (Marple, 1987). As a numerical technique for the harmonic inversion of a signal, i.e. a quantum spectrum, we apply the method of filter-diagonalization (Wall and Neuhauser, 1995) which allows to extract the frequencies $\tilde{S}_k$ in any given interval of interest. The method was recently improved in that it allows to significantly reduce the required length of the analyzed signal (Mandelshtam and Taylor, 1997a). The method has also been applied successfully for periodic orbit quantization, i.e., to calculate individual eigenenergies and resonances of classically chaotic systems by analytic continuation of Gutzwiller's divergent periodic orbit sum (Main et al., 1997b).

Operationally, harmonic inversion is performed by setting up a small generalized eigenvalue problem. To extract the frequencies w_k and amplitudes d_k in the small interval $[w_{\min}, w_{\max}]$ from a given signal c_n, which is assumed to have the form

$$c_n \equiv C(n\tau) = \sum_k d_k e^{-in\tau w_k} \;, \quad n = 0, 1, 2, \ldots N, \tag{3}$$

one has to carry out the following steps.
(i) Define a small grid $\{\varphi_j\}$, $j = 1, 2, \ldots, N_{win}$, in the interval $[w_{\min}, w_{\max}]$. The density of the grid has to be bigger than the density of the frequencies w_k in this interval. This grid corresponds to the set of N_{win} complex numbers $z_j \equiv e^{-i\varphi_j}$ on the unit circle.
(ii) Evaluate three complex symmetric matrices $\mathbf{U}^{(p)}$ of the size $N_{win} \times N_{win}$ with, e.g., $p = 0, 1, 2$, using

$$\begin{aligned} U^{(p)}_{jj'} &= \frac{1}{z_j - z_{j'}} \Bigg[z_j \sum_{l=0}^{M} c_{l+p} z_{j'}^{-l} - z_j^{-M} \sum_{l=M+1}^{2M} c_{l+p} z_{j'}^{M-l+1} \\ &\quad - z_{j'} \sum_{l=0}^{M} c_{l+p} z_j^{-l} + z_{j'}^{-M} \sum_{l=M+1}^{2M} c_{l+p} z_j^{M-l+1} \Bigg] \;, \; j \neq j' \\ U^{(p)}_{jj} &= \sum_{l=0}^{2M} (M - |M - l| + 1) c_{l+p} z_j^{-l} . \end{aligned} \tag{4}$$

(Note, that evaluation of $\mathbf{U}^{(p)}$ requires knowledge of c_n for $n = p, p{+}1, \ldots, N = 2M{+}p$.)
(iii) Solve the small generalized eigenvalue problem:

$$\mathbf{U}^{(p)} \mathbf{B}_k = u_k{}^p \mathbf{U}^{(0)} \mathbf{B}_k . \tag{5}$$

for the eigenvalues $u_k^{\,p} = \exp\left[-ipw_k/\tau\right]$ and eigenvectors $\mathbf{B}_k$.
(iv) Each value of p yields a set of frequencies w_k and amplitudes,

$$d_k = \left(\sum_j B_{jk} \sum_{n=0}^{M} c_n z_j^{-n} \right)^2 . \tag{6}$$

The converged w_k and d_k should not depend on p. This condition allows to identify spurious or non-converged frequencies by comparing the results with different values of p (e.g., with $p = 1$ and $p = 2$). For more details of the harmonic inversion technique see Mandelshtam and Taylor (1997b).

The complex actions $\tilde{S}_k^{\rm qm}$ in the chosen spectral domain and amplitudes $A_k^{\rm qm}$ are obtained from the resulting eigenvalues and eigenvectors of Eq. (5). Thus, the recurrence spectrum is effectively discretized, the number of terms being the number of eigenvalues in the spectral domain. This method is a variational one (as opposed to the Fourier transform) and therefore has practically an infinite resolution once the amount of information contained in the signal $\varrho(w)$ is greater than the total number of unknowns A_k and S_k. The minimal length of the quantum spectra required for the convergence of individual closed or periodic orbit parameters depends linearly on the mean density of orbits at that period, i.e., for a given finite length of the quantum spectra the region of the high resolution recurrence spectra is limited by the proliferation of classical orbits with increasing period.

In quantum calculations the bound state spectrum is given as a sum of δ-functions. In order to obtain $\varrho(w)$ on an evenly spaced grid the spectrum is regularized by convoluting it with a narrow Gaussian function having the width $\sigma \ll 1/\tilde{S}_{\max}$, where

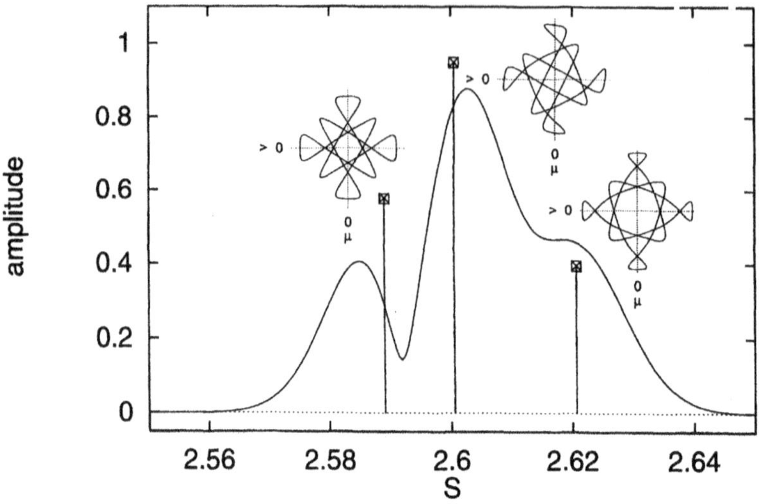

Figure 1. Recurrence spectrum for the density of states of the hydrogen atom in a magnetic field at scaled energy $\tilde{E} = E\gamma^{-2/3} = -0.1$. Smooth line: Conventional Fourier transform. Solid stick spectrum and crosses: High resolution quantum recurrence spectrum. Dashed sticks and squares: Recurrence spectrum from semiclassical *periodic orbit* theory. The recurrence peaks are identified by periodic orbits (drawn in semiparabolical coordinates).

$S_{\max}$ is the scaled action of the longest orbit of interest. The quantum spectrum reads $\varrho(w) = w^{-\alpha} \sum_n D_n \, \delta_\sigma(w - w_n)$, where the w_n are the eigenvalues of the scaling parameter w and the weights D_n are, e.g., the absolute value squared of transition matrix elements for excitations from a given initial state. The unweighted density of states is obtained with $\alpha = 0$ and $D_n = 1$.

We now demonstrate on the hydrogen atom in a magnetic field and in crossed magnetic and electric fields that physically interesting effects can be revealed in the quantum spectra by application of the harmonic inversion technique. We do not present the complete high resolution recurrence spectra here but concentrate on the physically interesting parts which can not be resolved by conventional Fourier transform.

Precision check of the periodic orbit theory

As a first example we apply the method to the density of states for the hydrogen atom in a magnetic field at scaled energy $\tilde{E} = -0.1$. At this energy the classical dynamics is completely chaotic and all periodic orbits are unstable. We calculated 9715 states in the region $w < 140$ by numerical diagonalization of the Hamiltonian matrix in a complete basis set. For details of the quantum calculations see, e.g., Main and Wunner (1994). The quantum density of states was analyzed both by conventional Fourier transform and by the high resolution method (Mandelshtam and Taylor, 1997). To get rid of unphysical sidepeaks in the conventional Fourier transform we multiply the spectra with a Gaussian window which is chosen in accordance with the total length of the quantum spectrum. A comparison between the Fourier transform, the high resolution quantum recurrence spectrum, and the semiclassical recurrence spectrum is presented in Fig. 1 around scaled action $\tilde{S} = 2.6$. The smooth line is the absolute value of the conventional Fourier transform. Its shape suggests the existence of at least three periodic orbits but obviously the recurrence spectrum is not completely resolved. The results of the high resolution spectral analysis are presented as sticks at the positions defined by the scaled actions $\tilde{S}_k^{\rm qm}$ with peak heights $|A_k^{\rm qm}|$. Note that the positions of the peaks are considerably shifted with respect to the maxima of the conventional Fourier transform. To compare the quantum recurrence spectrum with Gutzwiller's periodic orbit theory we calculated the classical scaled actions $\tilde{S}_k^{\rm cl}$ of the periodic orbits and their semiclassical amplitudes $|A_k^{\rm cl}| = \tilde{S}_k^{\rm cl}/\sqrt{|\det(M_k - I)|}$, with M_k being the monodromy matrix of orbit k. The semiclassical results are presented as dashed sticks and squares in Fig. 1. For illustration the shapes of periodic orbits are also shown (in semiparabolical coordinates $\mu = \sqrt{r+z}$, $\nu = \sqrt{r-z}$). For these three orbits the agreement between the semiclassical and the high resolution quantum recurrence spectrum is nearly perfect, deviations are within the stickwidths. For example, for the first peak in Fig. 1 we find $(\tilde{S}^{\rm cl} - \tilde{S}^{\rm qm})/\tilde{S}^{\rm qm} = 8 \times 10^{-6}$ and $(|A^{\rm qm}| - |A^{\rm cl}|)/|A^{\rm qm}| = 0.0043$. The deviation between the classical and quantum amplitude might be related to higher order $\hbar$ corrections of the semiclassical trace formula. In fact, Gutzwiller's trace formula is only the first term of the semiclassical series and the high resolution recurrence spectra might be a sensitive tool to identify these corrections quantitatively in quantum spectra.

Uncovering the "hidden" ghost orbits

In systems where classical orbits undergo bifurcations complex "ghost" orbits may be important for a complete understanding of quantum spectra, as was shown for the kicked top (Kuś et al., 1993) and for the hydrogen atom in a magnetic field (Main and Wunner, 1997). The contribution of a ghost orbit decreases exponentially with the energy distance from the bifurcation point and therefore ghosts can be detected in quantum spectra by conventional Fourier transform only if they are isolated from the other orbits, e.g., as a prebifurcation ghost of a saddle node bifurcation (Kuś et al., 1993). For other types of bifurcations, e.g., for the bifurcation of the perpendicular orbit of the hydrogen atom in a magnetic field ghost orbits are hidden behind a strong recurrence peak of a real orbit with nearly the same classical action (Main and Wunner, 1997). Such "hidden" ghosts can now be uncovered in high resolution quantum recurrence spectra. The key point is that the harmonic inversion of quantum spectra supplies complex frequencies $\tilde{S}_k^{\rm qm}$ in Eq. (2) which can be interpreted as complex actions $\tilde{S}_k^{\rm cl}$ of ghost orbits. In Fig. 2 we have analyzed the photoabsorption spectrum of the hydrogen atom in a magnetic field at scaled energy $\tilde{E} = -0.35$ in the region $w < 100$ (2823 transitions from the initial state $|2p0\rangle$ to final states with $m^{\pi_z} = 0^+$). The conventional Fourier transform (smooth line) has a maximum at $\tilde{S} = 1.895$ which is roughly twice the period of the perpendicular orbit but does not give any hint on the existence of a ghost orbit. The high resolution spectral analysis uncovers one real and two complex actions $\tilde{S}_k^{\rm qm}$ which are compared to the classical actions $\tilde{S}_k^{\rm cl}$ of the perpendicular orbit and a complex ghost (see crosses and squares in Fig. 2b respectively). For the frequency with the largest imaginary part no classical analogue has been found but its amplitude (see Fig. 2a) is rather small. For the two stronger contributions the agreement between quantum and classical recurrence spectrum is remarkably good. The shapes of the real and complex closed orbits are presented as insets in Fig. 2a (in semiparabolical coordinates μ, ν). The real part of the ghost orbit (solid line) is similar to the "Pacman" orbit R_2^1 (in the notation of Main and Wunner, 1997) which is created as a real orbit at much higher energy $\tilde{E} = -0.317$. This is the first observation of a hidden ghost orbit in quantum mechanical spectra.

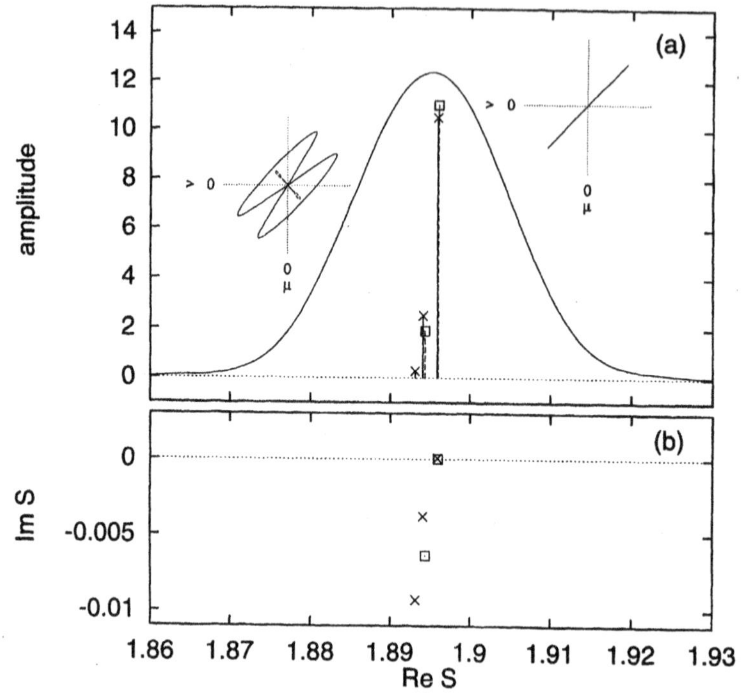

Figure 2. (a) Recurrence spectrum for the photoabsorption cross section of the hydrogen atom in a magnetic field at scaled energy $\tilde{E} = E\gamma^{-2/3} = -0.35$. Transition $|2p0\rangle \rightarrow |m^{\pi_z} = 0^+\rangle$. Smooth line: Conventional Fourier transform. Solid stick spectrum and crosses: High resolution quantum recurrence spectrum. Dashed sticks and squares: Recurrence spectrum from semiclassical *closed orbit* theory. The two strongest recurrence peaks are identified by a real and complex ghost orbit. The solid and dashed lines are the real and imaginary part (in semiparabolical coordinates). (b) Complex actions. Crosses and squares are the quantum and classical results respectively.

Symmetry breaking in crossed fields

The hydrogen atom in a magnetic field has a cylindrical symmetry around the magnetic-field axis, i.e., closed and periodic orbits do not depend on the azimuthal starting angle φ. This symmetry is broken in crossed magnetic and electric fields. Out of a manifold of closed orbits only two closed orbits with slightly different classical actions survive in the crossed fields (Neumann et al., 1997). As a result each recurrence peak splits into two. We investigate the symmetry breaking in the crossed field atom at scaled energy $\tilde{E} = -0.5$ and scaled field strength $\tilde{F} = 0.02$ where we calculated the photoabsorption spectrum (transitions from the initial state $|2p0\rangle$ to the final states with even z-parity) up to $w = 50$. Without electric field ($\tilde{F} = 0$) two recurrences occur at $\tilde{S} = 0.872$ for the perpendicular orbit and at $\tilde{S} = 1$ for the parallel orbit. With the weak electric field $\tilde{F} = 0.02$ the parallel orbit is not significantly affected but the perpendicular orbit splits into two closed orbits with slightly different classical actions $\tilde{S}^{\rm cl} = 0.864$ and $\tilde{S}^{\rm cl} = 0.881$. Their shapes are presented as insets in Fig. 3. As can be seen in the figure the two closed orbit contributions are not resolved by the conventional Fourier transform (smooth line) but they are clearly separated by the high resolution spectral analysis in agreement with the classical recurrence spectrum.

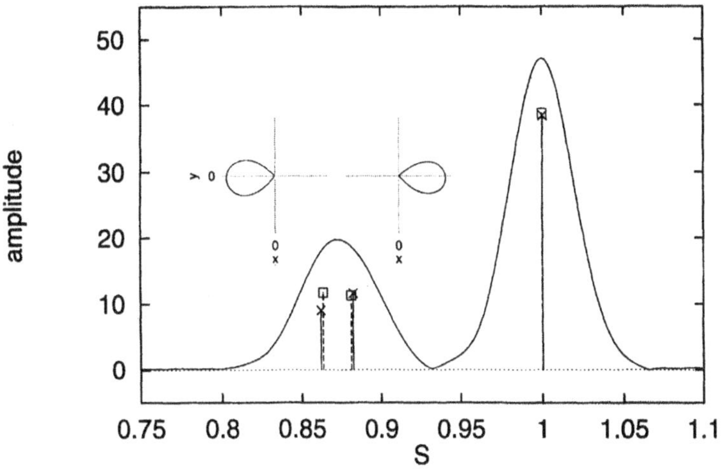

Figure 3. Recurrence spectrum for the photoabsorption cross section of the hydrogen atom in crossed magnetic and electric fields at scaled energy $\tilde{E} = E\gamma^{-2/3} = -0.5$ and field strength $\tilde{F} = F\gamma^{-4/3} = 0.02$. Transition $|2p0\rangle \rightarrow |\pi_z = +1\rangle$. Smooth line: Conventional Fourier transform. Solid stick spectrum and crosses: High resolution quantum recurrence spectrum. Dashed sticks and squares: Recurrence spectrum from semiclassical *closed orbit* theory. The two recurrence peaks around $\tilde{S} = 0.87$ are identified by closed orbits in the (x, y) plane.

CONCLUSION

We have shown that interesting physical phenomena can be revealed in high resolution quantum recurrence spectra by application of the harmonic inversion technique, thereby circumventing the restrictions imposed by the uncertainty principle of the conventional Fourier transform. The method allows, e.g., to test the validity of semiclassical theories, to identify hidden ghost orbits in the quantum spectra, and to observe the symmetry breaking in the spectra of the hydrogen atom in crossed magnetic and electric fields. The analysis has been demonstrated here on theoretically calculated quantum spectra but can be applied to experimental spectra as well.

REFERENCES

Bogomolny, E. B., 1989, *Sov. Phys. JETP* 69:275.
Du, M. L. and Delos, J. B., 1988, *Phys. Rev.* A 38:1896 and 38:1913.
Gutzwiller, M. C., 1990, "Chaos in Classical and Quantum Mechanics," Springer, New York.
Kuś, M., Haake, F., and Delande, D., 1993, *Phys. Rev. Lett.* 71:2167.
Main, J., Wiebusch, G., Welge, K. H., Shaw, J., and Delos, J. B., 1994, *Phys. Rev.* A 49:847.
Main, J. and Wunner, G., 1994, *J. Phys.* B 27:2835.
Main, J., Mandelshtam, V. A. and Taylor, H. S., 1997a, *Phys. Rev. Lett.* 78:4351.
Main, J., Mandelshtam, V. A. and Taylor, H. S., 1997b, *Phys. Rev. Lett.*, in press.
Main, J. and Wunner, G., 1997, *Phys. Rev.* A 55:1743.
Mandelshtam, V. A. and Taylor, H. S., 1997a, *Phys. Rev. Lett.* 78:3274.
Mandelshtam, V. A. and Taylor, H. S., 1997b, *J. Chem. Phys.*, submitted.
Marple, S., Jr., 1987, "Digital Spectral Analysis with Applications," Prentice-Hall, Englewood Cliffs.
Neumann, C., Ubert, R., Freund, S., Flöthmann, E., Sheehy, B., Welge, K. H., Haggerty, M. R., and Delos, J. B., 1997, *Phys. Rev. Lett.*, in press.
Wall, M. R. and Neuhauser, D., 1995, *J. Chem. Phys.* 102:8011.

ATOMS IN EXTERNAL FIELDS: GHOST ORBITS, CATASTROPHES, AND UNIFORM SEMICLASSICAL APPROXIMATIONS

Jörg Main and Günter Wunner

Institut für Theoretische Physik I, Ruhr-Universität Bochum,
D-44780 Bochum, Germany

INTRODUCTION

Highly excited atoms in external magnetic and electric fields are nontrivial systems possessing a classically chaotic counterpart, with at least two nonseparable, strongly coupled degrees of freedom. Atoms in magnetic fields have served as prototype systems for studying – both experimentally and theoretically – the quantum manifestations of classical chaos in real physical systems. A decisive advance for an interpretation of structures in the photoabsorption cross section was achieved by the development of *closed orbit theory* by Du and Delos (1988) and Bogomolny (1989). The method allows at least in low resolution a quantitative semiclassical calculation of spectra in terms of a few parameters of closed orbits starting at the nucleus and returning back to the nucleus. However, some features have been observed in quantum spectra which cannot be explained within the conventional closed orbit theory. An example is the observation of resonances in recurrence spectra of the hydrogen atom at positions where no closed classical orbits exist. These structures appear at energies near bifurcations of orbits, i.e., where orbits are born or vanish. Near bifurcations closed orbit theory fails, the semiclassical formulas diverge and are singular exactly at the bifurcation points. The purpose of this article is to calculate and interpret photoabsorption spectra of atoms in external fields in terms of closed classical orbits, in particular at energies close to bifurcations where the conventional theory fails. We discuss the classical dynamics, bifurcations, and ghost orbits, and construct the uniform approximations for various types of catastrophes.

CLASSICAL DYNAMICS

Search for real and "ghost" orbits

The nonrelativistic Hamiltonian for atoms in a magnetic field of strength B directed along the z axis reads (in atomic units, $\gamma = B/(2.35 \times 10^5\mathrm{T})$)

$$H = \frac{1}{2}\mathbf{p}^2 + V(r) + \frac{1}{2}\gamma L_z + \frac{1}{8}\gamma^2\varrho^2 = E\ . \qquad (1)$$

Atoms and Molecules in Strong External Fields
Edited by Schmelcher and Schweizer, Plenum Press, New York, 1998

The component of the angular momentum parallel to the field axis is conserved and we choose $L_z = m\hbar = 0$ in all classical calculations. The potential $V(r)$ is the Coulomb potential in case of the hydrogen atom and is modified by a short-ranged core potential modeling the non excited electrons of the ionic core in general Rydberg atoms (Hüpper et al., 1995).

A special feature of the Hamiltonian of the hydrogen atom is its scaling property with respect to the magnetic field strength. In scaled coordinates and momenta, $\tilde{\mathbf{r}} = \gamma^{2/3}\mathbf{r}$, $\tilde{\mathbf{p}} = \gamma^{-1/3}\mathbf{p}$ the classical Hamiltonian assumes the form

$$\tilde{H} = \gamma^{-2/3}H = \frac{1}{2}\tilde{\mathbf{p}}^2 - \frac{1}{\tilde{r}} + \frac{1}{8}\tilde{\varrho}^2 = \tilde{E} \, . \tag{2}$$

The classical trajectories obtained from the scaled equations of motion do not depend on both energy and magnetic field strength but only on one parameter, the scaled energy $\tilde{E} = E\gamma^{-2/3}$. Note that the classical action scales as

$$S = 2\pi\tilde{S}\gamma^{-1/3} \, . \tag{3}$$

The scaling property is lost, in general, for nonhydrogenic atoms in magnetic fields because the size of the ionic core does not depend on the magnetic field strength. However, for a limited range of the field strength the same scaling laws as for hydrogen can be approximately applied to nonhydrogenic atoms as well (Hüpper et al., 1995).

The Coulomb singularity presents an obstacle to the numerical integration of the equations of motion that follow from the Hamiltonian (2). The way out of this problem is a transformation of time $t \mapsto \tau$, with $dt = 2r d\tau$, called regularization (Stiefel and Scheifele, 1971), together with a coordinate transformation to semiparabolic coordinates $\mu = (r+z)^{1/2}$, $\nu = (r-z)^{1/2}$. The equations of motion obtained from the regularized Hamiltonian are free of singularities and allow efficient integration with the help of high-order numerical algorithms.

In a semiclassical approximation to photoabsorption spectra closed orbits which start at and return to the nucleus are of fundamental importance. Trajectories are started at the origin $\mathbf{r} = 0$ with an angle ϑ_i between the initial velocity and the magnetic field axis. The closed orbit search can be formulated as finding solutions (ϑ_i, T_f) so that the trajectory started with angle ϑ_i at time $t = 0$ returns back to the origin after period $t = T_f$. Numerically the problem is solved with the help of an iterative Newton algorithm and the roots (ϑ_i, T_f) are the *real* closed orbits when all parameters are defined real.

The analytic structure of the equations of motion allows a direct analytic continuation of the real phase space to the complex phase space. To search for complex closed orbits we again choose initial conditions $\mathbf{r} = 0$ but complex starting angle ϑ_i, i.e. the initial momenta become complex. With these complex initial conditions Hamilton's equations of motion can be numerically integrated from $t = 0$ to $t = T_f$. It may be noted that in general all closed orbit parameters become complex, in particular the recurrence time T_f, the classical action S, and the monodromy matrix $\mathbf{M}$, which is defined via

$$\begin{pmatrix} \delta q(T) \\ \delta p(T) \end{pmatrix} = \mathbf{M} \begin{pmatrix} \delta q(0) \\ \delta p(0) \end{pmatrix} = \begin{pmatrix} m_{11} & m_{12} \\ m_{21} & m_{22} \end{pmatrix} \begin{pmatrix} \delta q(0) \\ \delta p(0) \end{pmatrix} \tag{4}$$

where $\delta q(t)$ and $\delta p(t)$ are small deviations in coordinate and momentum space perpendicular to the direction of the orbit (Bogomolny, 1989). For classification of closed orbits we use the same nomenclature as in Main et al. (1991) and Main et al. (1994).

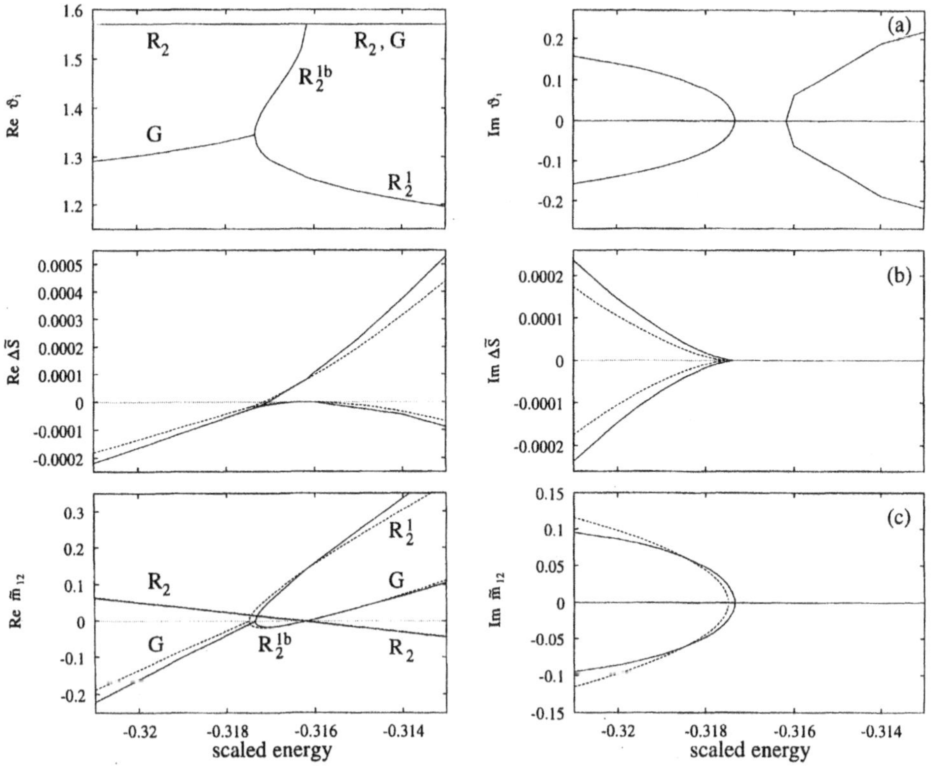

Figure 1. (*a*) Real and imaginary part of starting angle ϑ_i for closed orbits related to the bifurcating scenario of the (period doubled) perpendicular orbit R_2. (*b*) Difference $\Delta\tilde{S}$ between the classical action of the (period doubled) perpendicular orbit R_2 and real and ghost orbits bifurcating from it. (*c*) Monodromy matrix element $\tilde{m}_{12}$ of the perpendicular orbit R_2 and orbits bifurcating from it. Dashed lines: Analytical fits.

As an example we investigate here real and ghost orbits related to the period doubling of the perpendicular orbit R_1. This closed-orbit bifurcations is non elementary because various orbits with similar periods undergo two different elementary types of bifurcations at nearly the same energy. The structure of bifurcations and the appearance of ghost orbits can be seen clearly in the energy dependence of the starting angles ϑ_i in Fig. 1a. Orbits R_2^1 and R_2^{1b} are born in a saddle node bifurcation at $\tilde{E}_b^{(1)} = -0.31735345$, $\vartheta_i = 1.3465$. Below the bifurcation energy we find an associated ghost orbit and its complex conjugate. Orbit R_2^{1b} is real only in a very short energy interval ($\Delta\tilde{E} \approx 0.001$) and is then involved into the next bifurcation at $\tilde{E}_b^{(2)} = -0.31618537$, $\vartheta_i = \pi/2$. This is the period doubling bifurcation of the perpendicular orbit, R_1 which exists at all energies ($\vartheta_i = \pi/2$ in Fig. 1a). The real orbit R_2^{1b} separates from R_1 at energies *below* the bifurcation point, i.e. a real orbit vanishes with increasing energy. Consequently associated ghost orbits are expected at energies *above* the bifurcation, i.e. $\tilde{E} > \tilde{E}_b^{(2)}$, and indeed such "postbifurcation" ghosts have been found. Its complex starting angles are also shown in Fig. 1a. The energy dependence of scaled actions, more precisely the difference $\Delta\tilde{S}$ to the action of the period doubled perpendicular orbit R_2 is presented in Fig. 1b (solid lines) and the graph for the monodromy matrix element $\tilde{m}_{12}$ is given in Fig. 1c. As can be seen action and monodromy matrix element of ghost orbits related to the saddle node bifurcation of R_2^1 become complex at $\tilde{E} < \tilde{E}_b^{(1)}$ whereas these pa-

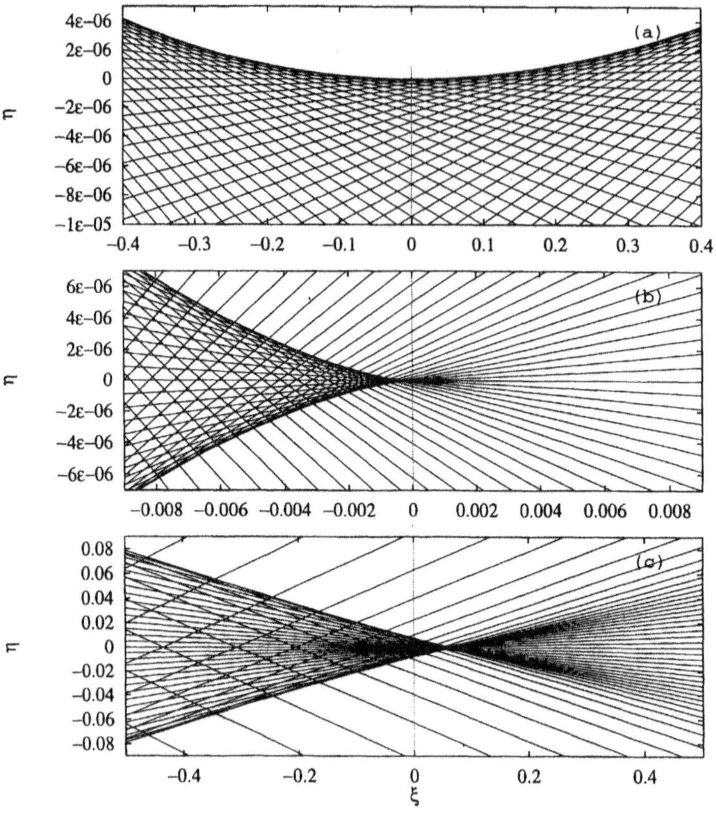

Figure 2. (a) Fold catastrophe of returning orbits (in rotated semiparabolic coordinates ξ, η) related to the saddle-node bifurcation of orbit X_1 at bifurcation energy $\tilde{E}_b = -0.11544216$. (b) Cusp catastrophe related to the period doubling bifurcation of the balloon orbit V_1^1 at bifurcation energy $\tilde{E}_b = -0.3420258$. (c) Butterfly catastrophe (bifurcation of the perpendicular orbit) at $\tilde{E}_b = -0.31618537$.

rameters remain real for the postbifurcation ghosts at $\tilde{E} > \tilde{E}_b^{(2)}$. Note that only ghost orbits with a positive imaginary part of the classical action are of physical relevance (Main and Wunner, 1997).

Bifurcations and catastrophes

We look for the manifold of trajectories returning back to the nucleus. Introducing rotated semiparabolic coordinates

$$\xi = \mu\cos(\vartheta_f/2) + \nu\sin(\vartheta_f/2) = \sqrt{2r}\cos((\vartheta - \vartheta_f)/2)$$
$$\eta = \nu\cos(\vartheta_f/2) - \mu\sin(\vartheta_f/2) = \sqrt{2r}\sin((\vartheta - \vartheta_f)/2)\,, \qquad (5)$$

the ξ and η axes are now parallel and perpendicular to the exactly returning orbit. For isolated closed orbits, i.e. orbits sufficiently far from bifurcations nearby returning orbits are approximately parallel to the ξ-axis. However, at bifurcations the returning orbits form *catastrophes* (Poston and Steward, 1978) as illustrated in Fig. 2. Fig. 2a shows a *fold* catastrophe related to the saddle-node bifurcation of orbit X_1 at bifurcation energy

$\tilde{E}_b = -0.11544216$. At energies $\tilde{E} > \tilde{E}_b$ the nucleus is located below the parabola type caustic in Fig. 2a and two different real closed orbits with slightly different shapes exist. With decreasing energy both orbits vanish exactly at the bifurcation point and below the bifurcation energy ($\tilde{E} < \tilde{E}_b$) there is no real closed orbit with similar shape. However, in the closed orbit search extended to the complex continuation of phase space the existence of prebifurcation ghost orbits could be revealed (Main and Wunner, 1997).

Fig. 2b presents a *cusp* catastrophe related to the period doubling bifurcation of the balloon orbit V_1^1 at bifurcation energy $\tilde{E}_b = -0.3420258$. The balloon orbit itself is already created at lower energy $\tilde{E} = -0.3913$ in a bifurcation from the parallel orbit. A special feature of its shape is the symmetry in the initial and final angle, i.e. $\vartheta_i = \vartheta_f$. It exists below and above the period doubling energy without any spectacular change of this shape. Above the period doubling energy a new orbit V_2^{2*} which breaks this symmetry ($\vartheta_i \neq \vartheta_f$) separates from V_1^1 and is closed roughly after two times the period of V_1^1.

The period doubling of the perpendicular orbit R_1 is a more complicated bifurcation scenario because various orbits with similar periods undergo two different elementary types of bifurcations at nearly the same energy (see Fig. 1). The returning trajectories form a *butterfly* catastrophe as illustrated in Fig. 2c. The catastrophes presented in Fig. 2 are three examples of the seven elementary catastrophes discussed in *catastrophe theory* (Poston and Steward, 1978). A more detailed analysis of these closed orbit bifurcations can be found in Main and Wunner (1997).

UNIFORM SEMICLASSICAL APPROXIMATIONS

The link between classical trajectories and photoabsorption spectra is made by semiclassical *closed-orbit* theory. The theory has been originally developed independently by Du and Delos (1988) and Bogomolny (1989) for the hydrogen atom in a magnetic field and has become a powerful tool for the interpretation of experimental spectra and exact quantum calculations as well. The agreement between quantum and semiclassical results serves as a measure for our understanding of atoms in external fields and deviations are usually an indication of new physical effects which are not yet adequately considered in the semiclassical theory. Examples are the observation of unidentified peaks in the recurrence spectra of hydrogenic and nonhydrogenic atoms which are related to complex "ghost" orbits (Main and Wunner, 1997) or core-scattered orbits (Hüpper et al., 1995). Indeed, these effects can be described and understood within a generalized closed-orbit theory. Our main interest is to construct uniform semiclassical approximations which are free of singularities at bifurcation points of closed orbits where the conventional closed-orbit theory breaks down. Here, we can not present the derivation of closed-orbit theory and its generalization to bifurcations and catastrophes which is discussed in detail in Main and Wunner (1997). Instead we directly present the results for the uniform semiclassical approximations and compare them to solutions of standard closed orbit theory.

In general the semiclassical oscillator strength can be written as

$$f = f^0 + 2(E - E_i)\,\mathrm{Im} \sum_{\mathrm{cl.o.}\ k} (\sin\vartheta_i^k \sin\vartheta_f^k)^{1/2}\, \mathcal{Y}_m^\delta(\vartheta_i^k)\tilde{\mathcal{Y}}_m^\delta(\vartheta_f^k) \times \mathcal{A}_k(\tilde{E},\gamma) \exp\left(i\left[S_m^k(\tilde{E},\gamma) - \frac{\pi}{2}\mu^k\right]\right) , \tag{6}$$

where f^0 is a smoothly varying continuous background, E_i and E are the energies of the initial and final state, and S_m and μ are the classical action and Maslov index of closed orbit k. The angular function $\mathcal{Y}_m^\delta(\vartheta)$ is defined in the Appendix and ϑ_i and ϑ_f are the initial and final angles of closed orbits starting at and returning to the nucleus. The non-universal part of Eq. 6 is the amplitude function $\mathcal{A}(\tilde{E},\gamma)$ which depends on the properties of the returning trajectories, i.e. the standard situation of isolated returning orbits or uniform approximations for the various types of catastrophes. In conventional closed orbit theory the amplitude function (in scaled variables) is given by

$$\mathcal{A}(\tilde{E},\gamma) = 2(2\pi)^{3/2}\gamma^{1/6}e^{i\pi/4}\left(\tilde{m}_{12}(\tilde{E})\right)^{-1/2} . \tag{7}$$

Note that the element $\tilde{m}_{12}$ of the monodromy matrix (4) is nonzero for isolated returning orbits but vanishes at the bifurcation points of closed orbits where the conventional semiclassical amplitude diverges. We will now discuss uniform approximations to the amplitude function $\mathcal{A}(\tilde{E},\gamma)$ for the three examples of bifurcations and catastrophes mentioned above, namely the fold, cusp, and butterfly catastrophe.

The fold catastrophe is related to a saddle node bifurcation. The amplitudes $A(\tilde{E},\gamma)$ around the bifurcation of orbit X_1 are presented in Fig. 3a (plotted is the absolute value of the amplitude) at constant magnetic field strength $\gamma = 4\cdot 10^{-6}$. The dashed line shows the superposition of the two real orbits X_1^a and X_1^b at energies $\tilde{E} > \tilde{E}_b$. The oscillatory structure of the amplitude is the result of the strong interference between these two orbits. At the bifurcation energy the amplitude diverges. Note that the amplitude at energies below the bifurcation point is zero in the standard formulation of closed orbit theory, i.e. when only real orbits are considered. The dashed dotted line is the extension when ghost orbit contributions are also included. We only consider the ghost with positive imaginary part of complex action because only this ghost has a physical meaning. The amplitude decreases exponentially in $-(\tilde{E}_b - \tilde{E})^{3/2}$ with decreasing energy but also exhibits an unphysical divergence at the bifurcation point. The solid line in Fig. 3a is the uniform approximation to the amplitude for which we obtain

$$A(\tilde{E},\gamma) = 2^{7/2}\pi^2\gamma^{1/9}\,(3\tilde{\sigma}/2)^{1/6}|\tilde{M}|^{-1/2}\,\mathrm{Ai}\left((3\tilde{\sigma}/2)^{2/3}\gamma^{-2/9}(\tilde{E}_b - \tilde{E})\right) \tag{8}$$

with $\tilde{E}_b$ the bifurcation energy and $\tilde{\sigma}$ and $\tilde{M}$ parameters which can be determined from classical trajectory calculations. Up to constant factors the uniform approximation to the amplitude is an Airy function of the energy difference $\tilde{E}_b - \tilde{E}$. Both the oscillatory structure of the amplitude at $\tilde{E} > \tilde{E}_b$ and the exponentially damped ghost orbit tail are well reproduced but in addition the singularity at the bifurcation point is now completely removed. Note that the maximum of amplitude is not located at the bifurcation point $\tilde{E} = \tilde{E}_b$ but is shifted to somewhat higher energy.

The cusp catastrophe is observed for example in the period doubling bifurcation of the balloon orbit V_1^1 and the uniform approximation to the amplitude is illustrated in Fig. 3b (magnetic field strength $\gamma = 10^{-7}$). The non uniform amplitude is plotted as a dashed line. Above the bifurcation energy three real orbits, the period doubled balloon, V_2^2, and the bifurcated orbit, V_2^{2*}, traversed in both directions are considered in Fig. 3b and the interference of these three orbits produces the oscillatory fluctuations of the amplitude. The non uniform solution is characterized by an unphysical divergence of the amplitude around the bifurcation energy but the singularity is removed in the uniform approximation (solid line in Fig. 3b)

$$A(\tilde{E},\gamma) = 8\pi\gamma^{1/12}\tilde{\sigma}^{1/4}|\tilde{M}|^{-1/2}\,\Phi\left(2\tilde{\sigma}^{1/2}\gamma^{-1/6}\left(\tilde{E}_b - \tilde{E}\right)\right) \tag{9}$$

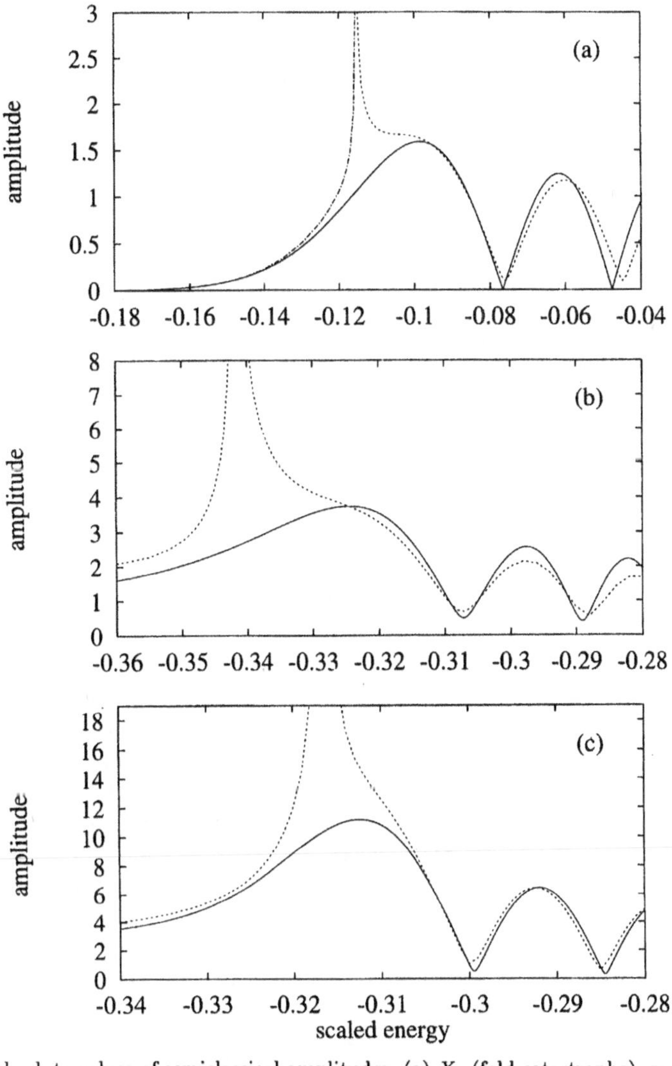

Figure 3. Absolute values of semiclassical amplitudes. (a) X_1 (fold catastrophe), magnetic field strength $\gamma = 4 \cdot 10^{-6}$ atomic units; (b) V_2^2 (cusp catastrophe), $\gamma = 10^{-7}$; (c) R_2 (butterfly catastrophe), $\gamma = 10^{-7}$. Dashed lines: Amplitudes in the standard model. Dashed dotted line (in (*a*)): Ghost orbit contribution. Solid lines: Uniform approximations.

with parameters $\tilde{E}_b$, $\tilde{\sigma}$, and $\tilde{M}$ completely determined by classical calculations. The energy dependence is given by a special case of Pearcey's integral (Pearcey, 1946),

$$\Phi(\pm x) = \int_{-\infty}^{+\infty} \exp\left[-i(\pm x t^2 + t^4)\right] dt = \frac{\pi}{2}\sqrt{\frac{x}{2}} \exp\left[i\frac{x^2}{8}\right]$$
$$\times \left\{ \exp\left[-i\frac{\pi}{8}\right] J_{-\frac{1}{4}}\left(\frac{x^2}{8}\right) \mp \exp\left[i\frac{\pi}{8}\right] J_{\frac{1}{4}}\left(\frac{x^2}{8}\right) \right\} . \tag{10}$$

In contrast to the fold catastrophe (Fig. 3a) there is a real instead of a ghost orbit below the bifurcation point and therefore the amplitude decreases $\sim (\tilde{E}_b - \tilde{E})^{-1/2}$ rather than exponentially with decreasing energy. Also the modulations of amplitudes at $\tilde{E} > \tilde{E}_b$ are less pronounced, particularly there are no energies with vanishing amplitude, i.e. complete destructive interference between all closed orbit contributions. Oscillations in periodic orbit contributions related to a cusp catastrophe of returning trajectories have been experimentally verified in Stark spectra of lithium atoms by Courtney et al. (1995).

The butterfly catastrophe is related to a more complicated bifurcation scenario as found e.g. around the period doubling of the perpendicular orbit. Results for $\gamma = 10^{-7}$ are presented in Fig. 3c. The solid line is the uniform approximation

$$A(\tilde{E}, \gamma) = \pi \gamma^{1/18} 3^{1/6} k^{1/3} (32\tilde{\sigma}/\tilde{M})^{1/2}$$
$$\times \Psi\left(-3^{1/3} k^{2/3} \gamma^{-2/9} \tilde{\sigma}(\tilde{E} - \tilde{E}_b^{(2)}), -(9k)^{1/3} \gamma^{-1/9}\right) \tag{11}$$

with k, $\tilde{M}$, and $\tilde{\sigma}$ obtained from classical calculations and

$$\Psi(x, y) \equiv \int_{-\infty}^{+\infty} \exp\left(-i\left[x t^2 + y t^4 + t^6\right]\right) dt \, . \tag{12}$$

For comparison the dashed line shows the non uniform solution of standard closed orbit theory. The modulations of amplitudes at energies above the bifurcation point are caused by the interference of three closed orbits, namely the period doubled perpendicular orbit, R_2, and the "pacman" orbit, R_2^1, traversed in both directions. Ghost orbits exist at these energies but have no physical meaning. Below the bifurcation point there are one real orbit, R_2, and in addition a ghost orbit (and its time reversal) contributing to the semiclassical photoabsorption spectrum. As in case of a fold catastrophe the amplitude of the ghost orbit is exponentially damped with decreasing energy but now the ghost is not isolated and its contribution is surpassed by that of the perpendicular orbit, i.e. the ghost is "hidden" behind the real orbit. The first direct observation of this hidden ghost orbit in high resolution quantum recurrence spectra is presented in Main et al. (1997).

CONCLUSION

Highly excited atoms in external fields are fundamental quantum systems which possess in many cases a classically chaotic analogue. The analysis of their quantum spectra is of fundamental importance for a deeper understanding of phenomena related to the so-called *quantum chaos.* The development and application of *closed-orbit* theory was the decisive step for a semiclassical calculation and interpretation of photoabsorption spectra. In its original version formulated by Du and Delos (1988) and Bogomolny (1989) closed-orbit theory was able to reproduce most structures in hydrogenic and even nonhydrogenic atoms but suffered from singularities at energies where orbits undergo bifurcations. Unphysical singularities at bifurcation energies of closed orbits are now removed in uniform semiclassical approximations and the analysis of these uniform solutions also reveals the role and physical importance of "ghost" orbits in complex phase space.

APPENDIX

Angular function $\mathcal{Y}_m^\delta(\vartheta)$

The angular functions $\mathcal{Y}_m^\delta(\vartheta)$ and $\tilde{\mathcal{Y}}_m^\delta(\vartheta)$ depend on the initial state ψ_i, dipole operator D, and phase shifts δ_l of the ionic core. They are linear superpositions of spherical harmonics

$$\mathcal{Y}_m^\delta(\vartheta) = \sum_{l=|m|}^{\infty} (-1)^l e^{i\delta_l} \mathcal{B}_{lm} Y_{lm}(\vartheta, 0) \tag{13}$$

$$\tilde{\mathcal{Y}}_m^\delta(\vartheta) = \sum_{l=|m|}^{\infty} (-1)^l e^{i\delta_l} \mathcal{B}_{lm}^* Y_{lm}(\vartheta, 0) , \tag{14}$$

with coefficients $\mathcal{B}_{lm}$ defined by the overlap integrals

$$\mathcal{B}_{lm} = \int d^3x \, (D\psi_i)(\mathbf{x}) \sqrt{2/r} \, \mathrm{J}_{2l+1}\left(\sqrt{8r}\right) Y_{lm}^*(\vartheta, \varphi) \tag{15}$$

where $\mathrm{J}_\nu(x)$ are Bessel functions. For more details see Gao and Delos (1992) and Hüpper et al. (1995).

REFERENCES

Bogomolny, E. B., 1989, *Sov. Phys. JETP* 69:275.

Courtney, M., Jiao, Hong, Spellmeyer, N., Kleppner, D., Gao, J., and Delos, J. B., 1995, *Phys. Rev. Lett.* 74:1538.

Du, M. L. and Delos, J. B., 1988, *Phys. Rev.* A 38:1896 and 38:1913.

Gao, J. and Delos, J. B., 1992, *Phys. Rev.* A 46:1455.

Hüpper, B., Main, J., and Wunner, G., 1995, *Phys. Rev. Lett.* 74:2650 and 1996, *Phys. Rev.* A 93:744.

Main, J., Wiebusch, G., and Welge, K. H., 1991, *Comm. At. Mol. Phys.* 25:233.

Main, J., Wiebusch, G., Welge, K. H., Shaw, J., and Delos, J. B., 1994, *Phys. Rev.* A 49:847.

Main, J. and Wunner, G., 1997, *Phys. Rev.* A 55:1743.

Main, J., Mandelshtam, V. A., and Taylor, H. S., 1997, *Phys. Rev. Lett.* 78:4351.

Pearcey, T., 1946, *Phil. Mag.* 37:311.

Poston, T. and Steward, I. N., 1978, "Catastrophe theory and its applications," Pitman, London.

Stiefel, E. L. and Scheifele, G., 1971, "Linear and Regular Celestial Mechanics," Springer.

QUADRATIC ZEEMAN SPLITTING OF HIGHLY EXCITED RELATIVISTIC ATOMIC HYDROGEN

D.A. Arbatsky, P.A. Braun

Department of Theoretical Physics, Institute of Physics,
St. Petersburg University, St. Petersburg 198904, Russia

In this communication the quadratic Zeeman splitting of highly excited atomic hydrogen is considered. The external magnetic field is assumed to be so weak that the second order splitting is comparable with the fine structure of the unperturbed atom. The Hamiltonian of the system can be written as

$$H = H_0 + \frac{\gamma}{2}(L_z + 2S_z) + V_d + V_{\text{rel}} \,. \tag{1}$$

Here $V_d = (\gamma^2/8)(x^2 + y^2)$ is the diamagnetic interaction; V_{rel} is the Pauli operator of the relativistic corrections comprising "the mass on speed dependence", the spin-orbit interaction and the Darwin contact interaction; γ is the magnetic field in atomic units ($\gamma = 1$ corresponds to $\mathbf{B} = 2.35 \times 10^5 T$).

There are three perturbations present in Eq.(1); their relative impact depends on the field strength and the extent of the atomic excitation. We will make the following assumptions:

- the spin and the orbital movement are uncoupled. This means that the first order Zeeman splitting induced by the operator $(\gamma/2)L_z$ is much larger than the fine structure, or

$$\frac{\gamma}{2} \gg \frac{\alpha^2}{n^3} \,,$$

  where α is the fine structure constant; n is the principal quantum number of the atomic state under consideration ($n \gg 1$);

- the first order (paramagnetic) Zeeman splitting is large compared with the second order (diamagnetic) one, or

$$\frac{\gamma}{2} \gg u_\gamma \equiv \frac{5\gamma^2 n^4}{16} \;;$$

Atoms and Molecules in Strong External Fields
Edited by Schmelcher and Schweizer, Plenum Press, New York, 1998

- the diamagnetic splitting is much smaller than the energy interval between adjacent shells in the unperturbed atom, or

$$u_\gamma \ll \frac{1}{n^3} ;$$

- the diamagnetic splitting is comparable with the fine structure, or

$$u_\gamma \sim \frac{\alpha^2}{n^3} . \tag{2}$$

The first and the second of these assumptions guarantee that the orbital angular momentum component L_z is conserved. The third assumption means that mixing between different shells of the atom can be neglected and n is a good quantum number. Finally Eq.(2) means that the magnetic fields are weaker than those usually considered in the theory of the second order Zeeman effect; in such fields the relativistic effects compete with the field-induced splitting completely modifying it in some cases (see below).

I. NUMERICAL CALCULATIONS AND SPECTRUM

Owing to the assumed hierarchy of the perturbation strengths calculation of the splitting can be performed in two stages. The paramagnetic operator splits the level $E_n^{(0)} = -1/2n^2$ of the non-relativistic unperturbed atom into a set of equidistant levels with the spacing $\gamma/2$,

$$E_{m\sigma}^{(1)} = \frac{\gamma}{2}(m+\sigma) , \tag{3}$$

where $\sigma \equiv \sigma_z = \pm 1$ is the spin quantum number. These levels are still degenerate. On the second stage we consider the lifting of this residual degeneracy. This is done by diagonalizing the sum of the operators of the relativistic corrections and the diamagnetic interaction in the basis set of the spin-orbitals belonging to the level (3). The respective basis set is formed by the functions

$$|nlm\sigma\rangle = R_{nl}(r) \cdot Y_{lm}(\theta, \phi) \cdot u_\sigma$$

with fixed n,m,σ and all possible l. Here R_{nl} stand for the non-relativistic radial wave functions of atomic hydrogen and u_σ is the spin function.

It may be argued that, since there is degeneracy of the first order levels with respect to the spin projection

$$E_{m\sigma}^{(1)} = E_{m+2\sigma,-\sigma}^{(1)} ,$$

mixing of states with $\sigma = 1$ and $\sigma = -1$ has to be taken into account. However, the relativistic correction operator can change the orbital angular momentum projection m by no more than 1; thus such mixing cannot take place in the first order by V_{rel}.

The only non-zero diamagnetic operator V_d matrix elements correspond to $\Delta m = 0$, $\Delta l = 0, \pm 2$:

$$w_l = \langle nlm| V_d |nlm\rangle , \qquad p_l = \langle n, l-2, m| V_d |nlm\rangle ;$$

their explicit expressions are well-known[see, e.g. 1, Eq.(1.5)]. The relativistic correction operator has non-zero matrix elements only for $\Delta m = 0, \pm 1$. The selection rules for the operator $V_{\rm rel}$ are $\Delta l = 0$, $\Delta\sigma = 0, \pm 2$, $\Delta m = 0, \pm 1$. However, as was pointed out, only the elements diagonal in m, σ are necessary:

$$V_{{\rm rel},l} \equiv \langle nlm\sigma | V_{\rm rel} | nlm\sigma \rangle = -\frac{\alpha^2}{2n^3}\left[\frac{2}{2l+1} - \frac{3}{4n} - \frac{m\sigma}{l(l+1)(2l+1)}\right] .$$

Thus the perturbation theory matrix is tridiagonal and the equation for its eigenvector $\mathbf{C} = \{C_l\}$ has the form of a three-term recurrence relation

$$p_l C_{l-2} + \left(w_l + V_{{\rm rel},l} - E^{(2)}\right) C_l + p_{l+2} C_{l+2} = 0 . \qquad (4)$$

Its eigenvalues $E^{(2)}$ determine the splitting of the degenerate sublevels (3).

The typical spectrum of the problem (4) is presented in Fig.1 as a function of the parameter

$$q \equiv 8\alpha^2/5n^8\gamma^2 .$$

Calculations for $n = 25, m = 5, \sigma = 1$ are shown. The depicted quantities are the eigenvalues $E^{(2)}$. Full and dashed lines refer to the even and odd states respectively. It is seen that the plot falls into four distinct regions denoted by Roman numbers. In the regions II and IV the spectrum consists of narrow $g-u$ doublets which split when the levels cross into the regions I and III. The non-relativistic limit is observed in the left hand side of the plot corresponding to $q = 0$ $(\gamma \to \infty)$, with its characteristic division of the spectrum into the vibrational and rotational part. The limit $q \to \infty$ corresponds to the trivial Paschen-Back case when the non-diagonal elements in Eq.(4) can be neglected and $E^{(2)}$ coincide with $w_l + V_{{\rm rel},l}$.

II. SEMICLASSICAL THEORY

Interpretation of the diamagnetic splitting in the non-relativistic case has been achieved by the classical perturbation theory and the correspondence principle[3,4]. We will use the same approach. The electron in the non-relativistic classical hydrogen atom moves along the Kepler orbit whose elements are determined by the angular momentum $\mathbf{L}$ and the Runge-Lenz vector $\mathbf{A}$. Owing to the magnetic field and the relativistic effects, $\mathbf{L}$ and $\mathbf{A}$ become slowly changing functions of time. The effective Hamiltonian determining their time dependence is one-dimensional. It can be obtained by averaging the perturbation over a period of the classical Kepler movement. A convenient choice for the canonical coordinate is the absolute value of the angular momentum $L = |\mathbf{L}|$. The conjugate momentum will then be the third Euler angle (eigenrotation) Φ of the frame of reference whose z and x axes are taken along the vectors $\mathbf{L}$ and $\mathbf{A}$ respectively.

The resulting classical Hamiltonian describing the evolution of the orbit of the atomic electron with the unperturbed energy $E^{(0)}$ and the angular momentum projection L_z under the influence of the second order effects can be written

$$H^{(2)}(L,\Phi) = H^{(2)}_{\rm dia}(L,\Phi) + W_{\rm rel}(L) .$$

The part describing the diamagnetic perturbation is well-known[1, Eq.(3.6); 3,4]

$$H^{(2)}_{\rm dia} = \frac{u_\gamma}{2}\left[\left(1 - \frac{3}{5}\frac{L^2}{n^2}\right)\left(1 + \frac{m^2}{L^2}\right) - \left(1 - \frac{L^2}{n^2}\right)\left(1 - \frac{m^2}{L^2}\right)\cos 2\Phi\right] .$$

Here $m = L_z$, $n = [-2E^{(0)}]^{-1/2}$. The function $W_{\rm rel}$ describes the relativistic effects and is obtained from the operator $V_{\rm rel}$ in the semiclassical limit $n \to \infty$ ($L/n \to const$, $|m|/n \to const \neq 0$):

$$W_{\rm rel} = -\frac{\alpha^2}{2n^3}\left(\frac{1}{L} - \frac{3}{4n}\right) .$$

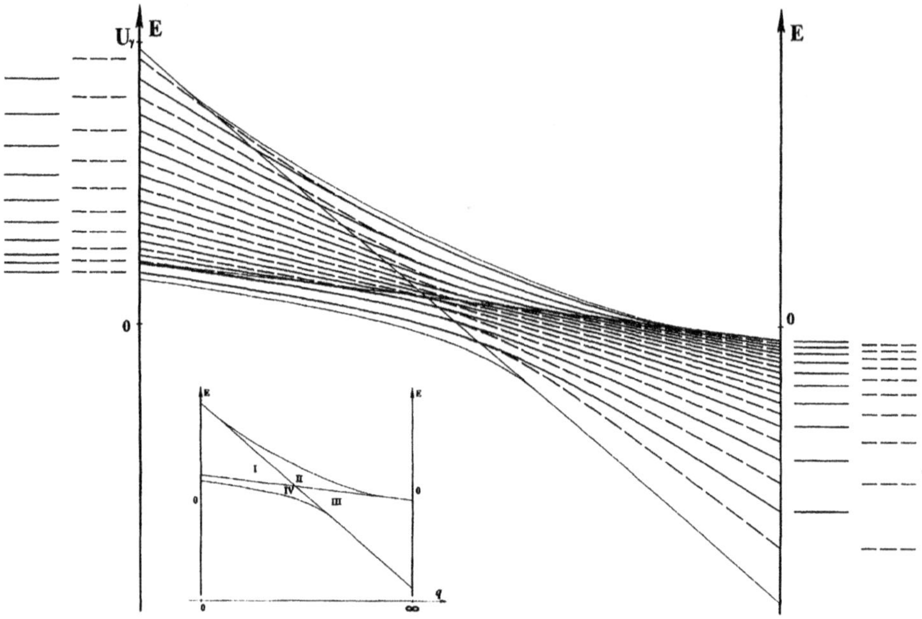

Figure 1. Spectrum of the Hamiltonian as a function of parameter q. Left hand side corresponds to $q=0$, right hand side corresponds to $q=+\infty$. ($n=25$, $m=5$, $\sigma=1$)

Note that the spin projection dependence vanishes in this limit.

The equation of the phase trajectory $L=L(\Phi)$ is given by the energy conservation law,

$$H^{(2)}(L,\Phi)=E^{(2)} . \tag{5}$$

The following symmetry property is of importance:

$$H^{(2)}(L,\Phi)=H^{(2)}(L,\Phi+\pi) . \tag{6}$$

It will be convenient to introduce two functions of the "coordinate L", which are in many respects similar to the potential energy:

$$U^{+}(L)\equiv\max_{\Phi}H^{(2)}(L,\Phi)=H^{(2)}\left(L,\frac{\pi}{2}\right),$$
$$U^{-}(L)\equiv\min_{\Phi}H^{(2)}(L,\Phi)=H^{(2)}(L,0) . \tag{7}$$

The extrema of these two functions with respect to L in the interval $|m|\le L\le n$ determine the upper and lower bound of the Hamiltonian $H^{(2)}$:

$$\min_{L}U^{-}(L)=\min_{L,\Phi}H^{(2)}(L,\Phi), \qquad \max_{L}U^{+}(L)=\max_{L,\Phi}H^{(2)}(L,\Phi). \tag{8}$$

Consequently they give bounds for the possible values of the energy correction under the diamagnetic and relativistic perturbation:

$$\min_{L}U^{-}(L)\le E^{(2)}\le\max_{L}U^{+}(L) .$$

For reasons based on the correspondence principle the extrema (8) also bound the spectrum of the secular problem (4). In Fig.1 the values of the expressions (8) calculated as functions of q are plotted by thin lines bounding the spectrum.

The angular momentum of the classical perturbed atom is an oscillating function of time. The range of this oscillation depends on the energy correction $E^{(2)}$ and is fixed by the inequalities following from Eqs.(5),(7)

$$U^-(L) \le E^{(2)} \le U^+(L) \tag{9}$$

The extrema of L (the turning points of its oscillation) are roots of the equations $E^{(2)} = U^+(L)$ and $E^{(2)} = U^-(L)$; Φ in the extrema is equal to $\pi/2 \pmod{\pi}$ or $0 \pmod{\pi}$ respectively.

The dependence of Φ on time can be either oscillation or unlimited growth corresponding to libration and rotation respectively. Libration takes place if the left and the right turning points of the L oscillation belong to the same potential curve. E.g., suppose that L attains both its minimal and maximal values when $E^{(2)} = U^+(L)$. Then Φ can never attain values $0, \pi, 2\pi \ldots$ (this would have been possible only in the non-existing turning point in which $E^{(2)} = U^-$). Therefore Φ cannot be an infinitely growing continuous function. According to Eq.(6) the phase trajectory in this case consists of two symmetric branches isolated from each other by the rays $\Phi = 0$ and $\Phi = \pi$. It can be proved in a similar way that when one of the turning points belongs to U^+ and another one to U^- the movement is rotation.

The character of the movement (libration or rotation) can change at certain values of $E^{(2)}$ which would mean the existence of a separatrix. Possible critical energies are the ordinates of crossings of the curves $U^+(L)$ and $U^-(L)$ which occur at $L = |m|$ and $L = n$:

$$E^{(2)}_{\text{crit},m} = u_\gamma \left[1 - \frac{3}{5}\frac{m^2}{n^2} - q\left(\frac{n}{|m|} - \frac{3}{4}\right)\right]; \qquad E^{(2)}_{\text{crit},n} = u_\gamma \left[\frac{1}{5}\left(1 + \frac{m^2}{n^2}\right) - \frac{q}{4}\right]. \tag{10}$$

However they determine separatrices only if they belong to the allowed energy correction range (9). There can be thus two, one or no separatrices. The energies (10) are linear functions of q. In Fig.1 they are depicted by two thin crossing straight lines dividing it into parts I...IV.

Consider now the classical movement and the spectrum of the problem for various values of q (assuming $|m|/n < 1/\sqrt{5}$, for example).

1. Case

$$q = 0$$

(the non-relativistic quadratic Zeeman effect). As seen in Fig.2a, the total range of $E^{(2)}$ is divided into two parts by the separatrix $E^{(2)}_{\text{crit},n}$. The energies in the upper part of the spectrum correspond to rotation since the left turning point belongs to U^- while the right one belongs to U^+. The energies lesser than $E^{(2)}_{\text{crit},n}$ correspond to vibration in the vicinity of $\Phi = 0$ and $\Phi = \pi$, with two isolated classical trajectories for each energy. This is well seen in Fig.3a where the phase portrait of the system is shown in the polar representation $L = f(\Phi)$ for the case $m/n = 0.1$.

The presence of two equivalent isolated phase trajectories corresponds in the quantum mechanical approach to the energy spectrum whose lower (vibrational) part consists of approximately degenerate pairs of levels. The levels in the upper (rotational) part are non-degenerate. These properties of the spectrum are, of course, well known[1-4].

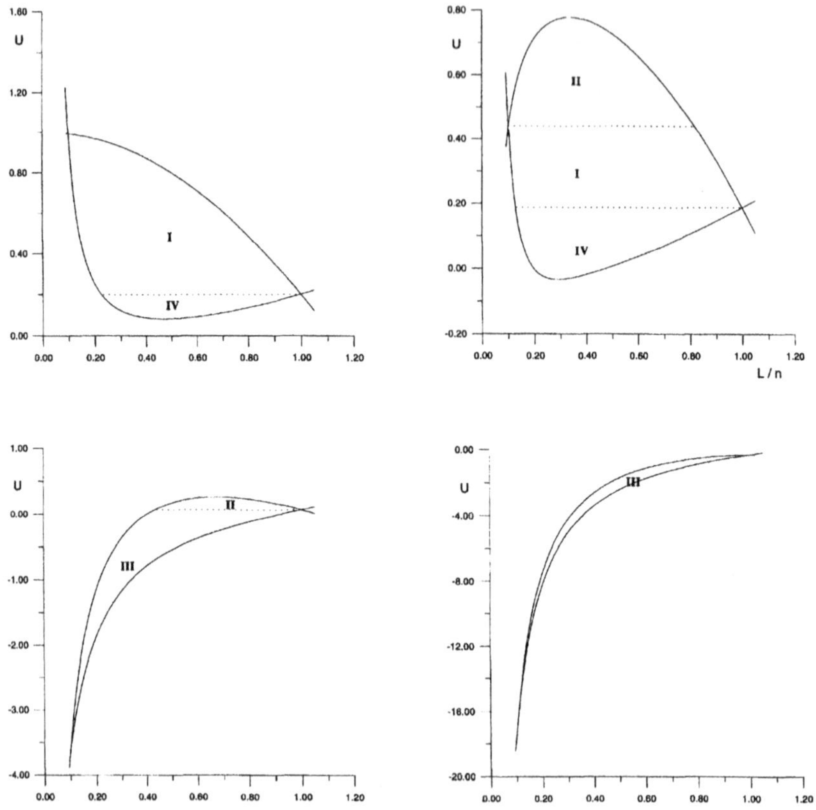

Figure 2. Potential functions $U^{\pm}(L)$. Classically allowed $E^{(2)}$-L combinations are in the area bordered by U^{+} from above and U^{-} from below. Critical $E^{(2)}$ values corresponding to the separatrices are shown by dashed horizontal lines. Case $m/n = 0.1$.
a) $q = 0$; b) $q = 0.06$; c) $q = 0.5$; d) $q = 2$.

2. Case

$$\frac{8}{5}\left(\frac{|m|}{n}\right)^3 < q < 2\frac{|m|}{n} - \frac{2}{5}\left(\frac{|m|}{n}\right)^3 .$$

Both of the critical energies fall now into the allowed range (9) of $E^{(2)}$ dividing it into three parts: the uppermost one corresponding to vibration in the vicinity of $\Phi = \pi/2$ and $\Phi = 3\pi/2$; the middle rotational one; the lower vibrational one with the vibration in the vicinity of $\Phi = 0$ and $\Phi = \pi$. This can be inferred from the plot of potential curves Fig.2b and the phase portrait Fig.3b. The corresponding quantum mechanical energy levels structure reflects these properties of the classical movement. The spectrum between the critical energies consists of singlets whereas the levels higher than $E^{(2)}_{\mathrm{crit},m}$ and lower than $E^{(2)}_{\mathrm{crit},n}$ are doublets.

3. Case

$$2\frac{|m|}{n} - \frac{2}{5}\left(\frac{|m|}{n}\right)^3 < q < \frac{8}{5} .$$

Only one of the separatrices remains within the range (9), namely $E^{(2)}_{\mathrm{crit},m}$. The classical movement is vibration around $\Phi = \pi/2$ or $\Phi = 3\pi/2$ and rotation for $E^{(2)}$ higher and lower than $E^{(2)}_{\mathrm{crit},m}$; see the plot of U (Fig.2c) and the phase portrait

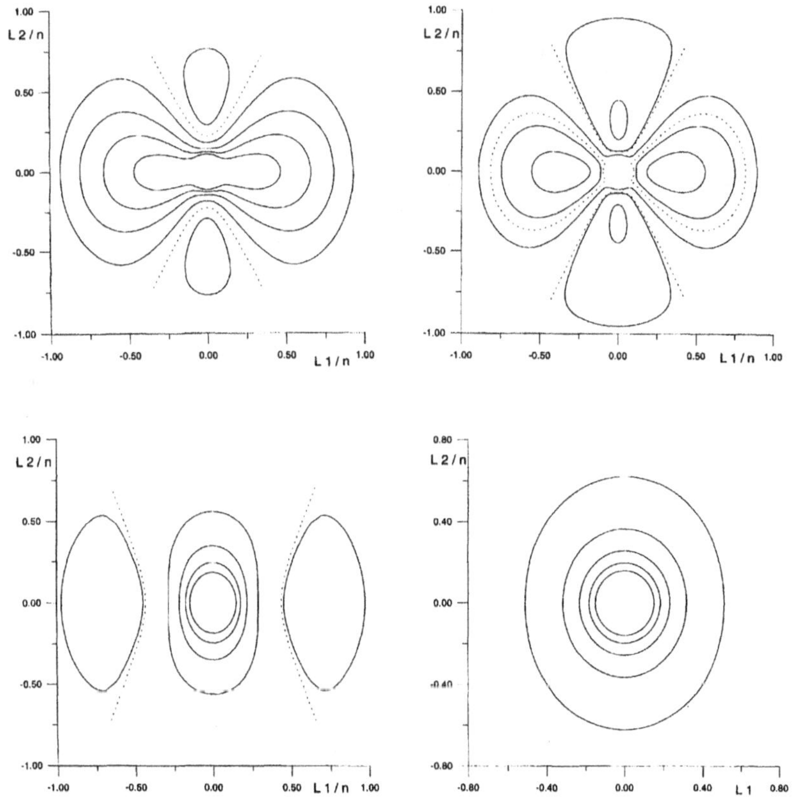

Figure 3. Phase portraits of the of the classical electron orbit evolution under the relativistic and diamagnetic perturbations. Angular momentum drawn as a function of the orbit eigenrotation Euler angle in the polar representation: $L_1 = L\cos\Phi$, $L_2 = L\sin\Phi$. The dotted lines are the separatrices of the movement. Case $m/n = 0.1$. a) $q = 0$; b) $q = 0.06$; c) $q = 0.5$; d) $q = 2$.

Fig.3c. The energy spectrum consists of doublets and singlets respectively and is thus turned "upside down" compared with the non-relativistic case.

4. Case

$$q > \frac{8}{5}\,.$$

Both of the critical energies leave the allowed range Eq.(9) so no separatrices exist in the movement (Fig.2d,3d). The only possible movement is rotation; the quantum mechanical spectrum consists of locally equidistant even and odd levels approaching to the Paschen-Back limit.

CONCLUSIONS

Using the classical mechanical perturbation approach we were able to explain the structure of the quadratic Zeeman effect spectrum of the atomic hydrogen Rydberg levels in weak magnetic fields in which the fine structure of the levels cannot be neglected. The general structure of the spectrum turns out to be very different from the non-relativistic scheme. In particular the division of the states into the upper rotational and the lower vibrational ones undergoes radical transformation. The role of the relativistic effects would become more pronounced and would be more important in much stronger field in the interesting case of the spectra of multiply charged ions in the magnetic fields possessing a much greater fine structure.

REFERENCES

1. P.A. Braun, Rev. Mod. Phys. 65:115 (1993).
2. J.B. Delos, S.K. Knudson, and D.W. Noid, Phys. Rev. A 28:7 (1983).
3. E.A. Solov'ev, Pis'ma Zh. Eksp. Teor. Fiz. 34:278 (1981) [JETP Lett. 34:265 (1981)].
4. E.A. Solov'ev, Zh. Eksp. Teor. Fiz. 82:1762 (1982) [Sov. Phys. JETP 55:1017 (1982)].

NEUTRAL TWO-BODY SYSTEMS OF CHARGED PARTICLES IN EXTERNAL FIELDS

L.S. Cederbaum and P. Schmelcher

Theoretische Chemie
Universität Heidelberg
Im Neuenheimer Feld 253
D-69120 Heidelberg
Federal Republic of Germany

1. FUNDAMENTAL ASPECTS

The general problem we are concerned with is a neutral system of two oppositely charged particles interacting via the Coulomb potential in the presence of an external homogeneous magnetic field and an additional electric field oriented perpendicular to the magnetic one. The corresponding Hamiltonian takes on the following appearance:

$$\mathcal{H}_L = \sum_{i=1}^{2} \left[\frac{1}{2m_i} (\mathbf{p}_i - e_i \mathbf{A}_i)^2 - e_i \mathbf{E} \mathbf{r}_i \right] + V(|\mathbf{r}_1 - \mathbf{r}_2|) \tag{1}$$

where $e_i, m_i, \mathbf{A}_i$ and $\mathbf{E}$ denote the charges, masses, vector potential and electric field vector, respectively. $\{\mathbf{r}_i, \mathbf{p}_i\}$ are the Cartesian coordinates and momenta in the laboratory coordinate system. In the absence of the external fields the total canonical momentum is conserved and equals the total kinetic momentum. The resulting symmetry group is formed by the translations in coordinate space. In center-of-mass (CM) and internal variables the CM motion then decouples, i.e. separates, completely from the relative motion.

In the presence of the external fields the vector potential appears in the Hamiltonian (1) and the space translation symmetry is, therefore, lost. However, there exists a generalization, i.e. the phase space translation group [1] which provides a symmetry associated with the CM motion of the system in the presence of the external fields. The new conserved quantity which is the corresponding generalization of the total canonical momentum of the field-free case is the so-called pseudomomentum $\mathbf{K}$ [1, 2]

$$\mathbf{K} = \sum_{i=1}^{2} (\mathbf{p}_i - e_i \mathbf{A}_i + e_i \mathbf{B} \times \mathbf{r}_i) - M \mathbf{v}_D \tag{2}$$

where M and $\mathbf{B}$ are the total mass and magnetic field vector, respectively and the term $M\mathbf{v}_D$ has been included in the pseudomomentum for convenience. $\mathbf{v}_D = \frac{(\mathbf{E} \times \mathbf{B})}{B^2}$ is the

Atoms and Molecules in Strong External Fields
Edited by Schmelcher and Schweizer, Plenum Press, New York, 1998

drift velocity of free charged particles in crossed fields. The latter is independent of the charge and mass of the particles [3]. Since the pseudomomentum is conserved its components commute with the above Hamiltonian (1):

$$[\mathbf{K}_\alpha, \mathcal{H}_L] = 0 \tag{3}$$

For a neutral system with $e_1 = -e_2 = e$, i.e. vanishing total charge, the components of $\mathbf{K}$ additionally commute among themselves:

$$[\mathbf{K}_\alpha, \mathbf{K}_\beta] = 0 \tag{4}$$

For charged two-body systems, where $e_1 + e_2 \neq 0$ holds, the components of $\mathbf{K}$ do not commute. This gives rise to a variety of interesting phenomena which are discussed elsewhere [4].

As a consequence of Eq. (4), the components of $\mathbf{K}$ can be made sharp simultaneously. This fact is the key ingredient for the so-called pseudoseparation of the CM motion which introduces the pseudomomentum as a canonical momentum and thereby eliminates the canonical CM coordinate. However, this does not mean that the CM and internal degrees of freedom decouple, i.e. are separated. The pseudoseparation merely uses the above exact constants of motion of the Hamiltonian in order to transform the coupling of the collective and internal degrees of freedom to a particularly simple form. The pseudoseparation transformation consists of a combined coordinate and unitary gauge transformation. In the literature [1, 2, 5] this transformation has been performed with the assumption of a fixed gauge for the vector potential in the Hamiltonian. However, fixing the gauge from the very beginning, i.e. already for the Hamiltonian in the laboratory coordinate system, possesses a serious drawback: It is not possible to discern between gauge dependent and gauge invariant terms in the transformed Hamiltonian. A gauge invariant pseudoseparation is, therefore, desirable and has been performed very recently [6]. As a result of this gauge independent pseudoseparation we obtain the following transformed Hamiltonian

$$\mathcal{H} = \mathcal{T} + \mathcal{V} \tag{5}$$

where

$$\mathcal{T} = \frac{1}{2\mu}\left(\mathbf{p} - q\mathbf{A}(\mathbf{r})\right)^2 \tag{6}$$

and

$$\mathcal{V} = \frac{1}{2M}\left(\mathbf{K} - e\mathbf{B} \times \mathbf{r}\right)^2 + V(r) + \frac{M}{2}\mathbf{v}_D^2 + \mathbf{K}\mathbf{v}_D \tag{7}$$

with the charge $q = \frac{e\mu}{\hat{\mu}}$, where $\mu = \frac{mM_0}{M}$ and $\hat{\mu} = \frac{mM_0}{M_0 - m}$ are different reduced masses. $\mathbf{K}$ is now the constant vector of the pseudomomentum and $\{\mathbf{r}, \mathbf{p}\}$ denote the canonical pair of variables for the internal relative motion. The Hamiltonian (5) is the sum of two terms: The kinetic energy $\mathcal{T}$ of the relative motion and the potential $\mathcal{V}$.

The explicite form of the kinetic energy $\mathcal{T}$ depends on the chosen gauge via the vector potential $\mathbf{A}$. The important novelty with respect to our Hamiltonian $\mathcal{H}$ is the appearance of the potential term $\mathcal{V}$. Apart from the Coulombpotential V and the constant term $\frac{M}{2}\mathbf{v}_D^2 + \mathbf{K}\mathbf{v}_D$ an additional potential term occurs in the total potential $\mathcal{V}$:

$$V_o = \frac{1}{2M}\left(\mathbf{K} - e\mathbf{B} \times \mathbf{r}\right)^2 \tag{8}$$

The latter term is gauge independent, i.e. does not contain the vector potential, and, therefore, fully deserves the interpretation of an additional potential term for the internal motion with the kinetic energy given in Eq. (6). Apart from the constant $\frac{\mathbf{K}^2}{2M}$

the potential V_o contains two coordinate dependent parts. The term linear in the coordinates $-\frac{e}{M}(\mathbf{K} \times \mathbf{B})\mathbf{r}$ consists of two Stark terms: one which is due to the external electric field $\mathbf{E}$ and a second one which is a motional Stark term with an induced constant electric field $\frac{1}{M}((\mathbf{K} + M\mathbf{v}_D) \times \mathbf{B})$. The latter electric field is oriented perpendicular to the magnetic one and arises due to the collective motion of the atom through the homogeneous magnetic field. Besides the linear terms there exists a quadratic, i.e. diamagnetic, term $\frac{e^2}{2M}(\mathbf{B} \times \mathbf{r})^2$ in the potential V_o.

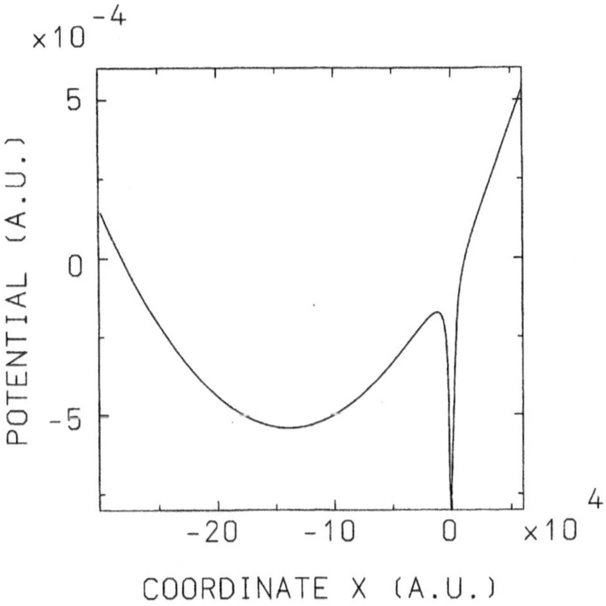

Figure 1. Intersection of the potential V (see Eq.(7)) along the direction of the motional electric field. Values of the pseudomomentum and magnetic field strength are $\mathbf{K} = (0, 1.4, 0)$ and $\mathbf{B} = (0, 0, 10^{-5})$ in atomic units ($B = 1a.u.$ corresponds to $2.35 \times 10^5 T$).

In the following we illustrate and discuss the qualitative properties of the potential $\mathcal{V}$. Fig. 1 shows an intersection of the potential $\mathcal{V}$ along the direction of the motional electric field ($y = z = 0$) for the choice $\mathbf{B} = (0, 0, B)$, $\mathbf{K} = (0, K, 0)$ and a vanishing external electric field. The Coulomb term $(\frac{1}{|x|})$ dominates for small values of the x coordinate. With increasing values of the coordinate x the Stark term $\left(\frac{-e}{M}\right) BKx$ increases and becomes comparable with the strength of the Coulomb potential. The diamagnetic term $(\frac{e^2}{2M})B^2x^2$ is, for our choice of the parameter values (see Sect. 2), in this coordinate region still negligible. Due to the competition of the Coulomb and Stark term a saddle point arises which in Fig. 1 is located at approximately $x \approx -1.1 \times 10^4$ a.u. For even larger absolute values of the of the x coordinate the Coulomb potential becomes small and the shape of the potential $\mathcal{V}$ is more and more determined by the diamagnetic quadratic potential term $(\frac{e^2}{2M})B^2x^2$. Due to the competition of the Stark and diamagnetic terms our potential $\mathcal{V}$ now develops a minimum which in Fig. 1 is located at approximately $x \approx -1.4 \times 10^5$ a.u. The existence of both the saddle point and the minimum depends, of course, on the values of the magnetic field strength and the pseudomomentum. For a derivation and discussion of the conditions for their existence we refer the reader to Sect. 2 and to the literature [6–9].

The above-discussed properties of our potential $\mathcal{V}$ have important implications on the dynamical behaviour of the atom. First of all, we observe that the ionization of the atom can take place only in the direction of the magnetic field axis: In the direction perpendicular to the magnetic field vector the diamagnetic term $\frac{e^2}{2M}(\mathbf{B} \times \mathbf{r})^2$ is dominating for large distances $\rho = (x^2 + y^2)^{1/2}$ and causes a confining behaviour of the potential $\mathcal{V}$. Therefore, ionization is not allowed perpendicular to the magnetic field. The second important observation is the fact that the presence of a minimum in the potential $\mathcal{V}$ leads to a potential well and consequently to new bound states or trajectories inside this well. These trajectories as well as the corresponding quantum mechanical states are extended objects in the sense that the electron and the nucleus are far from each other. The quantum dynamics inside this well will be discussed in Sect. 2 whereas the classical dynamics for energies beyond the saddle point energy is the subject of investigation of Sect. 4.

Finally we emphasize that the above potential V_o is inseparably connected with the finite nuclear mass. Assuming an infinite nuclear mass would simply yield $V_o \equiv 0$ and the above-discussed properties of the total potential $\mathcal{V}$ would disappear. In order to obtain the correct qualitative behaviour and properties of the atom in a strong magnetic field it is, therefore, necessary to treat the atom as an inseparable two-body system.

2. THE HYDROGEN ATOM IN CROSSED ELECTRIC AND MAGNETIC FIELDS

In Sect. 1 we showed that, in addition to the Coulomb singularity, there exists an outer potential well for the relative motion. This well is approximately an anisotropic harmonic potential in the vicinity of its minimum. Accordingly, we can use the results of Ref. [6] for the anisotropic harmonic potential and compare them to the numerically calculated exact eigenstates of the hydrogen atom in the well. Before doing this let us briefly discuss the conditions for the existence of the outer potential well [6–9].

Assuming the same orientations of the magnetic field vector and pseudomomentum as in Fig. 1 we obtain from the condition for a potential extremum $\frac{\partial \mathcal{V}}{\partial r_i} = 0$ (see Eq.(7)) for the y and z coordinate $y_0 = z_0 = 0$ and an equation for the x coordinate

$$x_0^3 + \frac{K}{B}x_0^2 - \frac{M}{B^2} = 0 \tag{9}$$

where $x_0 < 0$. In order to get both a minimum and a saddle point the cubic equation must have three real zeros. From the form of the discriminant we obtain the following inequality

$$K^3 > \frac{27}{4}BM \tag{10}$$

as a necessary condition for the existence of a minimum. In the literature [7] an explicit approximation formula has been given for the minimum coordinate: $x_0 \simeq -\frac{K}{B} + \frac{KM}{K^3 - 2MB}$. Hence, for laboratory field strengths ($B \sim 10^{-5}$a.u.) and pseudomomenta of the order of 1a.u. the minimum is located at a distance of about 10^5 a.u. from the Coulomb singularity. Therefore, for states in the well the electron and proton are separated about 100000 times as much as they are in the ground state of the hydrogen atom without external fields, i.e. we encounter a strongly delocalized atom of almost macroscopic size. Since the well exists only for negative values of x the separation is fixed in a certain direction of space. As a consequence, the atom possesses a large permanent dipole moment in contrast to the well known Rydberg states in a magnetic field which do not exhibit a permanent dipole moment for vanishing pseudomomentum.

In the following we will investigate the quantum mechanical states in the outer potential well [6–9] for laboratory field strengths. Many states up to a very high degree of excitation have been considered [6]. In order to interpret the full numerical results for the energies and wave functions of the hydrogen atom in a magnetic field (or alternatively crossed electric and magnetic fields) let us perform an expansion of the Coulomb potential $\frac{1}{|\mathbf{r}|}$ around the minimum of the potential well. Including only terms up to x_0^{-3} we get the approximated potential [6]

$$V_h = \frac{\mu}{2}\omega_x^2 x^2 + \frac{\mu}{2}\omega_y^2 y^2 + \frac{\mu}{2}\omega_z^2 z^2 + C_3 \tag{11}$$

where we used new coordinates with the origin at the minimum of the well. The frequencies are given by

$$\omega_{x,y} = \sqrt{\frac{2}{\mu}\left(\frac{B^2}{2M} \pm \frac{1}{x_0^3}\right)} \quad , \quad \omega_z = \sqrt{\frac{1}{\mu}\left(-\frac{1}{x_0^3}\right)} \tag{12}$$

and the constant reads $C_3 = \frac{2}{x_0} - \frac{B^2 x_0^2}{2M} + \frac{K^2}{2M} + \frac{ME^2}{2B^2} + \frac{KE}{B}$. With this potential the Hamiltonian of the hydrogen atom takes on the form of the Hamiltonian of a charged particle in a magnetic field with anisotropic harmonic interaction. Therefore, we can use the analytical results of Ref. [5] as an approximation for the low-lying energies and wave functions of the hydrogen atom in the well. The approximate energies are given by

$$E_{n_+ n_- n_z} = \left(n_+ + \frac{1}{2}\right)\omega_+ + \left(n_- + \frac{1}{2}\right)\omega_- + \left(n_z + \frac{1}{2}\right)\omega_z + C_3 \tag{13}$$

with the frequencies

$$\omega_{+,-} = \frac{1}{\sqrt{2}}\sqrt{\omega_x^2 + \omega_y^2 + \omega_c^2 \pm \sqrt{\left(\omega_x^2 + \omega_y^2 + \omega_c^2\right)^2 - 4\omega_x^2\omega_y^2}} \tag{14}$$

The quantum numbers n_+, n_- and n_z apply to the eigenstates of the Hamiltonian with the harmonic potential V_h only. However, we will use them as labels for the states of the hydrogen atom, too. For all calculated cases ω_+ will be much larger than ω_-, therefore n_+ will always be zero for the hundreds of states considered here. An increase of the value of the label n_- will correspond to an increase of the extension of the wave function in the x– and y–directions, and an increase of the value of the label n_z will correspond to an increase of the extension of the wave function in the z–direction.

In a second step, in order to determine the influence of the anharmonicity in the exact potential we will expand the $\frac{1}{|\mathbf{r}|}$-term up to higher powers of the components of $\mathbf{r}$ and treat them as small perturbations to the harmonic approximation of the Hamiltonian by means of first order perturbation theory. These perturbative calculations offer insight into the effects of the anharmonic parts of the potential onto the energies and the form of the wave functions. For details we refer the reader to Ref. [6].

In the following we discuss the results of these numerical calculations of the exact eigenenergies and wave functions and compare them to the corresponding quantities obtained by the harmonic approximation as well as the perturbation theoretical calculation.

Our choice of the parameter values is $K = 0.6$, $B = 10^{-5}$ and $E = 0$. The energy of the ground state in the potential well is $E_0 = -1.2499 \times 10^{-5}$ a.u. The binding energy of the ground state is $E_B = E_S - E_0 = 1.75 \times 10^{-5}$ a.u. where E_S is the threshold energy, i.e. the lowest energy of the ionized system. Even though the binding energies of the states in the well are relatively weak they should be stable as long as collisional interaction is prevented. The energy gap between the ground and first excited state corresponds to a frequency of the order of magnitude of a few tens of MHz.

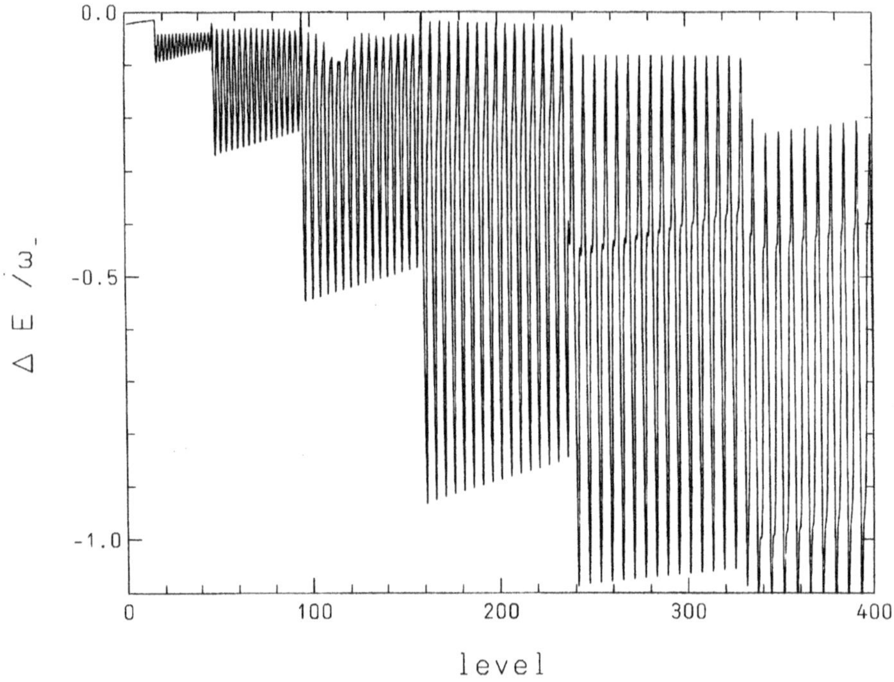

Figure 2. Difference between harmonic approximated and exact energies of a hydrogen atom in a magnetic field in units of the frequency ω_- as a function of the energy level. Parameters: $B = 10^{-5}$, $K = 0.6$ in atomic units.

In the following we consider the deviation of the exact energies from those of the harmonic approximation as a function of the level number. In Fig. 2 we have plotted the energy difference between the harmonic approximation and the exact energies of the hydrogen atom in units of ω_-. We see that the difference grows stepwise while neighboring states show very different deviations from the harmonic approximation. To explain these features let us look at the energy level number 162. We see that the difference between exact and approximated energies for this level is much larger than for the levels below 162. Level 162 has the quantum numbers $n_+ = n_- = 0$ and $n_z = 4$, i.e. the quantum number $n_z = 4$ appears for the first time. Looking at higher levels there are maxima of energy differences $\Delta E/\omega_-$ every 5th level above 162 up to level 237. For these levels (162,167, etc.) the quantum number n_z is 5 and $n_- = 1, 2, \ldots$. Between two levels with $n_z = 5$ there are levels with $n_z < 5$ and apparently the energy difference for these is smaller. That is, the difference between harmonic and exact energies is mostly determined by the quantum number n_z. Hence, the anharmonicity of the exact potential is most pronounced in z-direction. This can also be seen in perturbation theory for higher terms of the expansion of the Coulomb potential where the major contributions to the energy corrections are due to those terms containing high powers of z.

Next let us turn to the wave functions of the hydrogen atom in the well. For the states with either n_z or $n_- = 0$ the expectation values of the $y-$ and $z-$ coordinate are zero. For $n_z = 0$ the wave function is centered near the minimum of the well. With increasing quantum number n_z the center of the wave function is shifted towards

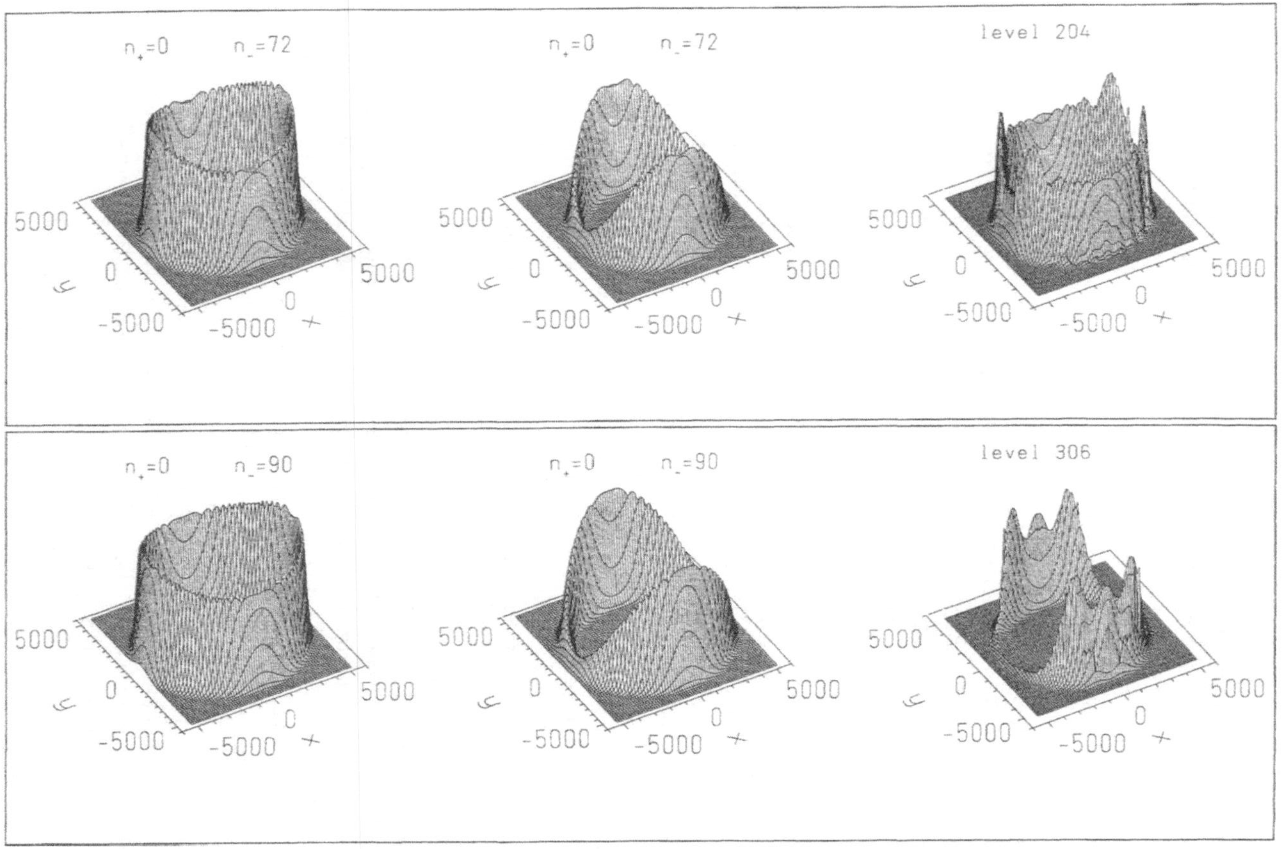

Figure 3. Square of the absolute value of the wave functions of the hydrogen atom in a magnetic field. Shown are intersections in the plane perpendicular to the magnetic field for the harmonic approximation, for perturbation theory, and for the exact Hamiltonian (from left to right). Parameters: $B = 10^{-5}$, $K = 0.6$ in atomic units.

negative values of the x coordinate. This is due to the increasing extension of the wave function in the z−coordinate which makes these states "feel" the potential in regions of larger z where the minimum of the well is shifted to negative x-values. We also observe that the extension of the wave functions in x and y is the same for states with the same n_+ and n_- and that the extension in z is the same for states with the same n_z. This shows that excitations in the plane perpendicular to the magnetic field are almost independent from those in z-direction. Furthermore, in general the extension of the wave functions in the x-y-plane is much smaller than parallel to the field. Note that the extension of the wave functions in the x-y-plane is also much smaller than it would be in the same potential without the presence of a magnetic field. Apparently, the form of the wave function is substantially determined by the field-dependent kinetic energy.

Looking at the form of the wave functions we restrict ourselves to intersections in the plane perpendicular to the magnetic field. In Fig. 3 we have plotted the square of the absolute value of the wave functions of two states that are highly excited in the plane perpendicular to the magnetic field. The ground state and less highly excited states are not shown since they do not exhibit major differences to the harmonic approximation. Shown are results in the harmonic approximation, this approximation plus corrections from first order perturbation theory for higher terms of the $\frac{1}{|\mathbf{r}|}$-expansion, and the numerical results for the exact wave functions. We see that the harmonic approximation leads to a wave function whose intensity is evenly distributed over an elliptic ring surrounding the minimum of the well. In contrast, due to corrections from anharmonic terms, the intensity of the wavefunctions of the hydrogen atom for states which are highly excited in the plane perpendicular to the magnetic field is largely reduced close to the x−axis both in the perturbation calculation and the exact results. Apparently this deviation from the wave functions of the harmonic approximation is due to the anharmonicity of the potential. In the perturbation calculation we included only a few terms of the $\frac{1}{|\mathbf{r}|}$-expansion. That is why the decrease of intensity of the wave functions appears on the level of perturbation theory for even smaller values of the label n_- than in the exact calculation.

Let us conclude with two remarks. First we would like to draw the reader's attention to the fact that the fundamental aspects discussed in Sect. 1 are valid for any neutral two-body system in a magnetic field. The existence of the outer potential well is a universal property of such systems provided that the pseudomomentum or alternatively the external electric field are sufficiently large. Hereby neutral particle hole systems, i.e. excitons, occuring for example in semiconductor bulk systems, are of particular interest. Since the Coulomb potential is screened by the ionic background and since the effective masses of the particle and hole are of comparable order of magnitude, laboratory fields are strong already for the ground state of the exciton. The extension of the bound states in the outer potential well is then of the same order of magnitude as the extension of the well-known states localized in the well due to the Coulomb singularity (see Ref. [10]).

Our second remark concerns the experimental observability of the above-discussed delocalized bound states of the hydrogen atom in crossed electric and magnetic fields. Direct state-to-state transitions for bound states in the outer well should in principle be observable in the radio-frequency regime. A second, probably more promising, approach to experimental verification of the existence of states in the well is the measurement of their dipole moment. There have been experiments that indicate the existence of atoms with very large dipole moments in magnetic fields [11, 12] for energies above the saddle point energy. In Ref. [11] results of experiments with Rydberg atoms in magnetic fields

($B \approx 0.1 \times 10^{-5}$ a.u.) were published. Their rough estimate for the dipole moment of these atoms is 1.9×10^4Debye. This value of the dipole moment is roughly of the order of magnitude to be expected.

3. CLASSICAL DIFFUSION OF THE CENTER OF MASS

The subject of investigation of the present section is the classical interaction of the CM and internal motion for neutral two-body systems and in particular for the hydrogen atom in a strong homogeneous magnetic field [13, 14]. For the present we concentrate on the case of a vanishing pseudomomentum $\mathbf{K}$ (the case $\mathbf{K} \neq 0$ will be discussed in Sect. 4). The relevant Hamiltonian is then given by Eq. (5) for $\mathbf{K} \equiv 0$

$$\mathcal{H}_o = \frac{1}{2\mu}\left(\mathbf{p} - e\frac{\mu}{\hat{\mu}}\mathbf{A}(\mathbf{r})\right)^2 + \frac{e^2}{2M}(\mathbf{B} \times \mathbf{r})^2 + V(r) \tag{15}$$

The component of the internal angular momentum parallel to the magnetic field is a conserved quantity for our case of a vanishing electric field and pseudomomentum $\mathbf{K} \equiv 0$. In spite of the fact that there appear no CM degrees of freedom in the Hamiltonian (15) the CM motion is by no means separated from the internal motion. This can be seen by establishing the equations of motion belonging to the Hamiltonian (5) and finally putting $\mathbf{K} = 0$. As a result, one obtains the following equation of motion for the CM

$$\dot{\mathbf{R}} = -\frac{e}{M}(\mathbf{B} \times \mathbf{r})) \tag{16}$$

where $\mathbf{R}$ is the CM coordinate vector. The motional Stark term $-\frac{e}{M}(\mathbf{K} \times \mathbf{B})\mathbf{r}$ in Eq. (5), therefore, intimately couples the CM and internal motion. The CM velocity (see Eq. (16)) is determined by the components of the relative coordinate perpendicular to the magnetic field. It is important to notice that the second quadratic term of the Hamiltonian $\mathcal{H}_o$ represents the kinetic energy of the CM according to Eq. (16), i.e. we have $\frac{M}{2}\dot{\mathbf{R}}^2 = \frac{e^2}{2M}(\mathbf{B} \times \mathbf{r})^2$. Since the dynamics of the cyclic CM coordinate is determined by the internal motion, the natural question arises how the transition from regularity to chaos in the internal motion reflects itself in the behaviour of the CM [13, 14]. The phase space of the internal motion is restricted to the energy shell whereas the phase space of the CM motion is, at least in principle, unbounded. Therefore, one may ask whether or not the phase space is filled out by the CM motion depending on the regularity or irregularity of the internal motion.

In the absence of a magnetic field, i.e. for $\mathbf{B} = 0$, the pseudomomentum coincides with the total canonical and kinetic momentum. In field-free space $\mathbf{K} \equiv 0$, therefore, means a vanishing CM velocity and the CM stands still. Let us begin our investigation of the CM motion in the presence of a magnetic field with the case of regular internal motion where the Coulomb potential dominates the dynamics. Fig. 4 shows a typical trajectory of the CM in the coordinate plane perpendicular to the magnetic field. Since the internal motion is quasiperiodic, the velocity as well as the coordinate of the CM are both also quasiperiodic. Only a bounded part of phase space is, therefore, filled out by the trajectory of the CM. The confinement of the CM trajectories to a circular bounded part of the phase space is a general feature of the deep regular regime. For a complete classification of the phase space in the regular regime accessible by low order classical perturbation theory we refer the reader to the literature (Ref. [14] contains a classification of the possibilities of the CM motion according to the classification of the internal motion given in Ref. [15]).

Fig. 4 also shows a typical center of mass trajectory for the case of a fully chaotic internal phase space. The eyecatching new feature is that the motion is no more restricted to some bounded volume of phase space. The trajectory of the CM motion of the hydrogen atom in the plane perpendicular to the magnetic field now closely resembles the random motion of a Brownian particle. In fact, the underlying equation of motion (16) for the CM motion is a Langevin-type equation without friction. The corresponding stochastic Langevin force is replaced by our intrinsic chaotic force $-e(\mathbf{B} \times \dot{\mathbf{r}})$. A main characteristic of the random Brownian motion is the diffusion law, i.e. the linear dependence of the travelled mean-square distance on time. The mean square distance $< \rho_s^2 >$ as a function of time for ensembles of trajectories for the hydrogen atom in the fully chaotic regime was investigated in Refs. [13,14]. Within statistical accuracy the dependence was shown to be linear. The mean square distance $< \rho_s^2 >$ of the CM after the time t, therefore, obeys the following diffusion equation

$$< \rho_s^2 >= D_s t \tag{17}$$

where D_s is the corresponding diffusion constant. Our intrinsic chaotic force, which is the kicker of the CM motion, possesses, therefore, the property of randomness in the sense that it provides the well-known diffusion law.

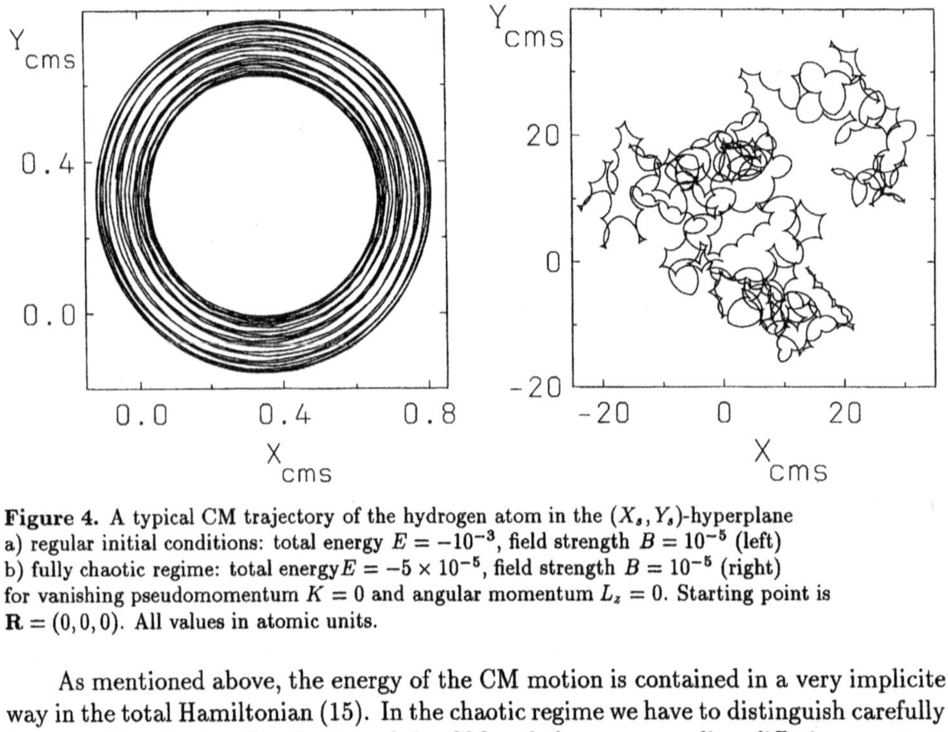

Figure 4. A typical CM trajectory of the hydrogen atom in the (X_s, Y_s)-hyperplane
a) regular initial conditions: total energy $E = -10^{-3}$, field strength $B = 10^{-5}$ (left)
b) fully chaotic regime: total energy $E = -5 \times 10^{-5}$, field strength $B = 10^{-5}$ (right)
for vanishing pseudomomentum $K = 0$ and angular momentum $L_z = 0$. Starting point is $\mathbf{R} = (0,0,0)$. All values in atomic units.

As mentioned above, the energy of the CM motion is contained in a very implicite way in the total Hamiltonian (15). In the chaotic regime we have to distinguish carefully between the velocity distribution of the CM and the corresponding diffusion constant resulting from the linear diffusion law. For a typical laboratory magnetic field strength $B = 10^{-5}$a.u. (≈ 2.35 Tesla) and a binding energy of 5×10^{-5}a.u. which is easily accessible experimentally, we arrive at a typical mean CM velocity of 10^{-5}a.u. ($\approx 20\frac{m}{s}$). The corresponding diffusion constant is of the order of magnitude of $1.5 \times 10^{-9}\frac{m^2}{s}$. We emphasize that these are results for a vanishing pseudomomentum $\mathbf{K} = 0$ which in the case of the presence of a magnetic field obviously does not mean that the CM velocity is equal to zero. Only in the field-free case $\mathbf{B} = 0$ implies that the CM stands still.

Other important examples which exhibit both confinement and diffusion in the classical dynamics of their cyclic collective coordinates are the positronium [16] and the excitonic atom [10]. Because of the comparable masses of the two particles in both cases the mean CM velocity as well as the diffusion constant are orders of magnitude larger than the corresponding values of the hydrogen atom.

For laboratory field strengths the chaotic regime corresponds to highly excited Rydberg states for which the quantum of action is a multiple of the elementary quantum of action $\hbar$. We, therefore, expect that an approach via classical dynamics is justified and gives some limited insight into the actual physical properties of the atom and, of course, is of interest by its own value. However, it is a priori not evident whether the above observed diffusion of the CM in the chaotic regime survives quantization or whether quantum interference effects will destroy the diffusion, i.e. localize the atom. In addition the effect of quantum localization might depend on the system under consideration (hydrogen atom, positronium, excitons).

4. INTERMITTENT DYNAMICS: A TYPICAL THRESHOLD PHENOMENON FOR FINITE PSEUDOMOMENTUM

In the present section we investigate the classical dynamics of a highly excited neutral two-body system and in particular of the hydrogen atom for a *nonvanishing pseudomomentum* $\mathbf{K}$ [17]. We will consider energies which are close to the ionization threshold and above the energy of the saddle point. The latter has been established in Sects. 1 and 2 within the context of our discussion of the potential picture for a neutral two-body system (see also Fig. 1). In Sect. 2 we derived an inequality for the absolute value of the pseudomomentum which has to be fulfilled in order to obtain an outer potential well. In the following we assume the existence of the outer potential well, i.e. we consider sufficiently large values of the pseudomomentum or the external electric field.

Let us begin our investigation of the classical dynamics by establishing the equations of motion belonging to the Hamiltonian (5)

$$\dot{\mathbf{R}} = \frac{1}{M}\mathbf{K} - \frac{e}{M}(\mathbf{B} \times \mathbf{r}) \tag{18}$$

$$\dot{\mathbf{r}} = \frac{1}{\mu}\mathbf{p} - \frac{e}{2\hat{\mu}}(\mathbf{B} \times \mathbf{r}) \tag{19}$$

$$\dot{\mathbf{p}} = -\frac{e}{M}(\mathbf{B} \times \mathbf{K}) - \frac{e}{2\hat{\mu}}(\mathbf{B} \times \mathbf{p}) + \frac{e^2}{4\mu}\mathbf{B} \times (\mathbf{B} \times \mathbf{r}) - e^2\frac{\mathbf{r}}{|\mathbf{r}|^3} \tag{20}$$

In contrast to the case of a vanishing pseudomomentum, which has been discussed in the previous section, the pseudomomentum now appears in both the equation for the CM (18) and the internal equation of motion (20). Apart from the purely translational term $(\frac{\mathbf{K}}{M})t$ the CM motion is again completely determined by the internal coordinate $\mathbf{r}$.

The typical new phenomenon for the trajectories of the highly excited hydrogen atom with nonvanishing pseudomomentum $\mathbf{K} \neq 0$ is their intermittent behaviour. Intermittency means that the trajectory alternately shows both quasiregular and chaotic phases. Fig. 5 shows for a typical trajectory the projection of the internal motion on a plane perpendicular to the magnetic field axis. One immediately realizes that there exist two alternating types of motion. During one phase of motion the electron and the nucleus are close together in the x, y plane and this shows up through the

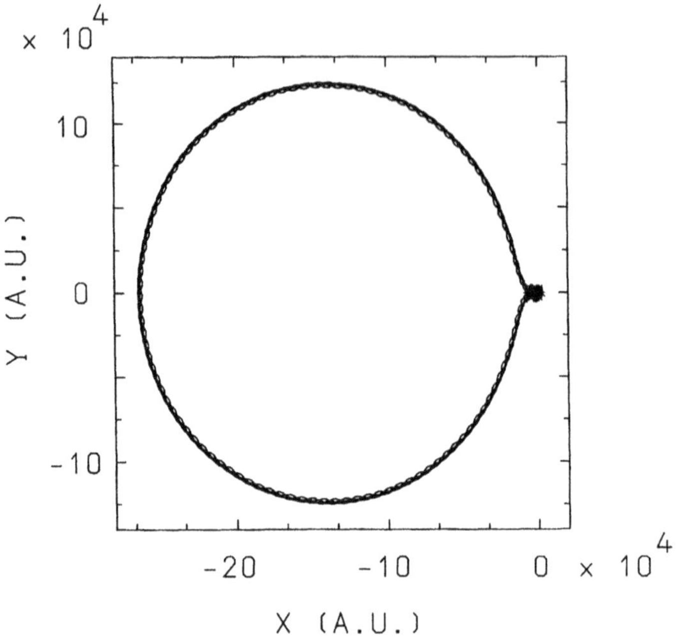

Figure 5. Typical intermittent trajectory of the internal motion of the hydrogen atom for nonvanishing pseudomomentum (projection onto the plane perpendicular to the magnetic field). Parameter values are: $B = 10^{-5}$, $E = 1.722 \times 10^{-4}$ and $\mathbf{K} = (0, 1, 0)$. All values in atomic units.

black bubble on the very rhs of Fig. 5. During this phase of motion the Coulomb and diamagnetic interactions are of comparable order of magnitude and the trajectory is, therefore, chaotic (see Refs. [18, 19] for a determination of local Ljapunov exponents). During the other regular looking phase the electron and the nucleus move far apart from each other. The relative motion in the x, y plane then approximately takes place on a circle with a large radius. Here the Coulomb energy provides only a small perturbation to the dominating magnetic interaction. The radius of the approximate circular large amplitude motion is given by

$$r = -\frac{1}{eB^2} |\mathbf{B} \times \mathbf{K}| \tag{21}$$

i.e. it is completely determined by the field strength and in particular the value of the pseudomomentum. On the other hand, we obtain a completely different interpretation of the pseudomomentum if the electron and the nucleus are very close together. In the latter case the Coulomb dominates the magnetic interaction and hence the pseudomomentum is approximately the linear kinetic momentum of the translational CM motion.

Fig. 6 shows the CM motion for the trajectory whose internal motion is given in Fig. 5. It consists of alternating phases of purely translational and circular motions. As already mentioned, the electron and the nucleus are strongly bound, i.e. close together, during the time interval of chaotic internal motion. This is precisely the time period during which the CM performs an almost purely translational motion. The time periods of quasiregular circular internal motion correspond to the periods of circular CM motion. Intermittency, therefore, shows up in the CM motion by alternating phases of more or less straightlined and circular motion.

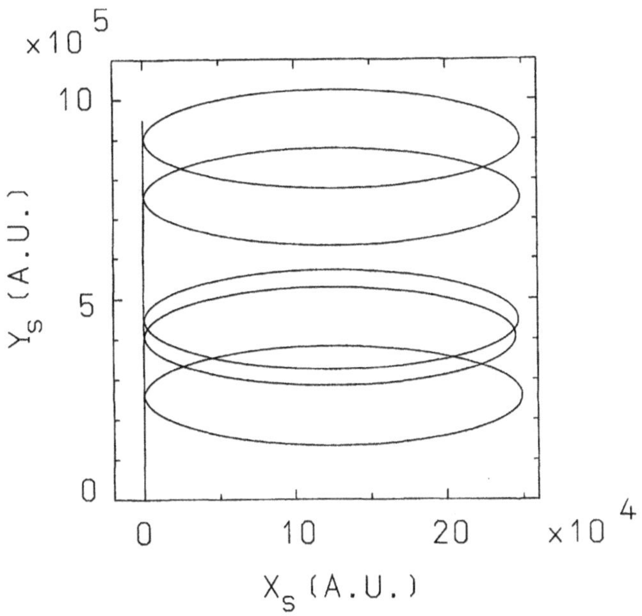

Figure 6. Center of mass motion of the hydrogen atom belonging to the internal motion shown in Fig. 5. The parameter values are the same as in Fig. 5 and the starting point is $\mathbf{R} = (0,0,0)$. All values in atomic units.

The sudden extension of phase space with increasing energy is a phenomenon which can be easily understood in the context of the potential picture discussed in Sects. 1 and 2. For energies below the saddle point energy the internal motion is confined either to the singular Coulomb well or to the outer potential well. For energies above the saddle point energy a sudden extension of the available coordinate space takes place and intermittency as a typical phenomenon occurs. The phases of strongly bound motion are located above the Coulomb well in the coordinate space whereas the large amplitude motion covers the upper part of the outer potential well.

If we consider the motion of the electron and the nucleus in a plane perpendicular to the magnetic field in the *laboratory coordinate system* we encounter an amazing phenomenon. During the quasiregular phases of motion the electron is localized in a small range of coordinate space whereas the nucleus performs the large amplitude motion on the circle shown in Fig. 5. At first glance this statement seems to contradict the traditional physical picture in which the light electron moves around the heavy nucleus. However, since the Coulomb energy provides only a small perturbation to the magnetic interaction during the quasiregular phase of motion, we expect the nucleus and the electron to perform their individual cyclotron motion which are more or less perturbed by the Coulomb interaction. The radius of the cyclotron motion of the nucleus is much larger than that of the electron due to its bigger mass. The strong localization of the electron and the large amplitude motion of the nucleus is, therefore, a characteristic feature for the case of a strongly dominating magnetic field. The electron performs a large amplitude in the direction parallel to the magnetic field.

For a detailed understanding of this phenomenon as well as the dynamical origin of the intermittent behaviour of the trajectories we refer the reader to the literature [17].

REFERENCES

1. J. Avron, I. Herbst, and B. Simon, Ann. Phys. (N.Y.) **114**, 431 (1978).
2. B. Johnson, J. Hirschfelder, and K. Yang, Rev. Mod. Phys. **55**, 109 (1983).
3. J. Jackson, *Klassische Elektrodynamik*, 2 ed. (de Gruyter, Berlin, 1983).
4. P. Schmelcher and L. Cederbaum, *Atoms and Molecules in Intense Fields, 28-62, Structure and Bonding 86* (Springer-Verlag, Berlin-Heidelberg, 1997).
5. P. Schmelcher, L. Cederbaum, and U. Kappes, *Conceptual Trends in Quantum Chemistry, 1-51* (Kluwer Academic, Dordrecht, 1994).
6. O. Dippel, P. Schmelcher, and L. Cederbaum, Phys. Rev. A **49**, 4415 (1994).
7. P. Schmelcher and L. Cederbaum, Chem. Phys. Lett. **208**, 548 (1994).
8. D. Baye, N. Clerbaux, and M. Vincke, Phys. Lett. A **166**, 135 (1992).
9. I. Dzyaloshinskii, Phys. Lett. A **165**, 69 (1992).
10. P. Schmelcher, Phys. Rev. B **48**, 14642 (1993).
11. M. Fauth, H. Walter, and E. Werner, Z. Phys. D. **7**, 293 (1987).
12. G. Raithel, M. Fauth, and H. Walter, Phys. Rev. A **47**, 419 (1993).
13. P. Schmelcher and L. Cederbaum, Phys. Lett. A **164**, 305 (1992).
14. P. Schmelcher and L. Cederbaum, Z. Phys. D. **24**, 311 (1992).
15. J. Delos, S. Knudson, and D. Noid, Phys. Rev. A **28**, 7 (1983).
16. P. Schmelcher, J. Phys. B **25**, 2697 (1992).
17. P. Schmelcher and L. Cederbaum, Phys. Rev. A **47**, 2634 (1993).
18. H. Fujisaka, Progr. Theor. Phys. **70**, 1264 (1983).
19. P. Grassberger and I. Procaccia, Physica (Amsterdam) **13D**, 34 (1984).

SEMICLASSICAL THEORY OF MULTIELECTRON ATOMS AND THE H_2^+ MOLECULAR ION IN INTENSE EXTERNAL FIELDS

N. H. March

Oxford University
Oxford
England

1. INTRODUCTION

The problem of atoms and molecules in intense external fields is, by now, embracing and interdisciplinary area of considerable importance. One, very academic, reason for this is the occurrence in white dwarfs, and in neutron stars, of very intense magnetic fields. Background to these areas is reviewed in some recent works [1-3] and will not therefore be elaborated further here. From a very practical standpoint, the advent of lasers, with associated intense electric fields, has greatly added to the urgency for understanding the non-linear response of atoms and molecules to such fields.

In the present study, we shall focus attention almost entirely on semiclassical approaches to treating atoms in such intense external fields. This way of tackling the problem has a long history, following the original Thomas-Fermi (TF) statistical method for treating the electronic structure of atoms in zero external fields. Some of the earliest references on the intense magnetic field problem in multielectron atoms are set out, for instance, in the paper by March and Tomishima [4], in which the scaling properties of the total ground-state energy $E(Z, N, B)$ of positive atomic ions with atomic number Z, number of electrons $N \leq Z$, in magnetic fields B which are extremely strong, are discussed. A more extensive treatment which delineates important regimes of magnetic field strength has been given much later by Lieb and coworkers [5] and will be referred to again below. This general area will be approached here, in section 2 below, starting again from the TF statistical theory. Because relativistic effects can play a significant role in intense magnetic fields, we shall present the theory first in relativistic form, but still place some emphasis on results in the non-relativistic limit. Since modern density functional theory has, as one of its prime aims, the construction of a differential equation from which to determine the ground-state electron density, some stress will be placed in the present account on such differential equations, both relativistic and non-relativistic, in the TF theory of multielectron atoms in intense magnetic fields **B**.

Following this discussion of strong **B** fields, section 4 will be concerned with static uniform electric fields E of arbitrary strength. Throughout the article, we shall focus

Atoms and Molecules in Strong External Fields
Edited by Schmelcher and Schweizer, Plenum Press, New York, 1998

on one-electron theory, described by a Hamiltonian H having the general form

$$H = \frac{1}{2m}\left(\mathbf{p} - \frac{e}{c}\mathbf{A}\right)^2 + V(\mathbf{r}) \tag{1}$$

where however in writing eq.(1) we have restricted ourselves for the moment, for simplicity, to the non-relativistic limit. In eq.(1) $\mathbf{A}$ represents the vector potential, defined by

$$\mathbf{B} = curl\ \mathbf{A} \tag{2}$$

which of course leaves open the choice of gauge. $V(\mathbf{r})$ represents the self-consistent potential energy in the atom or molecule under consideration, plus, when dealing with static uniform electric fields a term of the form

$$V_E(\mathbf{r}) = -eEz \tag{3}$$

corresponding to a uniform static field of strength E along the z-axis. Whereas in section 2, which is concerned purely with $\mathbf{B}$ fields, all attention will be focussed on the ground-state electron density $n(\mathbf{r}, B)$, in section 3 the main tool employed will be the Slater sum, denoted by $P(\mathbf{r}, \beta)$ and defined by

$$P(\mathbf{r}, \beta) = \sum_{all\ i} \Psi_i(\mathbf{r})\Psi_i^*(\mathbf{r})exp(-\beta\epsilon_i) \qquad : \quad \beta = (k_B T)^{-1} \tag{4}$$

where the Ψ_i and ϵ_i are the one-electron eigenfunctions and eigenvalues generated by the one-electron Hamiltonian H in eq.(1).

With this brief Introduction, let us turn to the problem of atoms in intense magnetic fields, using semiclassical TF theory.

2. RELATIVISTIC THOMAS-FERMI THEORY FOR MULTIELECTRON ATOMS IN INTENSE MAGNETIC FIELDS

The present writer [6], in considering the role of the virial, has conveniently summarized fully local relativistic Density Functional Theory (DFT). The Euler equation of the corresponding variational problem posed by minimizing the ground state energy with respect to the electron density n, subject to the normalization condition

$$\int n d\mathbf{r} = N, \tag{5}$$

is an equation for the chemical potential μ. This quantity plays the role of the Lagrange multiplier introduced to take care of the normalization requirement (5) as the density is varied. However, it is central to the theory to stress that the chemical potential μ is the same at every point in the inhomogeneous electronic charge cloud in the atom or molecule under consideration. In the extreme high field regime of TF theory, this equation for μ reads [7]

$$\mu = \sqrt{\left(\frac{c^2h^4n^2}{4e^2B^2} + m_0^2c^4\right)} - m_0c^2 + V(\mathbf{r}). \tag{6}$$

This equation relates ground-state electron density $n(\mathbf{r}, B)$ to the potential energy $V(\mathbf{r})$, which has to be determined self-consistently by relating $n(\mathbf{r}, B)$ to $V(\mathbf{r})$ via Poisson's equation of electrostatics.

2.1 Kinetic Energy

Let us note immediately that the non-relativistic (nr) kinetic energy per unit volume $t_{nr}(n)$ is readily extracted from eq.(5) as

$$t_{nr}(n) = \frac{h^4 n^3}{24 m_0 e^2 B^2} \tag{7}$$

where m_0 in the above eqs.(6,7) denotes the electron rest mass. It has been convenient in writing eq.(6) to subtract the energy $m_0 c^2$ associated with this mass. Returning to eq.(7) and noting that the total kinetic energy T_{nr} is given by

$$T_{nr} = \int t_{nr} d\mathbf{r} \tag{8}$$

one can evidently write the non-relativistic chemical potential equation in the form

$$\mu_{nr} = \frac{\delta T_{nr}}{\delta n(\mathbf{r})} + V(\mathbf{r}) = \frac{h^4 n^2}{8 m_0 e^2 B^2} + V(\mathbf{r}) \tag{9}$$

2.2 Self-Consistency

For the fully relativistic eq.(6), one can take the Laplacian ∇^2 acting on μ (equal to zero since μ is constant) to find

$$\frac{1}{2}\nabla\left(\frac{\frac{c^2 h^4 n}{2e^2 B^2}\nabla n}{\sqrt{\frac{c^2 h^4 n^2}{4e^2 B^2} + m_0^2 c^4}}\right) + \nabla^2 V = 0 \tag{10}$$

with the Poisson equation

$$\nabla^2 V = -4\pi n e^2 \tag{11}$$

representing the requirement of self consistency. After some manipulation, involving multiplying both sides of eq.(10) by $\left(\frac{c^2 h^4 n^2}{4e^2 B^2} + m_0^2 c^4\right)^{\frac{3}{2}}$ one is lead to the differential equation for the self-consistent ground-state density n as

$$\frac{1}{2}\left(\left(\frac{c^2 h^4 n^2}{4e^2 B^2} + m_0^2 c^4\right)\frac{c^2 h^4}{2e^2 B^2}\left(n\nabla^2 n\right) + \left(\frac{m_0^2 c^6 h^4}{2e^2 B^2}\right)(\nabla n)^2\right) \quad 4\pi n c^2 \left(\frac{c^2 h^4 n^2}{4e^2 B^2} + m_0^2 c^4\right)^{\frac{3}{2}} = 0 \tag{12}$$

We should add that eq.(12) is valid everywhere except at positions of nuclei, where there are delta functions representing the positively charged nuclei. Another way of viewing eq.(12) is to regard it as relating $n(\mathbf{r}, B)$ directly to the 'reduced' Laplacian $\frac{\nabla^2 n}{n}$ and the square of the so-called local wave number [8] $\left(\frac{\nabla n}{n}\right)^2$ (see also below).

2.3 Non-Relativistic Limit

The non-relativistic limit of eq.(12) can, of course, be taken directly. But perhaps the simplest procedure to obtain this limit is to return to eq.(9) and to form directly $\nabla^2 \mu_{nr}$ (equal to zero). The result

$$\nabla^2 \mu_{nr} = \frac{h^4}{4 m_0 e^2 B^2}\left(n\nabla^2 n + (\nabla n)^2\right) + \nabla^2 V(\mathbf{r}) \tag{13}$$

follows. Using eq.(11) in eq.(13), one readily obtains the desired differential equation for the self-consistent ground-state density $n(\mathbf{r}, B)$ in the non-relativistic limit as

$$\frac{\nabla^2 n}{n} + \left(\frac{\nabla n}{n}\right)^2 = \frac{1}{\pi^3 a_0 n l_B^4} \quad : \quad a_0 = \frac{\hbar^2}{m_0 e^2} \tag{14}$$

where in eq.(14) it has proved convenient to introduce the classical magnetic length l_B defined by

$$l_B = \sqrt{\frac{h}{2\pi eB}} \tag{15}$$

Eq.(14) is then evidently of dimensions $(length)^{-2}$. This eq.(14) makes, more simply, for the non-relativistic limit of eq.(12), the point stressed above that $n(\mathbf{r}, B)$ is related to $\frac{\nabla^2 n}{n}$ and $\left(\frac{\nabla n}{n}\right)^2$ at the same point $\mathbf{r}$. 'Local' density is thereby expressed in terms of density gradients quite precisely in this simplest form of DFT. The work of Hill et al. [9], in which the potential $V(\mathbf{r})$ was the tool employed, gave numerical solutions for atoms with different applied field strengths B. These solutions are readily translated into solutions of the density eq.(12) for the atomic electron density $n(\mathbf{r}, B)$.

Having dealt at some length with the semiclassical theory of the ground-state electron density in atoms and molecules in intense magnetic fields, let us consider briefly H_2^+ in a magnetic field.

3. H_2^+ MOLECULAR ION IN MAGNETIC FIELD: CURRENT DENSITY THEORY

Following the above semiclassical treatment of multielectron atoms, we shall quite briefly discuss the use of current density theory to treat the H_2^+ molecular ion in a magnetic field, following largely the work of Amovilli and March [10]. These authors considered only the case of magnetic field pointing along the internuclear axis in H_2^+, but in a recent study of current density theory, Holas and March [11] have shown that their final result is, in fact, quite general. However, for simplicity let us set out the case considered by Amovilli and March [10]. Then, brief reference will be made to the need to transcend oft-used hydrodynamic-like treatments [11], as clearly pointed out to the present writer by Dr. A. Holas.

Amovilli and March introduce the current density $\mathbf{j}$ in the usual way as

$$\mathbf{j} = n\mathbf{v} + \frac{e\mathbf{A}}{mc}n. \tag{16}$$

One then obtains for the ground state energy, $E^+(R, B)$ say :

$$E^+(R, B) = \frac{H\Psi}{\Psi} = \frac{(\nabla n)^2}{8n^2} - \frac{\nabla^2 n}{4n} + \frac{j^2}{2n^2} + \frac{i\nabla \cdot \mathbf{j}}{2} + V \tag{17}$$

with V simply accounting for potential energy. Her R denotes the internuclear separation, and evidently eq.(17) applies at any point $\mathbf{r}$. For stationary wave functions, and hence time-independent electron density $n(\mathbf{r}, R, B)$, it follows from the equation of continuity that

$$div\, \mathbf{j} = 0 \tag{18}$$

Hence one obtains the desired result for the ground-state energy in terms of electron density n and current density $\mathbf{j}$ as

$$E^+(R, B) = \frac{(\nabla n)^2}{8n^2} - \frac{\nabla^2 n}{4n} + \frac{j^2}{2n} + V \tag{19}$$

which is the result of Amovilli and March [10]: see also Holas and March [11].

One might note here that sometimes one writes a quantum fluid dynamics associated with the single-particle Schrödinger equation. This is done, for example, in March and Deb [12]. One is then led to an Euler-type equation of motion such as given, for example, in eq.(10.3) of the book by the present writer. Contact can then be established with the work of Amovilli and March, and it turns out that the comparison is only correct in detail if one neglects the gradient of the current density $\mathbf{j}$. Then the hydrodynamic result is recovered from the exact treatment of Amovilli and March [10]. Further discussion of this point is taken up by Holas and March [11].

Let us turn to the problem of such systems in strong static uniform electric fields of arbitrary strength E, but now in zero magnetic field. Hence the appropriate (now non-relativistic) Hamiltonian is obtained by putting $\mathbf{A} = 0$ in eq.(1) and incorporating the potential energy term V_ϵ in eq.(2) into the scalar potential energy $V(\mathbf{r})$.

4. ATOMS AND MOLECULES IN STATIC UNIFORM ELECTRIC FIELDS OF ARBITRARY STRENGTH

As mentioned in the Introduction, the main tool to be utilized in the discussion of atoms and molecules in static uniform electric fields of arbitrary strength E is the Slater sum defined in eq.(4). What is important for the semiclassical approach to be developed below is that this sum, say $P_{0E}(\mathbf{r},\beta)$, can be obtained analytically for free electrons in such an electric field $\mathbf{E}$. The result, as given, for example, by Jannussis [13] and by Harris and Cina [14] takes the form

$$P_{0E}(\mathbf{r},\beta) = (2\pi\beta)^{-\frac{3}{2}}\, exp\left(\beta E z + \frac{\beta^3 E^2}{24}\right) \tag{20}$$

Below, we shall also appeal to the off-diagonal generalization of eq.(4) to yield the so-called canonical (or Bloch) density matrix $C(r, r_0, \beta)$, namely

$$C(\mathbf{r},\mathbf{r}_0,\beta) = \sum_{all\ i} \Psi_i(\mathbf{r})\,\Psi_i^*(\mathbf{r}_0)\, exp(-\beta\epsilon_i) \tag{21}$$

which reduces to eq.(4) when one takes the diagonal element $\mathbf{r}_0 = \mathbf{r}$. Haris and Cina [19], for instance, give $C_{0E}(r, r_0, \beta)$ generalizing eq.(20) as

$$C_{0E}(\mathbf{r},\mathbf{r}_0,\beta) = C_{00}\, exp\left(\frac{\beta E}{2}(z+z_0) + \frac{\beta^3 E^2}{24}\right) \tag{22}$$

In eq.(22), C_{00} is the Bloch density matrix for free electrons in zero field, which was first calculated by Sondheimer and Wilson [15] as

$$C_{00}(\mathbf{r},\mathbf{r}_0,\beta) = (2\pi\beta)^{-\frac{3}{2}}\, exp\left(-\frac{|\mathbf{r}-\mathbf{r}_0|^2}{2\beta}\right) \tag{23}$$

It is also worthy of note at this point that Fallieros and Friar [16] have generalized the above result (22) to a time-dependent applied field $\mathbf{E}(\omega)$ with angular frequency ω. Taking the double Fourier transform $\mathbf{r} \to \mathbf{k}, \mathbf{r}_0 \to \mathbf{k}_0$, and using the Feynman propagator $K(t)$ instead of the Bloch matrix ($\beta \equiv it$) they obtain

$$K(\mathbf{k},\mathbf{k}_0,t) = (2\pi)^3\,\pi\pi\delta\left(\mathbf{k}-\mathbf{k}_0 - eE\frac{sin\omega t}{\hbar\omega}\right) exp(i\chi_\omega(\mathbf{k},t)) \tag{24}$$

where, as ω tends to zero, the 'phase' χ_ω tends to

$$\chi_0 = -\epsilon_k t - e\mathbf{E}\cdot\mathbf{k}\,\frac{t^2}{m} - \frac{1}{6}e^2\mathbf{E}^2\frac{t^2}{m\hbar} \tag{25}$$

with $\epsilon_k = \frac{\hbar^2 k^2}{2m}$. Taking the double Fourier transform of this limit recovers the static field result (22) when $i\cdot t$ is replaced by β.

We construct a semiclassical approach to the Slater sum $P(r,\beta,V,E)$ by 'switching on' an atomic or molecular potential $V(\mathbf{r})$ to the free electron form (20) above.

4.1 Semiclassical 'Switching On' of Potential Energy $V(\mathbf{r})$ to $P_{0E}(\mathbf{r},\beta)$

This is the point to make use of the Slater sum $P_{0E}(\mathbf{r},\beta)$ for free electrons in a uniform electric field of arbitrary strength to 'switch on' a suitable atomic or molecular potential $V(\mathbf{r})$. We then write the 'semiclassical' approximation $P(\mathbf{r},\beta,V,E)$ to the Slater sum as

$$P = P_{0E}(\mathbf{r},\beta)exp(-\beta U(\mathbf{r},\beta,E)) \tag{26}$$

where our task is to construct the effective potential $U(\mathbf{r},\beta,E)$ [17].

This we shall proceed to do via the integral form of the Bloch equation satisfied by the canonical matrix C of eq.(21). On the diagonal, this integral equation reads

$$P(\mathbf{r},\beta,E) = P_{0E}(\mathbf{r},\beta) - \int_0^\beta d\beta_1 \int d\mathbf{r}_1 C_{0E}(\mathbf{r},\mathbf{r}_1,\beta-\beta_1)\,V(\mathbf{r}_1)\,C(\mathbf{r}_1,\mathbf{r},\beta_1) \tag{27}$$

Since we are not able presently to solve for $C(\mathbf{r},\mathbf{r},\beta,E)$, we shall construct U in eq.(26) approximately by forcing agreement to first-order only in V. Thus, one can then write eq.(27) to yield P to first-order, with value P_1 say, as

$$P_1 = P_{0E} - \int_0^\beta d\beta_1 \int d\mathbf{r}_1 C_{0E}(\mathbf{r},\mathbf{r}_1,\beta-\beta_1)\,V(\mathbf{r}_1)\,C_{0E}(\mathbf{r}_1,\mathbf{r},\beta_1) \tag{28}$$

But the off-diagonal form of C_0 is known from eq.(22). We now expand eq.(26) to read

$$P_1 = P_{0E}\left(1-\beta U_1(r,\beta,E)\right) \tag{29}$$

and making comparison with eq.(28) yields then

$$P_{0E}(\beta U_1) = \int_0^\beta d\beta_1 \int d\mathbf{r}_1 C_{0E}(\mathbf{r},\mathbf{r}_1,\beta-\beta_1)\,V(\mathbf{r}_1)\,C_{0E}(\mathbf{r}_1,\mathbf{r},\beta_1) \tag{30}$$

In the 'fully semiclassical' limit where $V(\mathbf{r}_1)$ is so slowly varying in space that one can replace it by$V(\mathbf{r})$ and hence bring this outside the integral we find $U_1(\mathbf{r}) \approx V(\mathbf{r})$, i.e. independent of β and ϵ.

However, the present proposal is to retain the form (26) but with

$$U(\mathbf{r},\beta,E) \approx U_1 \tag{31}$$

where U_1 is to be obtained from eq.(3). Evidently this restores β and field E dependence to $U(\mathbf{r},\beta,E)$, though naturally only in an approximate manner motivated by semiclassical theory.

Inserting eq.(30) for U_1 into eq.(26) with U replaced by this approximation U_1, provides a way of incorporating an atomic or molecular potential $V(\mathbf{r})$ into the theory. Should a convergence factor be required in using eq.(30), a weak $\mathbf{B}$ field can be switched on perpendicular to E.

4.2 Proposed Use of Self-Consistent Thomas-Fermi Potential for Atoms

In the spirit of this approach, it would clearly be of interest to evaluate U_1 from eq.(3) with the self-consistent Thomas-Fermi potential for neutral atoms, defined by [11]

$$V(r) = -\frac{Ze^2}{r}\Phi(x) \tag{32}$$

where the 'shielding function' $\Phi(x)$ satisfies

$$\frac{d^2\Phi}{dx^2} = \frac{\Phi^{\frac{3}{2}}}{x^{\frac{1}{2}}} \tag{33}$$

with boundary conditions $\Phi(0) = 1$, Φ tends to zero at infinity. Both Φ and x are here dimensionless variables, with the independent variable x related to the radial distance from the atomic nucleus by

$$r = bx \quad : \quad b = \frac{0.88534a_0}{Z^{\frac{1}{3}}} \quad : \quad a_0 = \frac{\hbar^2}{me^2} \tag{34}$$

for an atom having atomic number Z. However no numerical calculations of U_1 in eq.(30) with $V(r)$ given by eqs.(32,33) are available at the time of writing.

5. SUMMARY AND FUTURE DIRECTIONS

For multielectron atoms in intense magnetic fields, semiclassical theory has been used to set up explicit differential equations determining the ground-state electronic density. Eq.(12) is the result for the relativistic Thomas-Fermi theory, which reduces to the much simpler form (14) in the non-relativistic limit.

The case of atoms and molecules in static uniform electric fields has also been considered in semiclassical theory. It would seem well worthwhile in future studies in this area to utilize eq.(26) for the Slater sum in conjunction with the approximation to the effective potential U denoted by U_1 in eq.(31). For atoms, the potential $V(r)$ appearing in this latter equation could be chosen as the Thomas-Fermi atom potential in eq.(33). Time-dependent generalizations based on eq.(24) as starting point are also of obvious interest for future work in the semiclassical treatment of atoms in laser fields.

Acknowledgment

Partial financial support from ONR for work on atoms and molecules in intense fields is acknowledged. The writer is especially indebted to Dr.P. Schmidt of ONR for much encouragement and for very stimulating discussions.

APPENDIX: ATOMS AND MOLECULES IN SUPER- AND HYPER-STRONG MAGNETIC FIELDS

In suitable units, the Thomas-Fermi theory presented in the text is valid when $\frac{B}{Z^3}$ is very small, even though both B and Z are large. In the opposite limit, when $\frac{B}{Z^3}$ is greater than a certain critical value depending on the ratio $\frac{N}{Z}$, N being the total number of electrons, Johnsen and Yngvason [19] derive the functional

$$E^{SS}[\rho] = \int \left[\frac{\partial}{\partial z}\{\rho\}^{\frac{1}{2}}\right]^2 d\mathbf{r} - Z\int \frac{\rho(\mathbf{r})}{r} d\mathbf{r} + \frac{1}{2}\int\int \frac{\rho(\mathbf{r})\rho(\mathbf{r}\prime)}{|\mathbf{r}-\mathbf{r}\prime|} d\mathbf{r}d\mathbf{r}\prime \tag{A.1}$$

where SS denotes, following these workers, the super-strong field regime. Atoms in this SS regime have the form of a thin cylinder with its axis in the direction of the magnetic field. The radius is of order $\sqrt{\frac{Z}{B}}$.

Lieb et al [5] have earlier discussed the hyperstrong regime in which the cylinder collapses to a needle. Then the functional (A.1) takes the simpler form, with HS denoting hyperstrong,

$$E^{HS}[\rho] = \int \left[\frac{\partial}{\partial z}\{\rho\}^{\frac{1}{2}}\right]^2 dz - \rho(0) + \int \rho(z)^2 dz \tag{A.2}$$

where $\rho^{\frac{1}{2}}$ denotes the density amplitude. Lehmann and March [20] have related this equation to the differential equation for the density amplitude, now fully quantal, and involving the so-called Pauli potential. Introducing

$$\lambda = \frac{N}{Z} \tag{A.3}$$

one can solve the Euler equation corresponding to the energy functional (A.2) to find [5,19]

$$\left(\rho^{HS}(z)\right)^{\frac{1}{2}} = \frac{2^{\frac{1}{2}}(2-\lambda)}{4sinh\left[\frac{1}{4}(2-\lambda)|z|+c\right]} \qquad \lambda < 2 \tag{A.4}$$

and

$$\left(\rho^{HS}(z)\right)^{\frac{1}{2}} = 2^{\frac{1}{2}}(2+|z|)^{-1} \qquad \lambda \geq 2 \tag{A.5}$$

where $tanh\ c = \frac{(2-\lambda)}{2}$.

$$E^{HS}(\lambda) = -\frac{1}{4}\lambda + \frac{1}{8}\lambda^2 - \frac{1}{48}\lambda^3 \qquad \lambda \leq 2 \tag{A.6}$$

Not only can one now bind $2Z$ electrons in an atom of atomic number Z but, as Lieb et al [5] stress, the binding energy of the last Z electrons is of the same order of magnitude as the first Z electrons bound in a neutral atom. The chemistry in the field of neutron stars turns out also to be very different from terrestrial chemistry, as two identical heavy atoms now bind with a binding energy of the homonuclear molecule of the same order as the energy of the isolated atoms. For further details, reference can be made to the article by Freeman and March [21]. We note in closing this Appendix the very recent work of Jones et al [22], who study atoms in strong magnetic fields by the Hartree-Fock formalism.

REFERENCES

1. Atoms in Intense Laser Fields: Ed. M. Gavrila (Academic: San Diego, 1992).
2. Atoms and Molecules in Intense Fields: Structure and Bonding, Vol. **86**, Eds. L.S. Cederbaum, K.C. Kulander and N.H. March (1997).
3. J.Yngavson, Lett.Math.Phys. **31**, 127 (1994)
4. N.H. March and Y. Tomishima, Phys.Rev.D**19**, 449 (1979)
5. E.H. Lieb, J.P. Solovej and J.Yngvason, Phys.Rev.Lett.**69**, 749 (1992)
6. N.H. March, Phys.Rev.A**48**, 4778 (1993)
7. M.S. Vallarta and N. Rosen, Phys.Rev.**41**, 708 (1932)
8. A. Nagy and N.H. March, Mol.Phys., in press (1997)
9. C. Amovilli and N.H. March, Chem.Phys.**146**, 207 (1990)
10. S.H. Hill, P.J. Grout and N.H. March, J.Phys.B**18**, 4665 (1985)
11. A. Holas and N.H. March, to be published
12. N.H. March and B.M. Deb, The Single Particle Density in Physics and Chemistry, Eds. Academic London (1987); see also N.H. March, Electron Density Theory of Atoms and Molecules, Academic New York (1992)
13. A.D. Jannussis, Phys.Stat.Solidi **36**, K17 (1969)
14. R.A. Harris and J. Cina, J.Chem.Phys.**79**, 1381 (1983)
15. E.H. Sondheimer and A.H. Wilson, Proc.Roy.Soc.A**210**, 173 (1951)
16. S. Fallieros and J.L. Friar, American J. Phys. **50**, 1001 (1982)
17. N.H. March and J.C. Stoddart, Rep.Prog.Phys.**31**, 533 (1968)
18. K.C. Kulander, F.H. Mies and K.J. Schafer, Phys.Rev.A**53**, 2562 (1996)
19. Johnsen and J.Yngvason, Phys.Rev. to appear (1996)
20. H. Lehmann and N.H. March, Pure and Appl.Chem.**67**, 457 (1995)
21. G.R. Freeman and N.H. March, J.Phys.Chem.**100**, 4331 (1996)
22. M.D. Jones, G. Ortiz and D.M. Ceperley, Phys.Rev.A**54**, 219 (1996)

ON THE GROUND STATE OF THE HYDROGEN MOLECULE IN A STRONG MAGNETIC FIELD

P. Schmelcher and T. Detmer

Theoretische Chemie
Universität Heidelberg
Im Neuenheimer Feld 253
D-69120 Heidelberg
Federal Republic of Germany

INTRODUCTION

The area of molecules in strong magnetic fields became during the past ten years a subject of increasing interest. This was on the o.h.s. motivated by the astrophysical discovery of huge field strengths in the vicinity of white dwarfs and neutron stars and on the o.h.s. by the possibility of studying strong field effects for highly excited Rydberg systems in laboratory fields.

Most of the existing investigations however deal with the H_2^+ ion (see Refs.[1-4] and references therein). In strong magnetic fields we encounter a variety of interesting new molecular phenomena. For the ground state of the H_2^+ ion an increase of the electron density between the nuclear charges leads to a contraction of the bond length. At the same time we observe an increase in the dissociation energy with increasing magnetic field strength. Moreover for the H_2^+ ion a class of states with purely repulsive potential energy curves (PECs) in field free space was shown to exhibit well-pronounced potential energy minima in a sufficiently strong magnetic field [2,3]. The topology of the electronic potential surfaces changes strongly with varying field strength. For intermediate field strengths it was shown, that the lowest-lying electronic states possess their global equilibrium configurations at positions corresponding to high symmetry, i.e. $\theta = 0^\circ$ or $\theta = 90^\circ$. However, for some excited states a global symmetry lowering occurs leading to global equilibrium configurations at $0^\circ < \theta < 90^\circ$ [3].

Only little is known concerning the electronic structure of the hydrogen molecule in a strong magnetic field [5-11]. Most of the investigations deal with the hydrogen molecule in superstrong fields ($B \geq 10^{11} G$). For intermediate field strengths there exist only two studies of qualitative character which investigate the PEC of the lowest $^1\Sigma_g$ state [6,7]. A detailed knowledge of the electronic structure of the hydrogen molecule is of particular relevance in astrophysics since it might lead to a better understanding of the spectra of white dwarfs and neutron stars. Hereby the ground state of the hydrogen molecule is of particular interest.

In the present investigation (see in particular also Ref.[12]) we perform a first step to elucidate the electronic ground state properties of the hydrogen molecule in a strong magnetic field. Particular emphasis is put on the intermediate regime which is of relevance to the physics of white dwarfs. We investigate the electronic structure of the lowest states of the Σ manifold, i.e. the lowest singlet and triplet states for a magnetic quantum number equal to zero. We therefore focus on the case of parallel internuclear and magnetic field axis. This configuration is distinct by its higher symmetry compared to the case of an arbitrary angle θ between the internuclear and magnetic field axis. Due to the efficiency of our method we are able to investigate in detail the electronic structure in the complete range of field strengths $B = 0\ au$ up to $B = 100\ au$.

THEORETICAL FRAMEWORK

We start with the total nonrelativistic molecular Hamiltonian in Cartesian coordinates. It is well known that the total pseudomomentum is a constant of motion [13,14]. For a neutral system like the hydrogen molecule the components of the pseudomomentum additionally commute with each other. Therefore, the Hamiltonian can be simplified by performing a so-called pseudoseparation of the center of mass motion [13,15,16]. Due to this pseudoseparation the center of mass coordinate and the conserved pseudomomentum are introduced as a pair of canonical conjugated variables. As a result, the center of mass coordinate does not appear in the transformed Hamiltonian. After applying this transformation, the exact Hamiltonian can be further simplified by a series of unitary transformations. For details of these transformations we refer the reader to the literature [15,16].

In order to separate the electronic and nuclear motion an adiabatic approximation has to be performed which means that we have to apply the Born-Oppenheimer approximation in the presence of a magnetic field. The validity of the Born-Oppenheimer approximation in the presence of a magnetic field has been studied in detail in refs [15-17] including all corrections due to the finite nuclear masses. In a first order approximation we choose the electronic Hamiltonian as the fixed nuclei Hamiltonian, i.e., we assume infinitely heavy masses for the nuclei. The origin of our coordinate system coincides with the midpoint of the internuclear axis of the hydrogen molecule and the protons are located on the z-axis. The magnetic field is chosen parallel to the z-axis of our coordinate system and we use the symmetric gauge for the vector potential. Finally our electronic Hamiltonian takes on the following appearance:

$$H = \sum_{i=1}^{2}\left\{\frac{1}{2}\mathbf{p}_i^2 + \frac{1}{8}(\mathbf{B}\times\mathbf{r}_i)^2 + \frac{1}{2}\mathbf{L}_i\mathbf{B} - \frac{1}{|\mathbf{r}_i - \mathbf{R}/2|} - \frac{1}{|\mathbf{r}_i + \mathbf{R}/2|}\right\} + \frac{1}{|\mathbf{r}_1 - \mathbf{r}_2|} + \frac{1}{R} + \mathbf{SB} \tag{1}$$

The symbols $\mathbf{r}_i$, $\mathbf{p}_i$ and $\mathbf{L}_i$ denote the position vectors, the canonical conjugated momenta and the angular momenta of the two electrons, respectively. $\mathbf{B}$ and $\mathbf{R}$ are the vectors of the magnetic field and internuclear distance, respectively and R denotes the absolute value of $\mathbf{R}$. With $\mathbf{S}$ we denote the vector of the total electronic spin. Since we deal with Σ states, the sum $\sum_{i=1}^{2}\mathbf{L}_i\mathbf{B}$ equals to zero. Throughout the paper we will use atomic units.

The Hamiltonian (1) commutes with the following operators:

a) The parity operator P due to the charge symmetry of the molecule . The corresponding eigenfunctions are labeled with the subscript g for gerade or u for ungerade parity.

b) The projection L_z of the electronic angular momentum onto the internuclear axis.

c) The square S^2 of the total electronic spin. The electronic functions are labeled with a left superscript describing the multiplicity $2S+1$ of the state.

d) The projection S_z of the total electronic spin on the internuclear axis. For the singlet states the only possibility is $M_s = 0$. For triplet states we are only interested in $M_s = -1$.

In the absence of a magnetic field we encounter an additional symmetry namely the reflections with respect to the electronic coördinates at the xz (σ_v) plane. The eigenfunctions possess the corresponding eigenvalues ± 1. This symmetry does not hold in the presence of a magnetic field! Therefore, the resulting symmetry groups for the hydrogen molecule are $D_{\infty h}$ in the case of field free space and $C_{\infty h}$ in the presence of a magnetic field [18].

In order to solve the fixed-nuclei electronic Schrödinger equation belonging to the Hamiltonian (1) we expand the electronic eigenfunctions in terms of molecular configurations. First of all we note that the total electronic wave function Ψ_{ges} can be written as a product of its spatial part Ψ and its spin part χ, i.e. we have $\Psi_{ges} = \Psi\chi$. For the spatial part Ψ of the wave function we use the LCAO-MO-ansatz, i.e. we decompose Ψ with respect to molecular orbital configurations ψ of H_2 which respect the corresponding symmetries (see above) and the Pauli principle, i.e.

$$\begin{aligned}\Psi &= \sum_{i,j} c_{ij}\left[\psi_{ij}(\mathbf{r}_1,\mathbf{r}_2) \pm \psi_{ij}(\mathbf{r}_2,\mathbf{r}_1)\right] \\ &= \sum_{i,j} c_{ij}\left[\Phi_i(\mathbf{r}_1)\,\Phi_j(\mathbf{r}_2) \pm \Phi_i(\mathbf{r}_2)\,\Phi_j(\mathbf{r}_1)\right]\end{aligned}$$

The molecular orbital configurations ψ_{ij} of H_2 are products of the corresponding one-electron H_2^+ molecular orbitals Φ_i and Φ_j. The H_2^+ molecular orbitals are built from atomic orbitals centered at each nucleus. A key ingredient of this procedure is a basis set of nonorthogonal optimized non-spherical Cartesian Gaussian atomic orbitals which has been established previously [3,19]. For a more detailed description of the construction of the molecular electronic wave function we refer the reader to Ref.[12]. In order to determine the molecular electronic wave function of H_2 we use the variation principle. That means we minimize the variational integral $\frac{\int \Psi^* H \Psi}{\int \Psi^* \Psi}$ by varying the coefficients c_{ij}. The resulting generalized eigenvalue problem reads as follows:

$$(\underline{H} - \epsilon \underline{S})\,\mathbf{c} = \mathbf{0} \tag{2}$$

In the present investigation for parallel internuclear axis and magnetic field axis, the Hamiltonian matrix $\underline{H}$ is real and symmetric and the overlap matrix is real, symmetric and positiv definite. The vector $\mathbf{c}$ contains the expansion coefficients. The matrix elements of the Hamiltonian matrix and the overlap matrix are certain linear combinations of matrix elements with respect to the optimized non-spherical Gaussian atomic orbitals. The latter matrix elements have already been calculated in Ref.[19]. However, the formulae for the electron-nucleus and in particular the electron-electron matrix elements given in Ref.[19] turned out to be not sufficiently efficient for numerical calculations with large basis sets. Both, the numerical stability as well as the efficient and fast computation of the matrix elements required a new approach to the integral evaluation within our basis set of atomic orbitals. In the new computational scheme the matrix elements are evaluated by a combination of a special quadrature method and a subsequent numerical integration. For each PEC about 300 points were calculated on an average. For each magnetic field strength we developed a basis set of atomic orbitals particularly adapted for minimizing the total energy. For further details of the computational procedure we refer the reader to Ref.[12].

RESULTS AND DISCUSSION

The $^1\Sigma_g$ state

Let us begin our investigation by considering the lowest $^1\Sigma_g$ state of the hydrogen molecule which is the ground state in field free space. This state has been extensively studied in field free space in the adiabatic approximation, both theoretically as well as experimentally. For the literature on theoretical investigations up to 1960 we refer the reader to the bibliography in Ref.[20]. Recently, several theoretical investigations have been performed to improve the overall energy values for the $^1\Sigma_g^+$ state. Kolos [21] as well as Wolniewicz [22] improved the electronic energy calculated in the Born-Oppenheimer approximation several times. As a reference in the case of field free space we use the energy values obtained by Wolniewicz in 1995 [22].

In field free space the energy curve of the $^1\Sigma_g^+$ state shows only one minimum at the equilibrium distance of 1.4 *au* with a total energy of -1.1744757 *au*. In the dissociation limit we have two H atoms in their ground states, i.e. $H_2 \rightarrow H\,(1s) + H\,(1s)$. Therefore, in the separated atom limit the total energy approaches -1.0 *au* which corresponds to the energy of two H atoms in the ground state. In the united atom limit we have a helium atom in the ^1S $1s^2$ state. For that reason the total energy without the nucleus-nucleus repulsion $\frac{1}{R}$ approaches an energy value of 2.90372 *au* with decreasing internuclear distance which is the ground state energy of the helium atom.

For the total energy at the equilibrium distance in the field free space we obtained a total energy of -1.173892 *au* which yields a dissociation energy of 0.173894 *au*. This corresponds to a relative accuracy in the total energy of about 6×10^{-4} compared to the bench mark result in Ref.[22]. This accuracy in the total energy even increases with increasing internuclear distance. For $R = 3$ *au* and $R = 4$ *au* we yield an error of 3×10^{-4} and 2×10^{-4}, respectively.

In contrast to the numerous investigations concerning the behavior and structure of the hydrogen molecule in field free space only a few studies deal with the hydrogen molecule in strong magnetic fields [5-11,23]. Most of them deal with the hydrogen molecule in superstrong magnetic fields larger than 10^{11} G or even 10^{12} G (in atomic units this corresponds to $B = 42.54414$ *au* and $B = 425.4414$ *au*, respectively).

First of all we mention that our computational method is by no means restricted to a special range of the magnetic field strength. We were therefore able to study the development of the total energy with respect to the field strength ranging from field free space up to a very strong field! Before entering into the discussion of our results let us introduce our notation for the united and separated atom limit in the presence of a magnetic field. Throughout the paper we will denote the atomic hydrogen states in the dissociation limit with $H\,(m_a^{\pi_a})$ where m_a denotes the atomic magnetic quantum number and π_a the atomic-z parity, respectively. The united atom limit is described by $^{2S+1}L_z^{\pi_z}$. Here $2S+1$ is the multiplicity, L_z is the projection of the electronic angular momentum onto the axis of the magnetic field and π_z is the z-parity. Now we are in the position to discuss the structure of the electronic PEC for the $^1\Sigma_g$ state in the presence of a magnetic field.

Figure 1 shows the energy curves of the $^1\Sigma_g$ state of the hydrogen molecule for different field strengths. In order to display electronic energies for varying magnetic

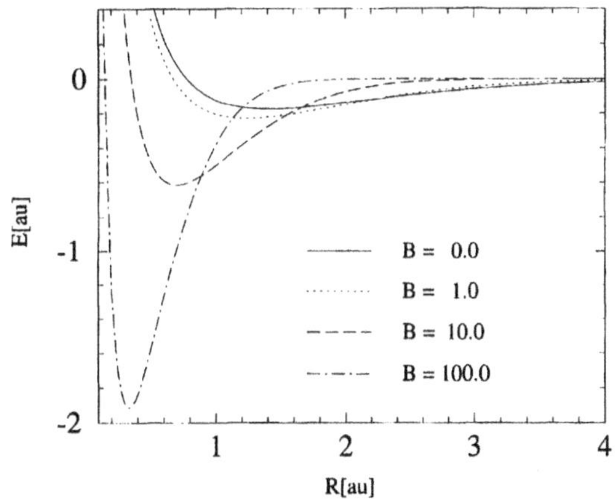

Figure 1. PECs for $B = 0.0$, 1.0, 10.0 and 100.0 au for the lowest ${}^1\Sigma_g$ state ; the energy is shown with respect to the dissociation limit i.e. $E(R) = E_t(R) - \lim_{R\to\infty} E_t(R)$

field strengths in the same viewgraph the total energy is substracted by the energy in the dissociation limit, i.e. we show the quantity $E(R) = E_t(R) - \lim_{R\to\infty} E_t(R)$. For the ${}^1\Sigma_g$ state the dissociation channel is $H_2 \to H(0^+) + H(0^+)$ which means that the energy in the dissociation limit corresponds to the energy of two hydrogen atoms in the lowest electronic state with positive z-parity. The appropriate electronic state in the united atom limit is the ${}^10^+$ helium state for any field strengths up to $B = 100au$.

In the following we discuss the changes in the dissociation energy and equilibrium distance with increasing strength of the magnetic field. The overall behavior we observe is a monotonously increasing total energy as well as dissociation energy and a monotonously decreasing equilibrium internuclear distance. The decrease in the equilibrium internuclear distance originates from the simoultaneous decrease of the electron cloud perpendicular and parallel to the magnetic field. Figure 1 illustrates the particularly drastical growth in the dissociation energy for magnetic field strengths $B \gtrsim 1$ au. At the same time the potential well becomes more and more pronounced, i.e. its width decreases strongly. Furthermore the asymptotic behavior of the PEC for large values of R changes with the magnetic field strength. With increasing value of B the dissociation limit is reached at much smaller values of the internuclear distance, i.e the onset of the asymptotic behavior can be observed for much smaller internuclear distances. Furthermore transition states appear in the PEC for magnetic field strengths $B \gtrsim 1$ au. For the position of the maximum we observe a strong decrease with increasing field strength. Simultaneously the height of the maximum with respect to the dissociation energy , i.e. $E_t - \lim_{R\to\infty} E_t$, increases from 1×10^{-6} at $B = 1.0$ au to 6×10^{-4} at $B = 100.0$ au.

We emphasize that the ${}^1\Sigma_g$ state is not the ground state of the hydrogen molecule in a magnetic field of arbitrary strength! In superstrong magnetic fields with $B \gtrsim 10^3 au$ it is well known that the ${}^3\Pi_u$ state represents the ground state of the hydrogen molecule [5,9]. The ground state for magnetic field strengths in the intermediate regime has not been investigated up to now. For sufficiently weak magnetic fields the ground state has to be the ${}^1\Sigma_g$ state. However, we will show in the following that for magnetic field strengths larger or equal than 0.2 au the total energy of the ${}^3\Sigma_u$ state is lower than the total energy of the ${}^1\Sigma_g$ state! This means that for $B \gtrsim 0.2$ au the ${}^3\Sigma_u$ state is the ground state of the hydrogen molecule. The "crossing field strength" of this lowest Σ_u

state with the $^3\Pi_u$ state, which is the ground state in superstrong magnetic fields, is not known so far.

Finally let us investigate the question whether vibrational levels exist in the PEC discussed above. This question is of great importance to the existence of bound states. In the presence of a magnetic field the determination of vibrational levels is a much more complicated task than in field free space. First we note that the Born-Oppenheimer approximation known from field free space is not valid in a magnetic field. In the presence of a magnetic field the nuclear charges are partially screened by the electrons against the external field. In order to correctly describe the partial screening of the nuclear charges the diagonal term of the nonadiabatic coupling elements has to be included in the nuclear equation of motion. This leads to a new kind of adiabatic approximation, the partially screened Born-Oppenheimer approximation [15,16,24]. Furthermore the nuclear equation of motion explicitely depends not only on the internuclear distance but also on the angle between the internuclear axis and the magnetic field axis. The facts discussed above clearly show that the nuclear dynamics is in general very complex. Within the present framework of the parallel configuration we can therefore provide only estimations for the energy levels of the vibrational ground states. Nethertheless this allows us to decide whether or not we encounter physically bound states with respect to the vibrational mode R. A lower bound of the vibrational energy within our approximation is given by the lowest vibrational state in the corresponding PEC using the field free kinetic energy $\frac{P^2}{2\mu}$. The corresponding nuclear equation of motion in field free space for the given electronic PEC was solved with the help of a discrete variable representation [25]. The upper bound for the energy of the lowest vibrational state in the presence of a magnetic field was obtained by simply adding the Landau energy of the nuclear motion to the value of the energy obtained for the corresponding vibrational level in field free space. In this way we obtained upper as well as lower bounds for the vibrational levels. These estimations of the vibrational levels were performed for each PEC shown in Figure 1.

For the PEC of the $^1\Sigma_g$ state the procedure described above yields many, i.e. of the order of magnitude of a few dozens, of vibrational levels for the entire regime $B = 0 - 100$ *au* of field strengths. This means that the $^1\Sigma_g$ state is a bound state with respect to the internuclear distance R for this wide range of field strengths.

The $^3\Sigma_u$ state

In field free space the electronic PEC of the $^3\Sigma_u^+$ state is repulsive, i.e. does not exhibit a well pronounced potential well. The united atom limit of this state is the ^3S $1s2p$ helium state and the dissociation channel is $H_2 \rightarrow H\,(1s) + H\,(1s)$. The, to our knowledge, most accurate results were obtained using Hylleraas-type expansions [26,27], explicitly correlated Cartesian Gaussian basis functions [28] or elliptical basis functions [29].

As a reference for our calculations we used the data given in Ref.[29] for internuclear distances smaller than 4 *au* and Refs.[26,27] for $R \geq 4.0$ *au*. For $R \leq 4$ *au* we obtained an overall relative accuracy of 6×10^{-4}. For larger values of the internuclear distance this accuracy further increases and we obtain a relative accuracy of at least 3×10^{-6}! Our accurate results, in particular for the $^3\Sigma_u^+$ state in field free space demonstrate the usefulness of our basis set of non-spherical nonorthogonal Cartesian Gaussian basis functions. The PEC for the $^3\Sigma_u$ state for different magnetic field strengths is presented in Fig.2.

Despite the fact that the PEC of the $^3\Sigma_u^+$ state is predominantly repulsive it exhibits a very shallow van-der-Waals minimum around $R \sim 8$ *au*. Due to the dissociation

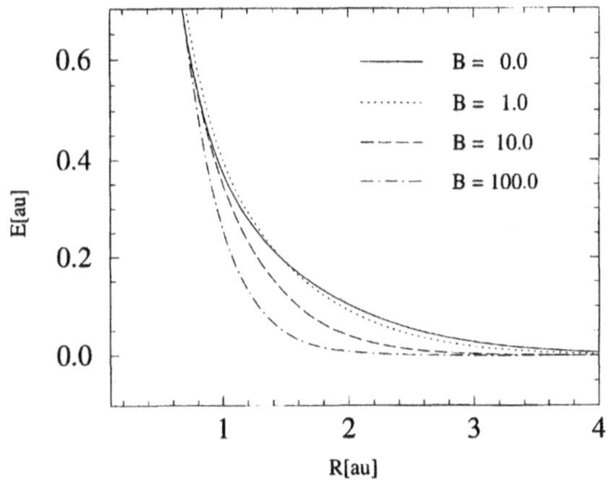

Figure 2. PECs for $B = 0.0, 1.0, 10.0$ and $100.0\ au$ for the lowest $^3\Sigma_u$ state; the energy is shown with respect to the dissociation limit i.e. $E(R) = E_t(R) - \lim_{R\to\infty} E_t(R)$

of the $^3\Sigma_u^+$ state into $H(1s) + H(1s)$ we encounter a dipole-dipole interaction of induced dipol moments for large internuclear distances which is proportional to $\frac{-1}{R^6}$. For the van-der-Waals minimum we obtained a dissociation energy of $1.885 \times 10^{-5}\ au$ at an internuclear distance of $7.9\ au$.

In the following we discuss the development of the PEC for the $^3\Sigma_u$ state depending on the magnetic field strength. First, we focus on the global structure which is shown in Fig.2. In the presence of a magnetic field the separated atom limit is given by $H_2 \rightarrow H(0^+) + H(0^+)$, i.e. the molecule dissociates into two hydrogen atoms in their ground states with positive z-parity. The corresponding united atom state in the presence of a magnetic field is the $^30^-$ helium state. First of all we mention that the onset of the asymptotic behavior with respect to the dissociation occurs for increasingly smaller internuclear distances with increasing magnetic field strength. A closer look at the two states $^3\Sigma_u, ^1\Sigma_g$ reveals that for magnetic field strengths larger than $0.2\ au$ the $^3\Sigma_u$ state is lower in energy than the $^1\Sigma_g$ state! Therefore, the crossing between these two states happens between the two field strengths 0.1 and $0.2\ au$. In Fig.3 and Fig.4 we illustrate the crossing of these two states. In Fig.3, which shows the total energy of the $^1\Sigma_g$ and $^3\Sigma_u$ states at a magnetic field strength of $B = 0.0\ au$, we can see the $^3\Sigma_u$ state being higher in the total energy as the $^1\Sigma_g$ state. Figure 4 shows the same states at a magnetic field strength of $B = 0.5\ au$. In this figure we observe the drastical decrease in the total energy of the $^3\Sigma_u$ state. The crossing of these two states has an important concequence for the stability of molecular hydrogen in astrophysics. Beyond $B \sim 0.2\ au$ the ground state of the hydrogen molecule is the $^3\Sigma_u$ state which is an unbound electronic state at least for the parallel configuration. It is a challenging task to clarify whether this is true for any angle of the internuclear axis with respect to the magnetic field axis. In principle it is possible that a potential well might develop if the internuclear and magnetic field axis does not coincide. The investigation of such configurations is an important task in the future. In addition we mention that beyond $B \gtrsim 3 \times 10^3\ au$ the global ground state of H_2 is the $^3\Pi_u$ state [5,9].

In the following we investigate the development of the van-der-Waals minimum depending on the magnetic field strength which is of particular interest for the determination of the global ground state of the hydrogen molecule in a magnetic field.

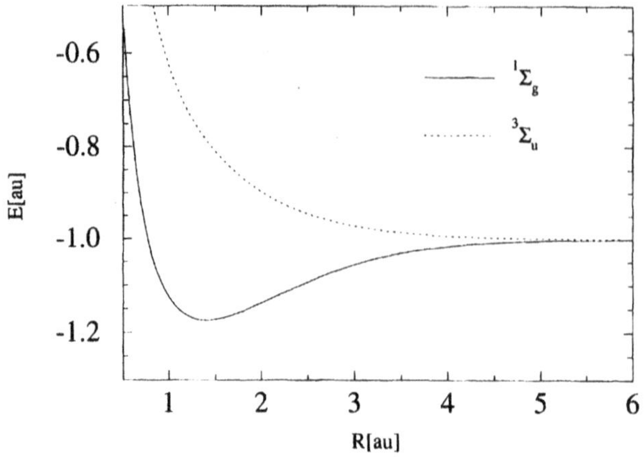

Figure 3. Total energy of the $^1\Sigma_g$ and $^3\Sigma_u$ state for $B = 0.0\ au$ (all quantities are given in atomic units)

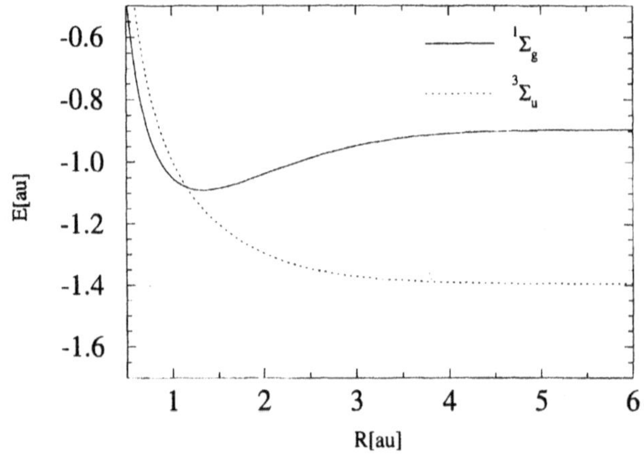

Figure 4. Total energy of the $^1\Sigma_g$ and $^3\Sigma_u$ state for $B = 0.5\ au$ (all quantities are given in atomic units)

In field free space the van-der-Waals potential is given by a law proportional $\frac{1}{R^6}$ due to the dipole-dipole interaction of induced dipoles in first order perturbation theory. In the presence of a magnetic field we have to pay attention to another interaction between atoms in s states. In first order perturbation theory two atoms in a magnetic field interact like two permanent quadrupoles. Therefore the leading expression in first order perturbation theory is proportional to $\frac{1}{R^5}$. The development of the van-der-Waals minimum with increasing magnetic field strength is as follows. The dissociation energy increases by $0.175 \times 10^{-5} au$ if we increase the field strength from zero to $0.2\ au$. At the same time we observe a monotonous decrease of the internuclear distance corresponding to the minimum from 7.9 to 7.7 au. We observe that the shape of the energy curve for $B = 0.2\ au$ differs only sligthly from that in field free space. For magnetic field strengths larger than $B = 0.2\ au$ the dissociation energy decreases drastically with increasing magnetic field strength down to 0.375×10^{-5} for $B = 1.0\ au$. Simultaneously the energy curve changes its shape. The gradient of the energy with respect to internuclear distances $R \leq R_{eq}$ becomes larger and a more and more shallow minimum can be observed. For magnetic field strengths larger than 1.0 *au* no minimum is observed! The gradient of the energy with respect to the internuclear distance increases further with increasing magnetic field strength changing the appearance of the PEC to an increasingly flatter curve. No vibrational levels were found for any field strength.

Finally we draw the readers attention to the fact that the calculations concerning the position and dissociation energy of the van-der-Waals minimum are close to the convergence limit of our calculations. The disappearance of the van-der-Waals minimum within our calculations reflects to our opinion a real physical effect but needs further investigations for a definite clarification. In order to answer the question about the lowest bound state of the hydrogen molecule in the superstrong regime of field strengths $B \lesssim 10^3$ *au* a detailed and very accurate investigation of the $^3\Pi_u$ electronic state has to be performed.

SUMMARY AND CONCLUSIONS

In the present paper we investigated the electronic structure of the hydrogen molecule in a magnetic field. We hereby focused on the case of parallel internuclear axis and magnetic field axis for a magnetic quantum number equal to zero. The key ingredient for our CI calculations is a basis set of nonorthogonal non-spherical Gaussian atomic orbitals which was established previously. Our results for the PECs in field free space showed a high accuracy compared to the existing data in the literature. The non-sherical atomic orbitals may therefore be very useful for precision calculations concerning molecules in field free space.

First we investigated the lowest $^1\Sigma_g$ state which is the ground state of the hydrogen molecule in field free space. In the presence of a magnetic field we observed a monotonous increase in the total energy. At the same time the equilibrium distance decreases and the dissociation energy (chemical binding energy) increases rapidly. The few existing data concerning the total energy of the $^1\Sigma_g$ state in the presence of a magnetic field were significantly improved by our calculations. By calculating lower and upper bounds for the lowest vibrational energy the PEC of the $^1\Sigma_g$ state was shown to contain many vibrational levels for any magnetic field strength up to 100 *au*.

Next we investigated the lowest $^3\Sigma_u$ state of the hydrogen molecule which is known to be repulsive in field free space and possesses only a very shallow van-der-Waals minimum at $R \sim 8.0$ *au*. The repulsive character of the PEC of the $^3\Sigma_u$ state remains for arbitrary field strengths up to $B = 100.0$ *au*. Due to the Spin-Zeeman shift in a magnetic field a crossing occurs between the $^1\Sigma_g$ and $^3\Sigma_u$ state in the range $0.1 \geq B \geq 0.2$ *au*. Therefore the lowest state of the hydrogen molecule in the presence of a magnetic field is the $^1\Sigma_g$ state for $B \leq 0.1$ *au* and the $^3\Sigma_u$ state for $B \geq 0.2$ *au*. In superstrong magnetic fields $B \gtrsim 3 \times 10^3$ *au* the ground state is the $^3\Pi_u$ state. The determination of the crossing of the $^3\Sigma_u$ and $^3\Pi_u$ state is a task which is left to a future investigation. Furthermore the nonexistence of a strongly bound ground state of the hydrogen molecule has to be confirmed by investigations concerning arbitrary angles between the internuclear axis and magnetic field axis. After considering these general properties of the $^3\Sigma_u$ state we investigated the development of the van-der-Waals minimum depending on the field strength. For the van-der-Waals minimum we observe a monotonous decrease in the equilibrium internuclear distance with increasing field strength. First the dissociation energy increases with increasing field strength and for $B \geq 0.5$ *au* the dissociation energy drastically decreases. For $B \geq 1.0$ *au* no minimum has been found. No vibrational levels exist for any field strength up to 1 *au*.

The existence of a minimum which supports a vibrational frequency for the lowest electronic state of the hydrogen molecule for intermediate magnetic field strengths is of particular interest to astrophysics in order to determine whether hydrogen molecules exist in the vicinity of white dwarfs. According to our investigations for $B \gtrsim 0.2\ au$ the ground state of the hydrogen molecule is not strongly bound and exhibits only a weak minimum due to the van-der-Waals interaction. However, in superstrong magnetic fields the ground state is strongly bound again. Finally we emphasize that for drawing definite conclusions about the existence or nonexistence of the van-der-Waals minimum the case of nonparallel internuclear axis and magnetic field axis has to be investigated. The determination of the corresponding potential energy surfaces is a complicated task which is left to future investigations.

Acknowledgements

The Deutsche Forschungsgemeinschaft (T.D.) is gratefully acknowledged for financial support. Computer time has been generously provided by the Rechenzentrum Heidelberg and in particular by the Rechenzentrum Karlsruhe.

REFERENCES

1. U.Wille, Phys.Rev.A **38**, 3210 (1988)
2. U.Kappes, P.Schmelcher and T.Pacher, Phys.Rev.A **50**, 3775 (1994)
3. U.Kappes, P.Schmelcher, J.Chem.Phys.**100**, 2878 (1994); Phys.Rev.A **51**, 4542 (1995); Phys.Rev.A **53**, 3869 (1996); Phys.Rev.A **54**, 1313 (1996); Phys.Lett.A **210**, 409 (1996)
4. Y.P.Kravchenko and M.A.Liberman, Phys.Rev.A **55**, 2701 (1997)
5. D.Lai, E.E.Salpeter and S.L.Shapiro, Phys.Rev.A **45**, 4832 (1992)
6. A.V.Turbiner, Pis'ma Zh. Eksp. Teor. Fiz. **38**, 510 (1983)
7. S.Basile, F.Trombetta and G.Ferrante, Nuovo Cimento **9**, 457 (1987)
8. T.S.Monteiro and K.T.Taylor, J.Phys.B **23**, 427 (1990)
9. G.Ortiz, M.D.Jones and D.M.Ceperley, Phys.Rev.A **52**, R3405 (1995)
10. D.Lai and E.E.Salpeter, Phys.Rev.A **53**, 152 (1996)
11. A.V.Korolev and M.A.Liberman, Phys.Rev.A **45**, 1762 (1992)
12. T.Detmer, P.Schmelcher, F.K.Diakonos and L.S.Cederbaum, Phys.Rev.A, September 1997
13. B.Johnson, J.Hirschfelder and K.Yang, Rev.Mod.Phys.**55**, 109 (1983)
14. J.E.Avron, I.W.Herbst and B.Simon, Ann.Phys.**114**, 431 (1978)
15. P.Schmelcher, L.S.Cederbaum and H.-D.Meyer, Phys.Rev.A**38**, 6066 (1988)
16. P.Schmelcher, L.S.Cederbaum and U.Kappes, in 'Conceptual Trends in Quantum Chemistry' edited by Eugene S. Kryachko (Kluwer Academic Publishers, 1994)
17. P.Schmelcher, L.S.Cederbaum and H.-D.Meyer, J.Phys.B**21**, L445 (1988)
18. P.Schmelcher and L.S.Cederbaum, Phys.Rev.A**41**, 4936 (1990)
19. P.Schmelcher and L.S.Cederbaum, Phys.Rev.A**37**, 672 (1988)
20. A.D.McLean, A.Weiss and M.Yoshimine, Rev.Mod.Phys.**32**, 211 (1960)
21. W.Kolos, J.Chem.Phys.**101**, 1330 (1994)
22. L.Wolniewicz, J.Chem.Phys.**103**, 1792 (1995)
23. M.Demeur, P.H.Heenen and M.Godefroid, Phys.Rev.A**49**, 176 (1994)
24. T.Detmer, P.Schmelcher and L.S.Cederbaum, J.Phys.B**28**, 2903 (1995)
25. D.T.Colbert and W.H.Miller, J.Chem.Phys.**96**, 1982 (1992)
26. W.Kolos and L.Wolniewicz, J.Chem.Phys.**43**, 2429 (1965)
27. W.Kolos and L.Wolniewicz, Chem.Phys.Lett.**24**, 457 (1974)
28. D.Frye, G.C.Lie and E.Clementi, J.Chem.Phys.**91**, 2366 (1989)
29. J.W.Liu and S.Hagstrom, J.Phys.B**27**, L729 (1994)

HYDROGEN MOLECULE IN MAGNETIC FIELDS: ON EXCITED Σ STATES OF THE PARALLEL CONFIGURATION

T.Detmer, P. Schmelcher, F. K. Diakonos, and L. S. Cederbaum

Theoretische Chemie
Universität Heidelberg
Im Neuenheimer Feld 253
D-69120 Heidelberg
Federal Republic of Germany

INTRODUCTION

The electronic structure of the hydrogen molecule in field-free space has been extensively studied, both theoretically and experimentally. Very accurate theoretical potential energy curves (PECs) have been calculated not only within the scheme of the Born-Oppenheimer approximation, but also including nonadiabatic and relativistic corrections. Contrary to these large number of investigations little is known concerning the electronic structure of the hydrogen molecule in the presence of a strong magnetic field. For intermediate field strengths there exist only two studies of almost qualitative character which investigate the PEC of the lowest $^1\Sigma_g$ state [1, 2]. A few investigations were performed in the high field limit [3–7], where the magnetic forces dominate over the Coulomb forces and therefore several approximations can be performed. Many interesting phenomena can be observed concerning the electronic structure and behavior of a hydrogen molecule in the presence of a magnetic field. The work of Turbiner [2] indicates a decrease of the internuclear equilibrium distance and a simultaneous increase of the dissociation energy with increasing magnetic field strength for the $^1\Sigma_g$ state. Ortiz and coworkers [6] showed that in sufficiently strong fields ($B \gtrsim 3 \times 10^3$ $a.u.$) the $^3\Pi_u$ state is the global ground state for a H_2 molecule oriented parallel to the magnetic field.

The structure and properties of molecular systems in the presence of a magnetic field are of great importance for molecular physics, solid-state physics and astrophysics. In particular we mention the discovery of huge magnetic fields in the vicinity of white dwarfs and neutron stars [8–10]. A detailed knowledge of the electronic structure of the hydrogen molecule might lead to a better understanding of the spectra of such objects. Hereby not only the ground state but also excited states are of interest. In this article we focus on some excited states of the hydrogen molecule for the parallel configuration, namely the lowest $^1\Sigma_u$ and $^3\Sigma_g$ state.

Within the present investigation we obtained accurate Born-Oppenheimer energies for the lowest $^1\Sigma_u$ and $^3\Sigma_g$ state. The position of the maxima and minima of the PECs were determined with an accuracy of 0.01 *a.u.* in the internuclear distance and the overall accuracy of our results is estimated to be better than 10^{-3}. For detailed data concerning the PECs of these two states we refer the reader to the literature [11].

In the following we briefly present some theoretical aspects and then discuss the electronic structure of the lowest $^1\Sigma_u$ and $^3\Sigma_g$ state depending on the magnetic field strength.

THEORETICAL ASPECTS

Our starting point is the fixed nuclei Hamiltonian within the scheme of the Born-Oppenheimer approximation in the presence of a magnetic field [11–13]. The origin of our coordinate system coincides with the midpoint of the internuclear axis of the hydrogen molecule and the protons are located on the z axis. The magnetic field is chosen parallel to the z axis of our coordinate system and we use the symmetric gauge for the vector potential. Therefore, the molecular Hamiltonian reads as follows:

$$H = \sum_{i=1}^{2} \left\{ \frac{1}{2}\mathbf{p}_i^2 + \frac{1}{8}(\mathbf{B}\times\mathbf{r}_i)^2 - \frac{1}{|\mathbf{r}_i - \mathbf{R}/2|} - \frac{1}{|\mathbf{r}_i + \mathbf{R}/2|} \right\} + \frac{1}{|\mathbf{r}_1 - \mathbf{r}_2|} + \frac{1}{R} + \mathbf{S}\cdot\mathbf{B} \quad (1)$$

We hereby neclect relativistic effects like, e.g., the spin-orbit coupling and the gyromagnetic factor of the electron was chosen to be 2. The symbols $\mathbf{r}_i$ and $\mathbf{p}_i$ denote the position vectors and the canonical conjugated momenta of the electrons, respectively. $\mathbf{B}$ and $\mathbf{R}$ are the vectors of the magnetic field and internuclear distance, respectively and R denotes the magnitude of $\mathbf{R}$. With $\mathbf{S}$ we denote the vector of the total electronic spin. Since we deal with Σ states, the sum $\sum_{i=1}^{2} \mathbf{L}_i \cdot \mathbf{B}$ equals to zero and does not occur in the above Hamiltonian.

In order to solve the fixed-nuclei electronic Schrödinger equation belonging to the Hamiltonian (1) we expand the electronic eigenfunctions in terms of molecular configurations. First of all we note that the total electronic wave function Ψ_{ges} can be written as a product of its spatial part Ψ and its spin part χ, i.e. we have $\Psi_{ges} = \Psi\chi$. For the spatial part Ψ of the wave function we use the LCAO-MO-ansatz, i.e. we decompose Ψ with respect to molecular orbital configurations of H_2 which respect the corresponding symmetries and the Pauli principle. The molecular orbital configurations of H_2 are products of the corresponding one-electron H_2^+ molecular orbitals which are built up by atomic orbitals centered at each nucleus. A key ingredient of this procedure is a basis set of nonorthogonal optimized nonspherical Gaussian atomic orbitals which has been established previously [14, 15]. For the case of a H_2 molecule parallel to the magnetic field, these basis functions read as follows:

$$\phi_{kl}^{m}(\rho, z, \alpha, \beta, \pm R/2) = \rho^{|m|+2k}(z \mp R/2)^{l}\, exp\left\{-\alpha\rho^2 - \beta(z \mp R/2)^2\right\} exp\{im\phi\} \quad (2)$$

The symbols $\rho = x^2 + y^2$ and z denote the cylindrical one particle coordinates. The parameter m, k, and l depend on the subspace of the H-atom, for which the atomic basis functions have been optimized. α and β are variational parameters. For a detailed description of the construction of the molecular electronic wave function we refer the reader to Ref. [11].

In order to determine the molecular electronic wave function of H_2 we use the variation principle, which leads to a generalized eigenvalue problem:

$$(\underline{H} - \epsilon\underline{S})\,\mathbf{c} = \mathbf{0} \quad (3)$$

In the present investigation for parallel internuclear and magnetic field axes, the Hamiltonian matrix $\underline{H}$ is real and symmetric and the overlap matrix is real, symmetric and positiv definite. The vector $\mathbf{c}$ contains the expansion coefficients. The matrix elements of the Hamiltonian matrix and the overlap matrix are certain linear combinations of matrix elements with respect to the optimized nonspherical Gaussian atomic orbitals. For the numerical solution of the eigenvalue problem (3) we used the standard NAG library.

THE $^1\Sigma_u$ STATE

The electronic PEC for the $^1\Sigma_u$ state of the hydrogen molecule in field-free space has been calculated with high accuracy by Kolos and Wolniewicz [16,17]. The PEC for this state in field-free space possesses a minimum at an internuclear distance of $R = 2.43$ $a.u.$ A detailed analysis of the wave function [16] shows the predominantly ionic character of the wave function for $3 < R < 7$ $a.u.$. This ionic character has also been confirmed by the analysis of the corresponding rotation-vibration spectrum of the hydrogen molecule [18]. Since H^- possesses only one weakly bound state in field-free space, one expects the ionic character of the hydrogen molecule in the $^1\Sigma_u$ state to decrease with increasing internuclear distance. For large internuclear distances the wave function can be described as a mixture of $(1s\sigma 2s\sigma)$ and $(1s\sigma 2p\sigma)$ configurations where the $(1s\sigma 2p\sigma)$ configuration predominates for very large internuclear distances. Therefore, the dissociation channel is given by $H_2 \rightarrow H\,(1s) + H\,(2p)$. The corresponding state in the united atom limit is the electronic ^3P $1s2p$ helium state.

In the presence of a magnetic field we observe a monotonous increase in the total energy with increasing magnetic field strength. At the same time the dissociation energy increases monotonously (cf. Fig. 1). The value of the equilibrium internuclear distance exhibits a minor increase from 2.42 $a.u.$ in field-free space to 2.53 $a.u.$ for $B = 0.2$ $a.u.$. However, for $B \gtrsim 0.2$ $a.u.$ we observe a drastic decrease in the equilibrium internuclear distance with increasing magnetic field strength. For $B = 1.0$ and 100 $a.u.$ the corresponding values are $R_{eq} = 2.30$ and 0.490 $a.u.$, respectively. Figure 1 shows the well becoming more and more pronounced with increasing magnetic field strength while the onset of the asymptotic convergence behaviour is shifted to larger internuclear distances with increasing field strength.

Another important property of the $^1\Sigma_u$ state in a magnetic field is the change of the dissociation channel in sufficiently strong fields. As described above, the wave function in field-free space exhibits a partially ionic character for certain internuclear distances. The dissociation into $H^- + H^+$ is not possible due to the nonexistence of strongly bound electronic states of H^-. In a magnetic field, however, it is known that a negatively charged atomic ion possesses infinitely many bound states in a non-zero constant magnetic field [19,20]. This means we encounter an infinite number of bound states of H^- in the presence of a magnetic field even though H^- has only one bound state in field-free space. With increasing field strength the binding energy of the H^- ion increases monotonously and H^- becomes more and more strongly bound. A critical field strength exists, for which the total energy of the ground state of H^- and the total energy of $H\,(0^+) + H\,(0^-)$, which is the dissociation limit for weaker field strengths, are equal.

The above considerations help us to understand the asymptotic behavior in the dissociation limit of the $^1\Sigma_u$ state as well as the shape of the PEC: For magnetic field strengths $B \lesssim 10.0$ $a.u.$ the molecular $^1\Sigma_u$ state dissociates into $H\,(0^+) + H\,(0^-)$. Since the total energy of H^- is larger than the energy of $H\,(0^+) + H\,(0^-)$ and since no crossings are allowed between electronic states of the same symmetry, an avoided crossing between the ground state and the first excited state in the $^1\Sigma_u$ subspace occurs in this regime of field strengths. With increasing magnetic field strength $B = 0 \to 10$ *a.u.* the position of the avoided crossing is shifted to increasingly larger internuclear distances (for $B = 0.1$ and 10.0 *a.u.* the position of the crossing is at ~ 12 *a.u.* and 81.47 *a.u.*, respectively). Between $B = 10.0$ and 20.0 *a.u.* the dissociation channel changes from $\mathrm{H}\,(0^+) + \mathrm{H}\,(0^-)$ to $\mathrm{H}^-\,(0_s^+) + \mathrm{H}^+$, where the subscribt s denotes a singlet state. Due to the change in the dissociation channel the onset of the asymptotic behavior of the PEC is shifted to increasingly larger internuclear distances. This leads to a strongly changing shape of the PEC which can be seen in Fig. 1.

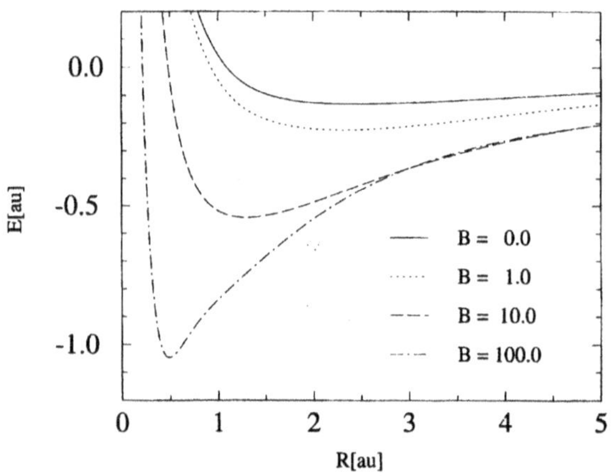

Figure 1. PECs for $B = 0.0$, 1.0, 10.0 and 100.0 *a.u.* for the lowest $^1\Sigma_u$ state; the energy is shown with respect to the dissociation limit, i.e., $E\,(R) = E_t\,(R) - \lim_{R\to\infty} E_t\,(R)$

As a further result we obtain the binding energy for H^- for magnetic field strengths of $B = 50.0$ and 100.0 *a.u.* The resulting total binding energies with respect to both electrons are 3.637999 *a.u.* and 4.561968 *a.u.*, respectively. The best available data for the ionization energy of the H^-–ion in a strong magnetic field $B \geq 100$ *a.u.* are given in Refs. [21, 22]. Compared to the value given in [21] for $B = 100.0$ *a.u.* our result of 4.561968 *a.u.* is more than one per cent lower in energy.

Finally let us comment on the existence of vibrational levels within the above PECs, which determines the existence of physically bound states. The determination of vibrational levels in the presence of a magnetic field is a more complicated task than in field-free space. The nuclear charges are partially screened against the magnetic field and this screening depends not only on the internuclear distance but also on the angle between the H_2 molecule and the magnetic field [12, 13, 23]. Within the present investigation of parallel internuclear and magnetic field axes, we can provide only lower and upper bounds for the vibrational energies. Hereby we can estimate whether a physically bound state may exist in the corresponding PEC. For the $^1\Sigma_u$ state we found that the number of levels increases with increasing field strength. This suggests that the $^1\Sigma_u$ state is a physically bound state in the entire range of field strengths $B = 0 \rightarrow 100$ $a.u.$

THE $^3\Sigma_g$ STATE

Very accurate Born-Oppenheimer energies for the lowest $^3\Sigma_g$ state of the H_2 molecule are given in Ref. [24]. Our CI calculation yields 0.737124 $a.u.$ for the total energy at the equilibrium distance of $R = 1.87$ $a.u.$ The relative accuracy compared to the result in Ref. [24] amounts to 4.4×10^{-5}. We emphasize that this high precision of our data is obtained for arbitrary internuclear distances. The dissociation channel for the $^3\Sigma_g$ state in field-free space is $H_2 \rightarrow H\,(1s) + H\,(2s)$, and the corresponding united atom state is the ^3S $1s2s$ helium state.

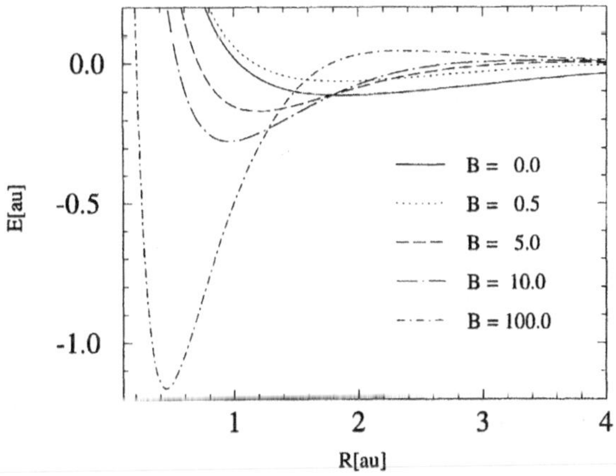

Figure 2. PECs for $B = 0.0$, 0.5, 5.0, 10.0 and 100.0 $a.u.$ in the range $0.1 < R < 4$ $a.u.$ for the lowest $^3\Sigma_g$ state showing the first minimum of this state ; the energy is shown with respect to the dissociation limit, i.e., $E\,(R) = E_t\,(R) - \lim_{R \rightarrow \infty} E_t\,(R)$

In the presence of a magnetic field the $^3\Sigma_g$ state dissociates into $H\,(0^+) + H\,(0^-)$. The corresponding united atom state is the $^30^+$ helium state. Due to the spin-Zeeman shift the total energy monotonously decreases with increasing magnetic field strength. For magnetic field strengths below 0.5 $a.u.$ we observe a monotonous decrease in the dissociation energy E_{d1} with increasing magnetic field strength from 0.112124 $a.u.$ to 0.062317 $a.u.$ for $B = 0.0$ and 0.5 $a.u.$, respectively. At the same time the equilibrium internuclear distance decreases slightly from 1.87 to 1.81 $a.u.$. For magnetic field strengths larger as 0.5 $a.u.$ we observe a drastical increase in the dissociation energy and a simultaneous decrease in the equilibrium internuclear distance. The development of the electronic PEC for internuclear distances smaller than 4.0 $a.u.$ is illustrated in Fig. 2. In Fig. 2 we can observe that the well of the PEC becomes increasingly more shallow for increasing field strengths in the range $B = 0.0 \rightarrow 0.5$ $a.u.$. With further increasing magnetic field strength the opposite behavior can be observed and the potential well becomes more and more pronounced and deeper.

A closer look at the PEC reveals that for magnetic field strengths between 0.05 and 50.0 $a.u.$ the $^3\Sigma_g$ state develops a second minimum which has no counterpart in field-free space. However, this minimum is very shallow and the maximum dissociation energy amounts to only 1.91×10^{-3} $a.u.$ at a magnetic field strength of 1.0 $a.u.$ The corresponding PECs are shown in Fig. 3. In this figure we observe a hump of the $^3\Sigma_g$ state at $B = 0.05$ $a.u.$ This hump also occurs for higher magnetic field strengths but is shifted to smaller internuclear distances and, therefore cannot be seen in Fig. 3. For the location of the maximum we observe an overall decrease in the internuclear distance with increasing magnetic field strength. Both, the maximum and second minimum of the PEC appear at the same magnetic field strength of 0.05 $a.u.$. However, the second minimum vanishes for magnetic field strengths larger than 50.0 $a.u.$ whereas the hump remains in the PEC.

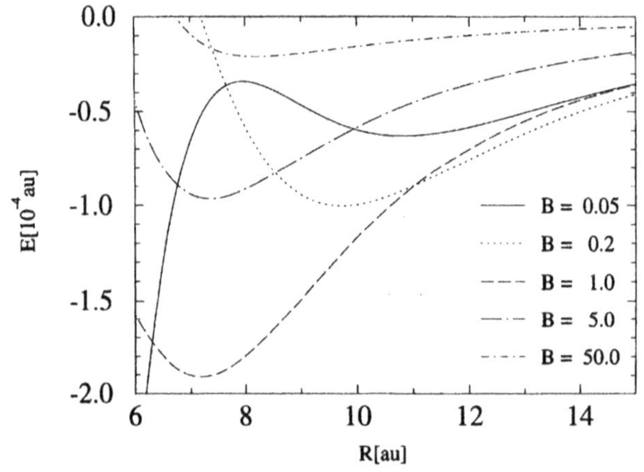

Figure 3. PECs for $B = 0.05, 0.2, 1.0, 5.0$ and 50.0 $a.u.$ in the range $6 < R < 15$ $a.u.$ illustrating the second minimum of the lowest $^3\Sigma_g$ state; the energy is shown with respect to the dissociation limit, i.e., $E(R) = E_t(R) - \lim_{R\to\infty} E_t(R)$

At the end of this section let us answer the question on whether vibrational levels may exist in the PEC of the $^3\Sigma_g$ state. For the first minimum we found in the order of ten vibrational levels. The number of levels slightly decreases from $B = 0.0$ to 0.5 $a.u.$ with decreasing depth of the well. For larger field strengths (5.0, 10.0, and 100.0 $a.u.$) the number of levels slightly increases. For the second, i.e., outer, minimum we found about five vibrational levels for the field strengths 0.05, 0.2, and 1.0 $a.u.$, respectively. For field strengths $B \gtrsim 5.0$ $a.u.$ the lower bound of the vibrational energy lies inside the potential well whereas the upper bound is above the well. Here the detailed dynamical behavior decides on the existence of vibrational levels, and no conclusions can be drawn from our estimations.

SUMMARY

In the present article we investigated the electronic structure of some excited states of the Σ manifold of the hydrogen molecule in the presence of a magnetic field. We hereby focused on the case of a parallel internuclear and magnetic-field axes. The key ingredient for our CI calculations is a basis set of nonorthogonal nonspherical Gaussian atomic orbitals which was established previously. Our results for the PECs in field-free space show a high accuracy compared to the existing data in the literature.

First we investigated the lowest $^1\Sigma_u$ state. With increasing magnetic field we observed a minor increase in the equilibrium internulear distance in the range $0.0 \leq B \leq 0.2$ $a.u.$ followed by a monotonous decrease with further increasing field strength. The dissociation and the total energy increases monotonously with increasing magnetic-field strength. Many vibrational levels were found for the PEC of the $^1\Sigma_u$ state in the entire regime $B = 0 - 100$ $a.u.$ The number of levels hereby increases with increasing field strength. An important feature of the $^1\Sigma_u$ state is the change in the dissociation channel with increasing magnetic-field strength. In field-free space we have $H_2 \rightarrow H(1s) + H(2p)$ in the separated atom limit. The wave function possesses a predominantly ionic character for large values of the internuclear distance. However, in field-free space the dissociation $H_2 \rightarrow H^- + H^+$ of the lowest $^1\Sigma_u$ state is not possible due to the nonexistence of strongly bound states for the H^-–ion in field-free space. In contrast to this strongly bound states of H^- exist in the presence of a magnetic field. Therefore a dissociation into $H^- + H^+$ is possible and we observe a change in the dissociation channel for magnetic-field strengths between $B = 10.0$ and 20.0 $a.u.$: For $B \lesssim 10.0$ $a.u.$ we have $H_2 \rightarrow H(0^+) + H(0^-)$. For field strengths slightly larger than 10.0 $a.u.$ the dissociation limit is given by $H_2 \rightarrow H^-(0_s^+) + H^+$. As a result of our calculations we therefore obtained the dissociation energy for H^- for $B = 50.0$ and 100.0 $a.u.$. Our result of 4.561968 $a.u.$ for the ionization energy of H^- at $B = 100.0$ $a.u.$ shows an improvement of more than 1 % compared to the best value given in the literature.

For the $^3\Sigma_g$ state we encounter a monotonous decrease in the total energy which is proportional to the field strength, and which arises due to the spin-Zeeman shift in the presence of a magnetic field. In the range $0.0 \leq B \leq 0.5$ $a.u.$ the dissociation energy decreases approximately by a factor of two compared to the dissociation energy in field-free space. Simultaneously the equilibrium internuclear distance decreases slightly from 1.87 to 1.81 $a.u.$. With further increasing magnetic-field strength we observe a drastical increase in the dissociation energy and a simultaneous decrease of the internuclear equilibrium distance. A more detailed investigation of the PEC for the $^3\Sigma_g$ state shows a second minimum for magnetic-field strengths between 0.05 and 50.0 $a.u.$ which has no counterpart in field-free space. However, this additional minimum is very shallow. Vibrational levels were found to exist within the first well of the PEC for the entire range of field strengths from $B = 0$ to 100 $a.u.$. For the second minimum for field strengths $B \gtrsim 5$ $a.u.$ the existence of vibrational levels depends on the detailed dynamical behavior and cannot be decided within the present approach. For $B \lesssim 5.0$ $a.u.$ a few vibrational levels exist.

REFERENCES

1. S. Basile, F. Trombetta, and G. Ferrante, Nuovo Cimento **9**, 457 (1987).
2. A. V. Turbiner, Pis'ma Zh. Eksp. Teor. Fiz. **38**, 510 (1983), [JETP Lett. **38**, 618 (1983)].
3. T. S. Monteiro and K. T. Taylor, J. Phys. B **23**, 427 (1990).
4. A. V. Korolev and M. A. Liberman, Phys. Rev. A **45**, 1762 (1992).
5. D. Lai, E. E. Salpeter, and S. L. Shapiro, Phys. Rev. A **45**, 4832 (1992).
6. G. Ortiz, M. D. Jones, and D. M. Ceperley, Phys. Rev. A **52**, R3405 (1995).
7. D. Lai and E. E. Salpeter, Phys. Rev. A **53**, 152 (1996).
8. J. P. Ostriker and F. D. A. Hartwick, Astrophys. J. **153**, 797 (1968).
9. J. Kemp, J. S. Swedlund, J. Landstreet, and J. Angel, Astrophys. J. **161**, L77 (1970).
10. J. Trümper *et al.*, Ann. N. Y. Acad. Sci. **302**, 538 (1977).
11. T. Detmer, P. Schmelcher, F. K. Diakonos, and L. S. Cederbaum, Phys. Rev. A (September 1997).
12. P. Schmelcher, L. S. Cederbaum, and H.-D. Meyer, Phys. Rev. A **38**, 6066 (1988).
13. P. Schmelcher, L. S. Cederbaum, and U. Kappes, *in "Conceptual Trends in Quantum Chemistry", edited by Eugene S. Kryachko* (Kluwer Academic Publishers, Dordrecht, 1994).
14. P. Schmelcher and L. S. Cederbaum, Phys. Rev. A **37**, 672 (1988).
15. U. Kappes and P. Schmelcher, J. Chem. Phys. **100**, 2878 (1994).
16. W. Kolos and L. Wolniewicz, J. Chem. Phys. **45**, 509 (1966).
17. W. Kolos and L. Wolniewicz, J. Chem. Phys. **48**, 3672 (1968).
18. G. Herzberg and L. L. Howe, Can. J. Phys. **37**, 636 (1959).
19. J. E. Avron, I. W. Herbst, and B. Simon, Commun. Math. Phys. **79**, 529 (1981).
20. J. Fröhlich, E. H. Lieb, and M. Loss, Commun. Math. Phys. **104**, 251 (1986).
21. M. Vincke and D. Baye, J. Phys. B **22**, 2089 (1989).
22. D. M. Larsen, Phys. Rev. B **20**, 5217 (1979).
23. T. Detmer, P. Schmelcher, and L. S. Cederbaum, J. Phys. B **28**, 2903 (1995).
24. D. M. Bishop and L. M. Cheung, Chem. Phys. Lett. **79**, 130 (1981).

ELECTRONIC PROPERTIES OF MOLECULES IN HIGH MAGNETIC FIELDS: HYPERMAGNETIZABILITIES OF H_2

Keith Runge and John R. Sabin

Quantum Theory Project
University of Florida
Gainesville, Florida 32611-8435
and
National High Magnetic Field Laboratory
Tallahassee, Florida 32310

ABSTRACT

Self-consistant field calculations using a Hamiltonian containing an external magnetic field and a basis of gauge origin independent atomic functions have been used to calculate some of the lower (< 5) magnetic hyperpolarizabilities of H_2. Comparison with other calculations gives good agreement.

INTRODUCTION

The study of the effect of large magnetic fields on the physical and electronic properties of molecules has traditionally been confined to the low field (≤ 50 T) regime, *i.e.* to those field strengths generally found in the laboratory. Recently, however, there has been a flurry of activity in the calculation of the properties of atoms and molecules in strong ($\sim 10^4$ T - 10^8 T) and extreme ($> 10^8$ T) fields, as these are the fields extant on the surfaces of white dwarf and neutron stars.[1] Futher, there has been growing interest in the geometrical and electronic properties of molecules in high fields (50 T - 10^4 T) - the upper limit of semicontinuous or pulsed fields, either presently or soon available in specialized laboratories, such as the National High Magnetic Field Laboratory (NHMFL) in Tallahassee, Florida.

The most noteworthy of the theoretical studies on molecules in this regime is reported in a series of papers from the Heidelberg group[2-7] and reviewed by them.[8] Their work indicates that changes in the electronic (electron density and spin distributions) and geometric (minimum energy bond lengths and angles, potential surfaces) structure of molecules and clusters might be expected in intense applied fields. These are of interest not only in their own right, but are important input information for other studies. For example, changes in the electronic structure of molecules leads inevitably to spectroscopic changes. Some evidence of this has been found in simple systems such as the Zeeman splitting in the H atom[9, 10] and the H_2 molecule.[11] A more complete spectroscopy of molecules and clusters in intense magnetic fields could yield information about the fundamental electronic and geometric structures involved. In addition, certain practical corollaries might emerge. Knowledge of the spectroscopy of simple molecules (e.g. carbon monoxide) as a function of field strength could also give an independent measurement of field strengths in situations where direct measurements of the magnetic field is not practical (e.g. in space or near rotating neutron stars).

In this contribution, we wish to concentrate on one aspect of the spectroscopy of molecules in high applied external fields: that portion of nonlinear optical response due to magnetic fields alone. The field of nonlinear optical effects is rich (*vide infra*), and deals with the effects of applied external electric, magnetic, and mixed fields on spectroscopic properties. The effects of electric fields are well documented (*cf. e.g.* Ref. 12) but there has been little investigation of those due to magnetic or mixed fields. We consider here the effects of static magnetic fields on H_2.

FORMALISM

Nonlinear Optical Properties

Consider the change of energy of a molecule when it is subjected to a static, uniform magnetic field **B** and/or electric field **E** . The lowering of the total energy due to the presence of the field(s) is usually expanded as:[12]

$$
\begin{aligned}
\Delta E = & -\mu_\alpha F_\alpha - \frac{1}{2}\alpha_{\alpha\beta}F_\alpha F_\beta - \frac{1}{6}\beta_{\alpha\beta\gamma}F_\alpha F_\beta F_\gamma \\
& -\frac{1}{24}\gamma_{\alpha\beta\gamma\delta}F_\alpha F_\beta F_\gamma F_\delta + \cdots \\
& - m_\alpha B_\alpha - \frac{1}{2}\chi_{\alpha\beta}B_\alpha B_\beta - \frac{1}{6}Y_{\alpha\beta\gamma}B_\alpha B_\beta B_\gamma \\
& -\frac{1}{24}\zeta_{\alpha\beta\gamma\delta}B_\alpha B_\beta B_\gamma B_\delta + \cdots \\
& - G_{\alpha\beta}F_\alpha B_\beta - \frac{1}{2}\xi_{\alpha\beta\gamma}F_\alpha B_\beta B_\gamma - \frac{1}{2}X_{\alpha\beta\gamma}F_\alpha F_\beta B_\gamma \\
& -\frac{1}{4}\eta_{\alpha\beta\gamma\delta}F_\alpha F_\beta B_\gamma B_\delta + \cdots
\end{aligned} \tag{1}
$$

where the tensorial componenets of the fields are indicated by Greek subscripts. In this preliminary work, we concern ourselves only with that part of eq. 1 which deals purely with the applied magnetic field. In this case,

$$
\begin{aligned}
\Delta E = & - m_\alpha B_\alpha - \frac{1}{2}\chi_{\alpha\beta}B_\alpha B_\beta - \frac{1}{6}Y_{\alpha\beta\gamma}B_\alpha B_\beta B_\gamma \\
& - \frac{1}{24}\zeta_{\alpha\beta\gamma\delta}B_\alpha B_\beta B_\gamma B_\delta + \cdots
\end{aligned} \tag{2}
$$

Here, m_α is the magnetic dipole function, $\chi_{\alpha\beta}$ is the magnetizability, and $Y_{\alpha\beta\gamma}$, $\zeta_{\alpha\beta\gamma\delta}$, *etc.* are the higher hypermagnetizabilities. Previously, the hypermagnetizabilities have not been of concern, as they are thought to be small and laboratory fields have not been large enough to measure them. With the high fields presently, or soon to be, available at the NHMFL, this is, hopefully, no longer the case.

The Hamiltonian

We restrict our considerations to fields less than $\sim 10^8$ T, where relativistic effects are negligible. Consider a set of N charged particles having masses $\{m_i\}$, charges $\{q_i\}$, and total nuclear charge Z. Thus $M = \sum_i m_i$ is the total mass of the system and $Q = \sum_i q_i$ is the total charge on the system. The nonrelativistic Hamiltonian for this system of particles in an external magnetic field $\mathbf{B}$ with the scalar potential $\mathcal{V}$ can be written

$$\mathcal{H} = \sum_i^N \frac{\pi_i^2}{2m_i} + \mathcal{V}[\{|\mathbf{r}_i - \mathbf{r}_j|\}] \tag{3}$$

The eigenfunctions of this Hamiltonian are, of course, functions of the coordinates of all particles; nuclei and electrons. Here

$$\pi_i = \mathbf{p}_i - q_i \mathbf{A}_i \tag{4}$$

is the mechanical momentum of the ith particle. We use the symmetric gauge, so that the vector potential is connected to the field *via*

$$\mathbf{A}_i = \frac{1}{2}[\mathbf{B} \times \mathbf{r}_i] \tag{5}$$

We now introduce the pseudomomentum of the ith particle

$$\mathbf{k}_i = \pi_i + q_i[\mathbf{B} \times \mathbf{r}_i] \tag{6}$$

and the total pseudomomentum

$$\mathbf{K} = \sum_i \mathbf{k}_i \tag{7}$$

Following Schmelcher *et al.* [8] (see also Ref. 13), one finds that for a neutral diatomic molecule, the Hamiltonian can be written in the Born-Oppenheimer approximation as

$$\begin{aligned}\mathcal{H}_{el} = {} & \frac{1}{2M}K^2 - \frac{e}{M}[\mathbf{K} \times \mathbf{B}](\sum_i \mathbf{r}_i - \frac{1}{M_0}\mathcal{M}\mathbf{R}) \\ & + \sum_i \frac{1}{2m}(\mathbf{p}_i - \frac{e}{2}[\mathbf{B} \times \mathbf{r}_i])^2 \\ & + \frac{1}{2M_0}(\sum_i \mathbf{p}_i + \frac{e}{2}\sum_i [\mathbf{B} \times \mathbf{r}_i] - \frac{e}{M_0}\mathcal{M}[\mathbf{B} \times \mathbf{R}])^2 + \mathcal{V}(\{\mathbf{r}_i\}, \mathbf{R})\end{aligned} \tag{8}$$

Gauge Origin Independent Atomic Functions

In order to use the many years of accumulated experience in calculation of molecular electronic structure, we would like to use Gaussian atomic basis functions. One encounters a problem, however, as the effect of the magnetic field on the electronic Hamiltonian is to alter the kinetic energy of the electron. This change in kinetic energy needs to be accounted for in the Gaussian atomic basis functions themselves. We achieve this end through the use of an exponential factor[13] which is introduced in order to aviod the gauge problem normally encountered in electronic structure calculations in high external magnetic fields.[2,4] Thus, we work in a basis of gauge origin independent atomic functions (GIAF's). These GIAF's are composed of the commonly chosen normalized Gaussian basis functions, χ, multiplied by gauge origin independence factors

$$\xi_\mu = \exp(-\frac{i}{2}[\mathbf{B} \times \mathbf{R}] \cdot \vec{r}_i)\chi_\mu \tag{9}$$

Here ξ_μ is the GIAF, **B** is the magnetic field, **R** is the location of the nucleus on which the GIAF is centered, $\vec{r_i}$ is the position of electron i, and χ_μ is a Gaussian atomic function. The effect of the exponential introduced into the GIAF is to boost the momentum of the electron to the mechanical momentum of the electron in the magnetic field.

The calculation of one and two electron integrals in the GIAF basis is not as common as the calculation of such integrals in the atomic Gaussian basis. For this reason, we choose to calculate these integrals in the Gaussian basis and then transform them to the GIAF basis. The overlap matrix is calculated in both Gaussian and GIAF bases.

We arrive at the transformation between integrals expressed in the Gaussian and the GIAF bases by beginning with the completeness relations as expressed in both the Gaussian and the GIAF bases

$$\begin{aligned} \hat{I} &= \sum_{\mu\nu} |\xi_\mu\rangle(\langle\xi_\mu|\xi_\nu\rangle)^{-1}\langle\xi_\nu| \\ &= \sum_{\mu\nu} |\chi_\mu\rangle(\langle\chi_\mu|\chi_\nu\rangle)^{-1}\langle\chi_\nu| \end{aligned} \tag{10}$$

Using the convention that a superscript denotes the basis in which a matrix for the operator $\hat{O}$ is written

$$\begin{aligned} \langle\xi_\mu|\hat{O}|\xi_\nu\rangle &= \mathbf{O}^{(\xi)} \\ \langle\chi_\mu|\hat{O}|\chi_\nu\rangle &= \mathbf{O}^{(\chi)} \end{aligned} \tag{11}$$

we obtain the transformation matrices **B** and **C**

$$\begin{aligned} \mathbf{O}^{(\xi)} &= \mathbf{B}^\dagger \mathbf{O}^{(\chi)} \mathbf{B} \\ \mathbf{O}^{(\chi)} &= \mathbf{C}^\dagger \mathbf{O}^{(\xi)} \mathbf{C} \end{aligned} \tag{12}$$

where

$$B_{\mu\nu} = \sum_k (S^{(\chi)})^{-1}_{\mu\kappa} \langle\chi_\kappa|\xi_\nu\rangle \tag{13}$$

takes one from the GIAF basis to the Gaussian basis and

$$C_{\mu\nu} = \sum_k (S^{(\xi)})^{-1}_{\mu\kappa} \langle\xi_\kappa|\chi_\nu\rangle \tag{14}$$

takes one in the opposite direction. For the purposes of the calculations reported here, we shall employ restricted Hartree-Fock theory. Extensions to correlated treatments of molecules are straight forward. As any implementation of this scheme will involve finite bases while eq. 12 presumes a complete basis, one must be aware of basis set effects.

We note that the two-electron integrals of conventional Hartree-Fock theory involve the four index quantities

$$\langle\kappa\lambda|r_{12}^{-1}|\mu\nu\rangle = \int d^3r_1 d^3r_2 \chi + \kappa(\vec{r_2})\chi_\lambda(\vec{r_2}) r_{12}^{-1} \chi_\mu(\vec{r_1})\chi_\nu(\vec{r_1}) \tag{15}$$

Transforming these two-electron integrals into the GIAF basis would require a four index transformation of the type shown above. However, the quantity that appears in

Table 1. Calculated Magnetic Moments of H_2 Parallel Configuration

moment	this work	CB[a]
m_z	4×10^{-5}	0^b
χ_{zz}	-0.7721 ± 0.0006	0.7769
Y_{zzz}	0.1192 ± 0.0102	
ζ_{zzzz}	3.3157 ± 0.0608	

[a]Ref. 21
[b]identically zero

the Fock matrix itself is a sum of products of two-electron integrals and density matrix elements

$$\begin{aligned} F_{\mu\nu} &= H^{(1)}_{\mu\nu} + G_{\mu\nu} \\ G_{\mu\nu} &= \sum_{\kappa\lambda} P_{\mu\nu}(2\langle\kappa\lambda|r_{12}^{-1}|\mu\nu\rangle - \langle\mu\lambda|r_{12}^{-1}|\kappa\nu\rangle) \end{aligned} \tag{16}$$

where $\mathbf{F}$ is the Fock matrix, $\mathbf{H}^{(1)}$ is the one-electron hamiltonian, $\mathbf{P}$ is the density matrix and $\mathbf{G}$ is the two-electron piece of the Fock matrix. $\mathbf{G}$ is itself a two index quantity, so that we transform the density matrix into the Gaussian basis and the two-electron piece of the Fock matrix is constructed in the Gaussian basis. The Fock matrix thus constructed in the Gaussian basis is then transformed into the GIAF basis. Hence, we replace one four-index transformation by two two-index transformations.

Calculational Details

In the calculations reported here, we have used bases consisting of segmented contracted Gaussian functions rich in tight functions. Such bases have previously been shown to give good results for molecular properties with magnetic field dependencies, such as NMR chemical shifts[14] and spin-spin coupling constants.[15]

For hydrogen, an (8s) contracted to {5s} primitive set,[16] augmented with a (3p) set of polarization functions[17] was used. The large basis allows us to reproduce previous results to at least microhartree accuracy. The iterative Hartree-Fock procedure is based on an energy convergence criterion which, for these pilot calculations, was set to a microhartree. It is clear that this criterion serves as a limit to the accuracy of these calculations rather than any insufficiency in the basis.

Results and Discussion

Calculations were carried out on the $^1\Sigma_g^+$ ground state of H_2 at a variety of external magnetic fields in the range $0 < B < 1$ au, [18] with the molecular bond axis parallel to the field direction. In each case the minimum energy geometry was determined. As expected, and in agreement with similar calculations,[19, 20] the equilibrium bond length decreases and the total energy rises.

The relative change in minimum energy $[\Delta E = E(B) - E(0)]$ with respect to the field free case as a function of applied field for H_2 in the parallel configuration is plotted in Fig. 1. The rise is attributable to the increase in kinetic energy of the electrons in the field. If this curve is fit to eq. 2, the various response strengths can be determined.

Using a simple least squares fit to a fourth order polynomial of the form of eq. 2, the diagonal components of the moment tensors along the field direction can be calculated. The results are presented in Table 1, where they are compared to a similar SCF calculation of Cybulski and Bishop (CB). [21]

As can be seen, the agreement is good: the magnetic dipole moment is vanishingly small, as it should be, and the agreement of the magnetizability with that calculated by CB is within a percent. To the best of our knowledge, the third and fourth order moments have not been previously reported. In this case, the third order contribution to the energy is indeed small, but the fourth order term appears large enough so that it might be measured at high field.

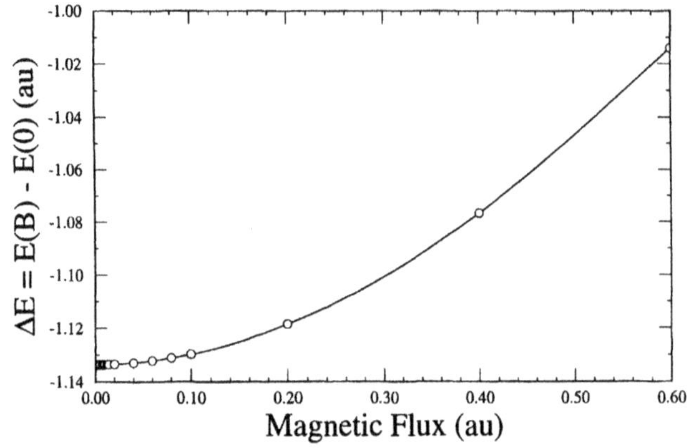

Figure 1. Relative change of minimum energy as a function of applied field for H_2.

Acknowledgments

This work was supported in part by grant #DAAH04-95-1-0326 from the U.S. Army Research Office. We thank Prof. David A. Micha for his support of this project.

REFERENCES

1. See, for example, other papers in this Proceedings.
2. P. Schmelcher and L. S. Cederbaum, Phys. Rev. A **37**,672 (1988).
3. P. Schmelcher, L. S. Cederbaum and H.-D. Meyer, Phys Rev. A **38**, 6066 (1988).
4. U. Kappes and P. Schmelcher, J. Chem. Phys. **100**, 2878 (1994).
5. U. Kappes, P. Schmelcher and T. Pacher, Phys. Rev. A **50**, 1775 (1994) .
6. U. Kappes and P. Schmelcher, Phys. Rev. A **51**, 4542 (1995).
7. U. Kappes and P. Schmelcher, Phys. Lett. A, **210**, 409 (1996).
8. P. Schmelcher, L. S. Cederbaum and U. Kappes, Molecules in magnetic fields: Fundamental aspects, in *Conceptual Trends in Quantum Chemistry*, pages 1-51, Kluwer Academic Publishers, Dordrecht, 1994.
9. K. Runge, Presentation at the *1000 Tesla Workshop*, Tallahassee, April, 1995.
10. Y. P. Kravchenko, M. A. Liberman, and B. Johansson, Bull. A.P.S. **40**, 1284 (1995).
11. T. S. Monteiro and K. T. Taylor, J. Phys. B: At. Mol. Opt. Phys. **23**,427 (1990).

12. D. Bishop, Adv. Quant. Chem. **25**, 1 (1994).

13. K. Runge and J. R. Sabin, Int. J. Quantum Chem. **00**, 000 (1997).

14. J. Gauss, J. Chem. Phys. **99**, 3629 (1993).

15. S. A. Perera, M. Nooijen, and R. J. Bartlett, J. Chem. Phys. **104**, 3290 (1996).

16. The s orbital exponents and their contraction coeffiicients are [41856108023, 0.0026559; 62.745732323, 0.0205875; 14.279648916, 0.1069490; 4.041829912, 0.4333780; 1.3163925229, 1.0000000; 0.47346270504, 1.0000000; 0.18280016580, 1.0000000; 0.07288443779, 1.0000000].

17. Polarization function exponents of 2.2920, 0.8380, and 0.2920 were obtained from the Alrichs group by anonymous FTP at tchibm3.chemie.uni-karlsruhe.de.

18. For magnetic field, 1 $au = 2.35 \times 10^5$T for $\hbar = m_e = e = 1$.

19. T. Detmer, P. Schmelcher, F. K. Diakonos and L. S. Cederbaum, Phys. Rev. A Sept. 1997.

20. Y. P. Kravchenko, private communication.

21. S. M. Cybulski and D. M. Bishop, J. Chem. Phys. **101**, 424 (1994).

SHALLOW DONOR STATES IN A MAGNETIC FIELD

Tjeerd O. Klaassen,[1] Janette L. Dunn,[2] and Colin A. Bates[2]

[1]Faculty of Applied Physics / Delft Institute of Microelectronics and Submicrontechnology (DIMES), Delft University of Technology, P.O. Box 5046, 2600 GA Delft, The Netherlands
[2]Physics Department, University of Nottingham, University Park, Nottingham NG7 2RD, UK

INTRODUCTION

The central question in the subject of atoms and molecules in strong external fields concerns the influence of external electromagnetic fields on the electronic properties of an atom or a molecule. Electric and magnetic fields can be considered to be "strong" if their action results in sizable modifications of the electronic structure, i.e. if the interaction strength of the electrons with those fields is comparable to their structural binding forces in the atom.
Donor impurities in III-V semiconductors like GaAs appear ideal subjects for such studies. The Coulomb interaction that causes an electron to be bound to the positively charged donor nucleus is very small as the result of the large dielectric constant of the semiconductor and the small "effective" mass of the electron. Consequently, magnetic (and electric) fields of moderate strength, and which are thus relatively easy to realize in the laboratory, do have a profound influence on the electron wave functions. In many aspects, the "donor + electron" system is closely related to the hydrogen atom. This "mimicked hydrogen atom" can be studied under the condition that the interaction with the external magnetic field is comparable to, or even outweighs, the nucleus-electron Coulomb interaction. Such a situation is very interesting, as it is similar to that of genuine hydrogen atoms in the extreme intense magnetic fields (10^5-10^6 T) present in the atmospheres of white dwarf and neutron stars. The donor can thus serve as a perfect experimental testing ground for theoretical models on properties of the hydrogen atom under these extreme conditions.
In this paper, we will present the results of magneto-optical experiments on donors in GaAs and compare the data with predictions from recent theoretical work on the hydrogen atom. Our emphasis will be on the many "Lyman" transitions observed involving Landau-like states. We will discuss the deviations from the ideal model system due to band non-parabolicity, electron-phonon interactions and the presence of ionized impurities. The possibilities of using this shallow donor system to study

Atoms and Molecules in Strong External Fields
Edited by Schmelcher and Schweizer, Plenum Press, New York, 1998

Table 1. Energy and length scales for the H - atom and the donor in GaAs

Quantity	Expression	units	Hydrogen	Donor
Electron mass	m^*	m_e	1	0.0665
Dielectric constant	ϵ_r		1	12.56
Rydberg	$\mathcal{R} = m^* e^4/32\pi^2\epsilon_0^2\epsilon_r^2\hbar^2$	cm^{-1}	11×10^4	46.1
Bohr radius	$a_0 = 4\pi\epsilon_0\epsilon_r\hbar^2/m^*e^2$	nm	0.0529	9.96
Larmor energy	$\hbar\omega_L = eB/2m^*$	cm^{-1}	$0.467\times$B(T)	$7.02\times$B(T)
Larmor radius	$(\hbar/eB)^{1/2}$	nm	$25.53/\text{B(T)}^{1/2}$	$25.53/\text{B(T)}^{1/2}$
γ	$\hbar\omega_L/\mathcal{R}$		$\text{B(T)}/2.35\times10^5$	B(T)/6.57

the influence of strong electric fields on electron states of the hydrogen atom will be illustrated also, and some preliminary experimental results given.

DONOR IN A MAGNETIC FIELD

The properties of electrons in a conduction band minimum of a semiconductor are usually treated within an effective mass approximation: the influence of the crystal potential is described in terms of an *effective* electron mass $m^* \neq m_e$, the true electron mass. In GaAs, the electron bound to a positively charged donor is characterized by an effective mass $m^* = 0.0665m_e$ which, together with the large value of the relative dielectric constant (ϵ_r=12.56) results in a strongly reduced Coulomb interaction. The radius of the electron orbit therefore is a factor $\epsilon_r m_e/m^* \approx 200$ larger than that of the Bohr orbit of the H-atom. Hence in first order the electron does not feel the details of the donor chemical binding, but only the Coulomb interaction with its nucleus. It is thus a hydrogen-like centre. The mimicked hydrogen atom and the genuine one differ only in the magnitudes of their electric- and magnetic interactions; these are compared in Table 1. The influence of the magnetic field on the electron states is determined by the ratio of the electric- and magnetic interactions $\gamma = \hbar\omega_L/\mathcal{R}$. Whereas $\gamma = 1$ for the H-atom is reached at a field of 2.35×10^5T, for the donor electron a field of only 6.57T is needed; this is easily achievable in the laboratory. Because of the small value of the Rydberg constant, the optical transitions for the donor occur in the 30 to 300 μm (far infrared) region of the optical spectrum.
The Schrödinger equation describing the electron bound to a donor nucleus, subject to a magnetic field parallel to the z-axis, is given by

$$\mathcal{H}\Psi = \left(-\frac{\hbar^2}{2m^*}\Delta - \frac{e^2}{4\pi\epsilon_0\epsilon_r r} + \frac{1}{2}\hbar\omega_{cr}l_z + \frac{1}{8}m^*\omega_{cr}^2(x^2+y^2)\right)\Psi = E\Psi \qquad (1)$$

Here Δ is the Laplace operator, l_z is the z-component of the angular momentum, and $\omega_{cr} = 2\omega_L$ is the cyclotron frequency. To obtain formally the complete set of solutions for finite magnetic field, one needs to expand the wavefunctions in Landau states $\phi_{N,m}(\rho)$ [1]:

$$\Psi_{Nm\nu}(\rho,\phi,z) = \sum_{N'\geq 0} \phi_{N',m}(\rho)e^{im\phi}\mathcal{F}_{N'm\nu}(z). \qquad (2)$$

Inserting eq.(2) in eq.(1) yields the matrix equation:

$$\left[-\frac{\hbar^2}{2m^*}\frac{\partial^2}{\partial z^2} + \left(N+\frac{1}{2}\right)\hbar\omega_{cr} - E\right]\mathcal{F}_{N'm\nu}(z) + \sum_{N'} V_{N,N'}^m(z)\mathcal{F}_{N'm\nu}(z) = 0 \qquad (3)$$

Here N is the Landau level index and m the z component of the angular momentum. The first term contains the effect of the magnetic field and the last term the attractive

potential of the donor. Eq.(3) has two sets of solutions; the first set contains the free electron states, characterized by the set of quantum numbers $(N, m, \hbar k_z)$, with $\hbar k_z$ the z component of momentum and $m \leq N$. The free electron states form a continuum above each Landau level N, with energies $E = (N + \frac{1}{2})\hbar\omega_{cr} + \hbar^2 k_z^2/2m^*$.
The second set represents the bound states. For these states N is *not* a good quantum number because the matrix elements $V^m_{N,N'}(z)$ mix Landau states with different N and equal m. Fortunately, for $\gamma \geq 0.1$ the contribution of only one Landau level is found to dominate, thus enabling the bound states to be characterized by an unambiguous label N and by the good quantum numbers m and ν (where ν is the number of nodes of the wavefunction along the z-axis). For these bound states the total energy is $E_{N,m,\nu} = (N + \frac{1}{2})\hbar\omega_{cr} - E(N, m, \nu)$, with $E(N, m, \nu) > 0$ the Coulomb binding energy. Landau level N is the ionization threshold. For every N an infinite number of bound states (N, m, ν) exists with $m = N, N-1, N-2, ..$ and $\nu = 0, 1, 2, ...$ The overall parity of a state is $\pi = (-1)^{(m+\nu)}$.
The binding energy as a function of N, m, ν and γ can be estimated without actually solving the above eigenvalue equation. For Landau states, the expectation value of ρ^2 can be shown to equal $2(2N - m+1)/\gamma$. So, with increasing field the average radius of these orbits in the xy-plane will decrease, leading to an increase of the Coulomb binding. At fixed field the average radius will be larger for a larger N and/or a smaller m; consequently the Coulomb binding will be smaller. A similar argument holds for ν: if the number of nodes of the wavefunction along the z-axis is large, the average z-axis distance between electron and nucleus is large, and thus the Coulomb interaction is small. In addition, the presence of the l_z term in the Hamiltonian (eq.(1)) leads to the equality $E(N, m, \nu) = E(N - m, -m, \nu)$.

Fig. 1 shows the order of the bound states for $\gamma \approx 1$ according to numerical calculations (see below); apart from some small deviations this order is consistent with the considerations given above[2].
It must be noted that the Coulomb potential $V^m_{N,N'}(z)$ of Eq.(3) mixes states with equal m. Inspection of Fig. 1 shows that a state (N, m, ν) with $N \geq 1$ and $m < N$ can mix with free electron states with equal m and equal energy. An electron in such a bound state can thus perform a (very fast) radiationless - energy conserving - transition towards a continuum state. These states are therefore called autoionizing-, metastable- or Landau-like states [1]. Apart from a lifetime broadening, a transition to such a state will also show the asymmetric "Fano resonance" shape [3]. For all N=0 states and states with $m = N \geq 1$ such a transition is not possible, because no free electron states with the same m at equal energy exist. Only these states extrapolate for B $\rightarrow$ 0 to the well known zero field hydrogen states. In Fig. 1, the zero field quantum numbers (n, l, m) are also given for the "hydrogen-like" states.

FAR-INFRARED MAGNETO-SPECTROSCOPY

In general, magneto-spectroscopy experiments are carried out at liquid Helium temperature (T=4.2K) to ensure that the electron is bound to the donor. A molecular gas laser, optically pumped by a tunable CO_2 laser, is often used as a source of monochromatic far-infrared radiation. Although not continuously wavelength tunable, these lasers can easily be made to operate at a large number of different wavelengths in the 30 μm - 2 mm. range. As the energies of the electron states depend strongly on the magnetic field, an optical spectrum can be measured by scanning the magnetic field at a fixed laser wavelength. Experiments are performed typically on 10μm thick epitaxial layers grown on semi-insulating GaAs substrates; donor concentrations have to be below 10^{15}cm^{-3} in order to have "isolated" centres.

When an electron is optically excited, the donor can subsequently be ionized easily through emission (or absorption) of an acoustic phonon. In that way, some of the excited electrons end up in the conduction band and the optical absorption can be measured simply by monitoring the photo-induced conductivity of the sample. In Fig. 2 the results of a field scan at T=4.2K and λ=70.5 μm (142 cm^{-1}) is shown (lower trace). As only the 1s^0 ground state is populated, all transitions shown belong to the "Lyman series". To emphasize the oscillatory behaviour of the photoconductivity at the low field (or high energy-) side of the spectrum, a spectrum at λ=53.9 μm (186 cm^{-1}) is shown also. Making use of the selection rules for electric dipole transitions and the parity of the states, the final states of the transitions have been identified. From field scans at a different wavelengths the transition energy *versus* field curves as shown in Fig. 3 have been obtained [4,5]. Figs. 2 and 3 summarise the essential features of the optical spectrum of the donor in GaAs (and consequently also of the hydrogen atom in very large magnetic fields) for energies up to 270 cm^{-1}.

The spectrum (for transitions to states above the N=0 level) is dominated by a repeating sequence of 4 transitions. Between Landau levels N and (N-1) the main features of the spectrum consist of the transitions to the states $(N,1,0)$, $(N,-1,0)$, $(N,0,1)$, and $(N,1,2)$, with the $(N,1,0)$ transition having the largest intensity. Towards higher N the intensities decrease. For N=1 essentially the same pattern can be recognized, but, due to the overall higher intensities some additional, weaker transitions

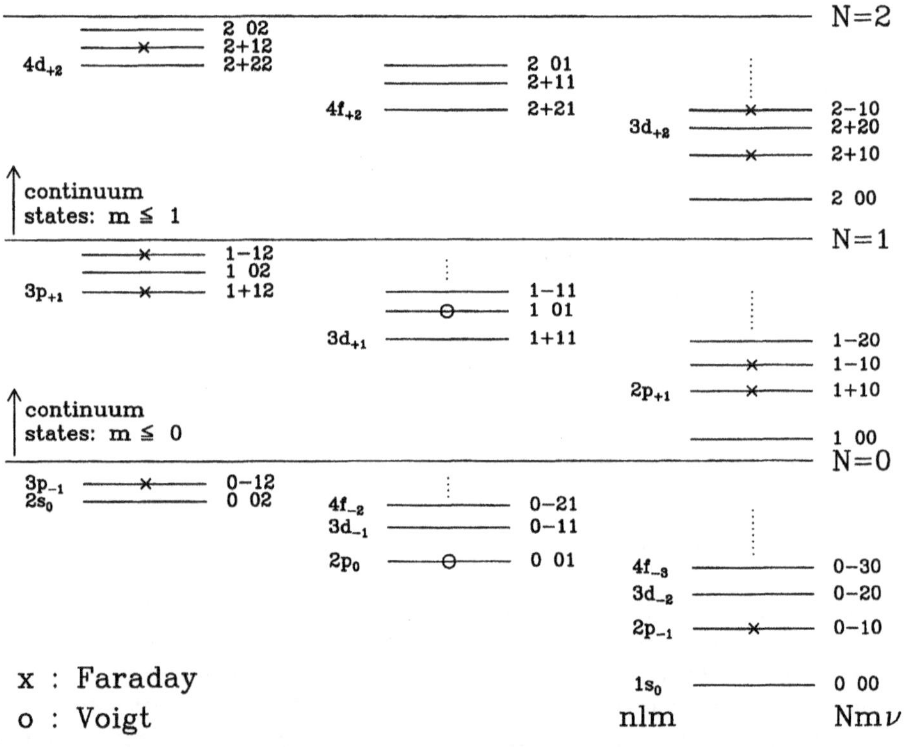

Figure 1. Energy level diagram for the electron states in an intermediate magnetic field. Final states for (allowed) "Lyman" transitions are indicated by ×/o for the Faraday/Voigt configuration. For the true hydrogen-like states also the zero field (n, l, m) notation is given.

are visible too. The lines through the experimental points in Fig. 3 drawn as a guide to the eye can not be distinguished on that scale from the theoretical fits (see below) for energies below 250 cm^{-1}.

The transitions towards Landau-like states clearly show the asymmetric shape (dip at the low field side), characteristic for Fano-resonances[3] . The transitions to the hydrogen-like $(1,1,0)$ and $(1,1,2)$ states have a notably smaller linewidth than all other transitions, and therefore reveal a fine structure, which is obscured in the other transitions.

INTERACTION OF DONOR ELECTRON WITH HOST

In order to compare experimental data with predictions for the hydrogen model, the influence of the GaAs host on the donor spectrum has to be examined. A detailed inspection of the experimental data shows the presence of small as well as large deviations from the behaviour expected for the genuine hydrogen atom.

Finestructure at Transitions

The first reason for the fine structure observed on the transitions to the $(1,1,0)$ and $(1,1,2)$ states, clearly seen in the 70.5 μm spectrum in Fig.2, is the presence of chemically different donor species in the sample. The details of the chemical bond of the donor influence to a small extend the binding of the electron in an s-orbit and lead to a finestructure at transitions out of the $1s^0$ state. The magnitude of this so-called central cell shift (≤ 1 cm^{-1} for B=0) increases slightly with increasing field due to contraction

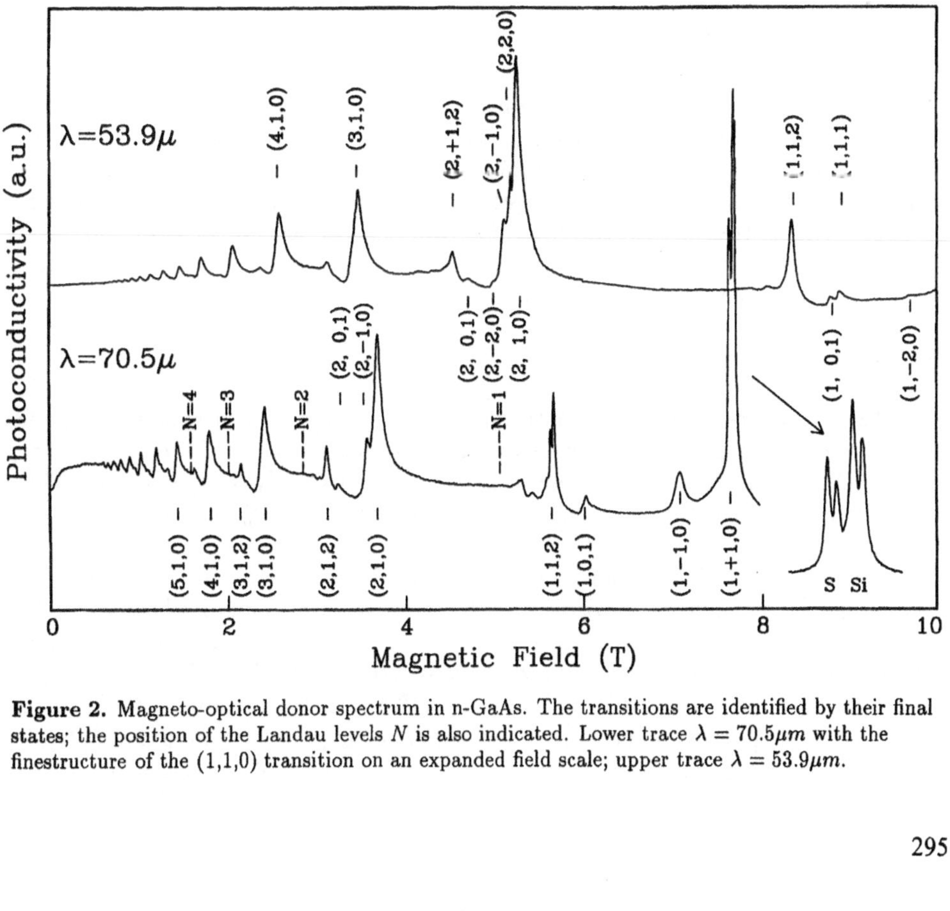

Figure 2. Magneto-optical donor spectrum in n-GaAs. The transitions are identified by their final states; the position of the Landau levels N is also indicated. Lower trace $\lambda = 70.5\mu m$ with the finestructure of the (1,1,0) transition on an expanded field scale; upper trace $\lambda = 53.9\mu m$.

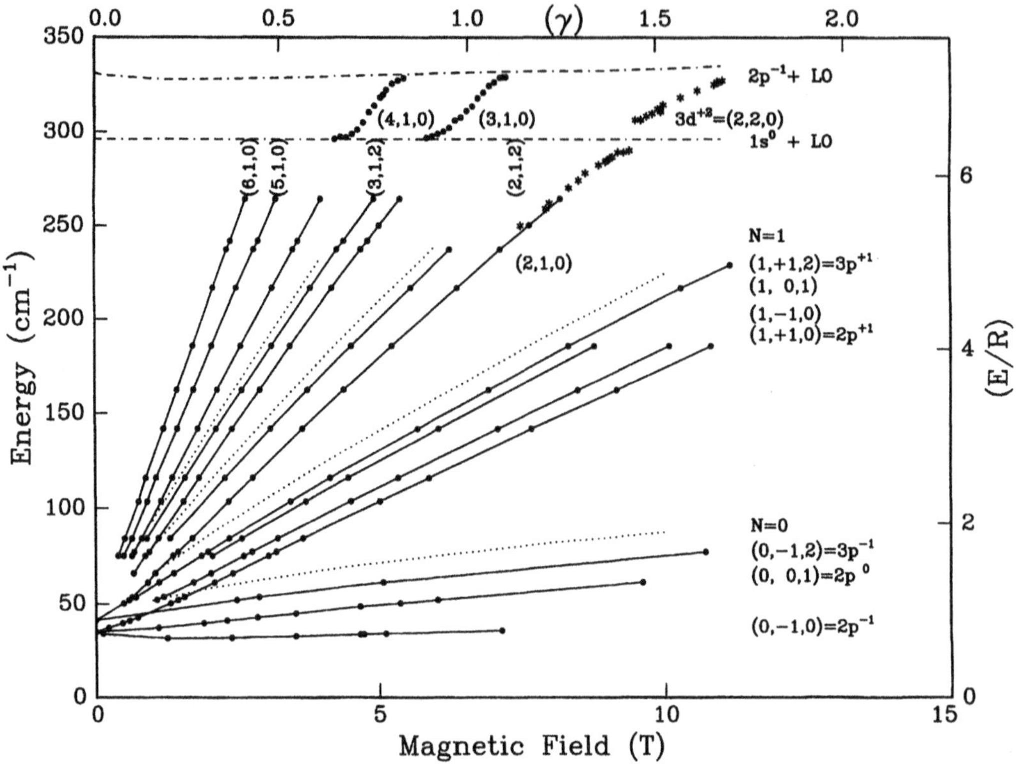

Figure 3. Field dependence of the energies for some of the stronger Lyman transitions. Apart from the energy- and field scales that apply for the donor in GaAs, also reduced scales with general applicability are given. The data near the LO-phonon energy (296 cm^{-1}) show the avoided crossing behaviour resulting from the electron-phonon interaction.

of the electron wavefunction [6]. Both transitions show a two-fold structure originating from the presence of Si and S donors in the sample. The additional splitting of the (1,1,0) transition is due to the presence of the electron spin. In general the electron spin causes all electron states to have an equal two- fold splitting, which is *not* observable in electric dipole transitions. However, due to mixing of energy bands in GaAs, this splitting becomes energy dependent, and therefore becomes observable as a two-fold, magnetic field dependent, splitting of all transitions[7].

Ionized Impurities

Besides donors, a finite concentration of acceptors is also present in a sample. At low temperatures, these acceptors capture electrons from neutral donors, thereby creating N_a negatively charged acceptors and as many positively charged donors. These ionized impurities create electric field gradients at the neutral donor sites, which cause mixing of donor states with different quantum numbers m and ν, resulting in the observation of "forbidden" transitions. Moreover, it leads to an inhomogeneous line broadening [8]. If radiation across the band gap of the semiconductor host is applied, the ionized impurities are neutralised and the above mentioned effects are suppressed. The two spectra in Fig. 2 clearly show the influence of this effect. The lower trace is from a sample with $N_d = 10^{14}cm^{-3}$ and $N_a = 2.10^{13}cm^{-3}$, measured *with* band gap radiation. The upper trace has been taken *without* bandgap radiation in a sample with higher impurity concentrations. Here the $(1,1,2)$ and $(1,1,0)$ transitions (the latter not shown) are broadened, thus obscuring the fine structure. Apart from the allowed transitions, forbidden transitions to the $(2,\pm2,0)$, $(1,1,1)$, and $(1,-2,0)$ states are also visible in the upper trace.

Non-Parabolicity of the Conduction Band

The conduction band in the Γ-point of GaAs is not perfectly parabolic: this leads to an increase of the value of m^* with energy. As a result the experimental transition energies are somewhat lower than expected for constant m^*. For the data in Fig. 2 these deviations are not more than a few cm^{-1} for the highest transition energies and can easily be calculated [4].

Electron-Phonon Interaction

As GaAs is a polar semiconductor, the electrons interact with the optical phonons. This leads to a change in the energy of the electron state (N,m,ν) given by [9] :

$$\Delta E_{Nm\nu} = -\sum_{N'm'\nu'}\sum_{q}\frac{|\langle N'm'\nu',\vec{q}|\mathcal{H}_{ep}|Nm\nu,0\rangle|^2}{E_{N'm'\nu'}+\hbar\omega_{LO}-E_{Nm\nu}-\Delta E_{Nm\nu}} \tag{4}$$

where $|Nm\nu\rangle$ is the wavefunction of the unperturbed state of energy $E_{Nm\nu}$ with no LO-phonons, and $\langle N'm'\nu',\vec{q}|$ is the wave function of an unperturbed electronic state of energy $E_{N'm'\nu'}$ with one LO-phonon with wave vector $\vec{q}$ and energy $\hbar\omega_q \simeq \hbar\omega_{LO} = 296$ cm^{-1} (phonon replicated state). $\mathcal{H}_{ep}$ is the electron-phonon interaction hamiltonian. Whereas for energies far below $\hbar\omega_{LO}$ this interaction merely leads to a slight increase of the effective mass, with similar effects as mentioned above, at higher energies drastic effects are observed. As soon as the energy difference between states $|Nm\nu\rangle$ and $|N'm'\nu'\rangle$ reaches the LO-phonon energy, $\Delta E_{Nm\nu}$ will resonantly increase, resulting in an anticrossing between states $|Nm\nu\rangle$ and $\langle N'm'\nu',\vec{q}|$. Cheng et al. [9] measured in GaAs the behaviour of the $1s^0 \rightarrow (1,1,0)$ transition. The upper part of Fig. 3 shows the results of a similar experiment, but now on the transitions to the $(3,1,0)$ and $(4,1,0)$ metastable states [10,11] Data on the transition to the $(2,2,0)$ state is also given [12]. The two-photon excitation technique employed by the latter authors enabled also the observation of transitions in the energy region between the LO- phonon and the TO-phonon at 270 cm^{-1} (Reststrahlen band) where GaAs is completely opaque. The observed anti-crossing behaviour of these transitions has been described properly on the basis of Eq. 4, using wavefunctions obtained from variational calculations.

Electron Dynamics

An essential difference between donors and hydrogen atoms is found in the dynamics of excited electrons. As the donor excitation energies are small, the probability of relaxation of excited electrons through spontaneous emission is very low. Therefore, electrons excited into bound states return to the groundstate, mainly through non-radiative processes. Experiments on optical saturation [13] and time resolved photoconductivity [14,15] show that electrons excited into hydrogen-like (resp. metastable) states decay on typically a hundred (resp. ten) picosecond time scale into the conduction band through acoustic phonon emission (resp. autoionisation). After 10-100 ns the electrons are then captured by ionized donors, and decay subsequently in about 100 ns to the groundstate under acoustic phonon emission.

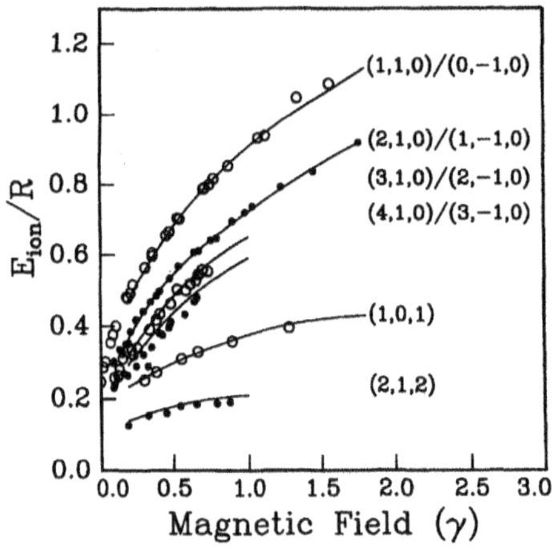

Figure 4. Experimental ionization energies of some dominant transitions compared to numerical results from variational calculations. Field and energy are given on reduced scales.

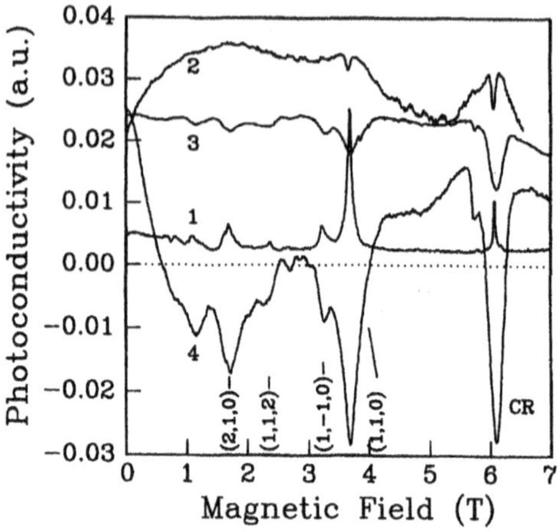

Figure 5. Photoconductive response as a function of magnetic field for λ=118.8 μm at different dc- and ac- (2.54 Thz) electric field intensities; see the text for explanation.

COMPARISON WITH THE HYDROGENIC MODEL

Many authors have used a variety of numerical techniques to obtain accurate data on energies of hydrogen-like states in finite magnetic field. A comprehensive set of calculations on field dependent energies of many states has been given by Rösner et al. [16]. Among others, Friedrich et al.[17] obtained results on energies and - lifetime determined - linewidths of some low lying metastable states, that agree very well with experiments on GaAs [18]. Recently Barmby et al. [5] used a variational method to derive accurate data on energies of a large number of those metastable- as well as hydrogenic states that are observed experimentally in GaAs. For the z dependent part of the electron wavefunctions $\mathcal{F}_{Nm\nu}(z)$ the following functions for ν = 0,1,2 are used:

$$\mathcal{F}_0(z) = (\frac{2\alpha_0}{\pi})^{1/4} e^{-\alpha_0 z^2} \tag{5}$$

$$\mathcal{F}_1(z) = (\frac{32\alpha_1^3}{\pi})^{1/4} z e^{-\alpha_1 z^2} \tag{6}$$

$$\mathcal{F}_2(z) = (\frac{2\alpha_2}{\pi})^{1/4} \frac{2\alpha_2}{(3\alpha_0^2 + 2\alpha_0\alpha_2 + 3\alpha_2^2)^{1/2}} \times [2(\alpha_0 + \alpha_2)z^2 - 1] e^{-\alpha_2 z^2} \tag{7}$$

with α_ν the N and m dependent variational parameters. In Fig. 4 the experimental ionization energies $E(N, m, \nu)$ of some states, corrected for the above mentioned deviations, are seen to compare very well with the results of these variational calculations. Very recently, Barmby et al. [19] used a finite difference approach to obtain analytical electron wavefunctions, allowing for mixing between different Landau levels (see Eq.(2)). Also those results on transition energies agree very well with experiment.

DONOR IN AN ELECTRIC FIELD

Apart from intense magnetic fields, also (random) electric fields up to 10^{10}Vm^{-1} are present in the atmospheres of white dwarfs and neutron stars. Is it possible to use the donor system also to study experimentally the influence of *electric* fields on the

optical spectrum? The essential quantity that determines that influence is the ratio of the electric potential across a Bohr orbit and the Rydberg constant. Using the quantities given in Table 1, it is easily shown that a field of $10^{10}Vm^{-1}$ will have about the same impact on a H-atom as a field of 2.10^4Vm^{-1} has on a donor. The latter field is equal to 0.4 mV $\approx 0.07\mathcal{R}$ across 20 nm, the diameter of the donor Bohr orbit.
Random electric fields of that amplitude are present in GaAs. In a sample with an impurity concentration of $3.10^{14}cm^{-3}$ the mean inter- impurity distance is about 140 nm. A donor at such a distance from an ionized impurity experiences a potential of $V = 0.12$ mV over 20 nm! The effects of such random fields on the shape of some donor transitions has been studied by Larsen [8].
In a preliminary experiment we inspected the influence of intense external electric fields on the shape of the donor spectrum. A planar log-periodic Thz antenna on top of an epitaxial GaAs sample was used together with a Si hyperhemispherical lens to collimate 0.5 W pulses of λ=118 μm radiation. Even assuming an antenna efficiency of only 1%, this would result in an optical intensity of about $100kWcm^{-2}$ at the antenna apex, which is equivalent to a Thz field of about 6 mV over 20 nm. The dc bias current, used to detect the induced photoconductivity, created dc fields as high as 0.5 mV over 20 nm.
In Fig. 5 the influence of the intensity of the optical and dc field on the induced photoconductivity is illustrated. Trace 1 shows a normal field scan at relatively low intensity of the optical and bias fields. Apart from transitions to donor states, also at B=6.08T the cyclotron transition (CR) between the N=0 and the N=1 Landau levels is observed. Trace 2 is taken at low bias voltage, but large optical intensity. The photoconductivity has strongly increased and, due to severe saturation, only the prominent transitions - (1,1,0) and C.R.- are visible, but very broad and with a dip in the centre of the transitions. With increasing bias voltage (trace 3) the spectral features reappear, be it with an "inverse" photoconductive response. With maximum bias voltage and optical intensity the resonant photoconductive response even becomes negative; the conductance is *smaller* with than without optical excitation (trace 4).
Probably deformation of electron states, as well as the specific response of the "hot" electrons in the semiconductor host are responsible for these complicated effects. It is clear that these high electric fields strongly influence the donor spectrum. Further experimental and theoretical work has to be done to understand the physical processes underlying these effects.

SUMMARY

The spectrum of an electron bound to a donor in GaAs has been shown to behave as a very good hydrogenic center. Its far-infrared magneto-optical spectrum provides much experimental information, not otherwise available, to verify theoretical models of the hydrogen atom in very intense magnetic fields. Deviations from the theoretical predictions are observed, and can be related to the interaction with the semiconductor host. Preliminary experiments show that also the influence of additional intense dc- and optical electric fields on the properties of the H-atom can be studied using this model system.

REFERENCES

1. J. Simola, J. Virtamo, Energy levels of hydrogen atoms in a strong magnetic field, *J. Phys. B: At. Mol. Phys.* 19:3309 (1978).

2. T.O. Klaassen, Magneto-optical spectroscopy of hydrogenic centres in semiconductors, *C.R. Acad.Sci. Paris* 323, Série IIb:187 (1996).
3. U. Fano, Effects of configuration interaction on intensities and phase shifts, *Phys. Rev.* 124:1866 (1961).
4. A. van Klarenbosch, T.O. Klaassen, W.Th. Wenckebach, C.T. Foxon, Identification and ionization energies of the shallow donor metastable states in GaAs:Si, *J. Appl. Phys.* 67:6323 (1990).
5. P.W. Barmby, J.L. Dunn, C.A.Bates, T.O. Klaassen, Landau- like states in the magneto-optical spectrum of a shallow donor impurity: theory versus experiment, *Phys. Rev. B.* 54:8566 (1996).
6. H.R. Fetterman, D.M. Larsen, G.E. Stillman, P.E. Tannenwald, J. Waldman, Field-dependent central-cell corrections in GaAs by laser spectroscopy, *Phys. Rev. Lett.* 26:975 (1971).
7. W. Zawadski, P. Pfeffer, H. Sigg, Five-level k.p. model for conduction electrons in GaAs. Description of cyclotron resonance experiments, *Solid State Commun.* 53:777 (1985).
8. D.M. Larsen, Inhomogeneous broadening of the Lyman-series absorption of simple hydrogenic donors, *Phys. Rev.B* 13:1681 (1976).
9. J.-P. Cheng, B.D. McCombe, J.M. Shi, F.M. Peeters, J.T. Devreese, Magnetopolaron effect on shallow donors in GaAs, *Phys. Rev. B* 48:7910 (1993).
10. A.J. van der Sluijs, K.K. Geerinck, T.O. Klaassen, W.Th. Wenckebach, C.T. Foxon, Resonant polaron effect on impurity metastable states in n-GaAs, *J.Appl. Phys.* 75:3698 (1994).
11. P.W. Barmby, J.L. Dunn, C.A. Bates, E.P. Pearl, C.T. Foxon, A.J. van der Sluijs, K.K. Geerinck, T.O. Klaassen, A. van Klarenbosch, C.J.M.G. Langerak, Donor metastable states and resonant electron-phonon interaction in n-GaAs, *J.Phys.: Condens. Matter* 6:7867 (1994).
12. P.C.M. Planken, H.P.M. Pellemans, P.C. van Son, J.N. Hovenier, T.O. Klaassen, W.Th. Wenckebach, P.W. Barmby, J.L. Dunn, C.A. Bates, C.T. Foxon, C.J.G.M. Langerak, Using far-infrared two-photon excitation to measure the resonant-polaron effect in the Reststrahlen band of GaAs:Si, *Opt. Commun.* 124:258 (1996).
13. A.J. Kalkman, H.P.M. Pellemans, T.O. Klaassen, W.Th. Wenckebach, Far-infrared pump-probe measurement of the lifetime of the $2p^{-1}$ shallow donor level in n-GaAs, *Int.J. of Infrared and Millimeter Waves* 17:569 (1996).
14. J. Burghoorn, T.O. Klaassen, W.Th. Wenckebach, The dynamics of shallow donor ionization in n-GaAs studied with subnanosecond FIR-induced photoconductivity, *Semicond. Sci. Technol.* 9:30 (1994).
15. P.C.M. Planken, P.C. van Son, J.N. Hovenier, T.O. Klaassen, W.Th. Wenckebach, B.N. Murdin, G.M.H. Knippels, Preliminary experiments on short pulse/high power free electron laser spectroscopy in n-GaAs, *Infrared Phys. Technol.* 36:333 (1995).
16. W. Rösner, G. Wunner, H. Herold, H. Ruder, Hydrogen atoms in arbitrary magnetic fields: I. Energy levels and wavefunctions, *J.Phys. B: At. Mol.Phys.* 17:29 (1984).
17. H. Friedrich, M. Chu, Autoionizing states of the hydrogen atom in strong magnetic fields, *Phys. Rev. A* 28:1423 (1983).
18. A. van Klarenbosch, K.K. Geerinck, T.O. Klaassen, W.Th. Wenckebach, C.T. Foxon, Ionization energies and lifetime broadening of autoionizing states of the hydrogen atom in strong magnetic fields: theory vs. experiment, *Europhys.Lett.* 13:237 (1990).
19. P.W. Barmby, J.L. Dunn, C.A.Bates, T.O. Klaassen, High field calculations for shallow donor impurities in bulk GaAs: a finite difference approach, unpublished, (1997).

QUANTUM DOTS IN STRONG MAGNETIC FIELDS

P. A. Maksym

Department of Physics and Astronomy
University of Leicester
Leicester LE1 7RH
UK

INTRODUCTION

Quantum dots are semiconductor nanostructures that can confine very small numbers of electrons. Depending on the method of fabrication, typical dot dimensions are in the range 10-100nm and electron numbers range from one to several hundred. Most dots are fabricated by applying an electrostatic confining force to electrons in the plane of a quasi-two dimensional system and this review is particularly concerned with these dots. The aim is to focus on the properties of quantum states of interacting electrons with particular emphasis on the strong magnetic field limit. In addition, a brief discussion of dot fabrication is given to make this review self-contained and motivate the theoretical model used to study the quantum states. More general reviews of dot physics and technology can be found in the literature [1, 2, 3, 4, 5, 6].

ELECTROSTATIC CONFINEMENT

Dot Technology

Generally speaking, dots can be made in several ways. The most common method is to constrain the motion of electrons in a quasi-two dimensional system by applying a confining force in the plane of the system. Early work used both mesa-etching [7] and electrostatics [8] to produce confinement but electrostatic confinement has been most widely used. Very recent work, however, has used a combination of mesa-etching and electrostatic confinement to produce very high quality dots [9]. The possibility of confinement by localised magnetic fields has also been discussed [10, 11]. Another approach to dot fabrication is to use a technique called self-organised growth which enables dots to be grown directly on a semiconductor substrate. This leads to dots as small as 10nm which have potential applications to semiconductor lasers [12].

A typical electrostatically confined system is shown in figure 1. The quasi-two dimensional electronic system in this case occurs at a GaAs-AlGaAs heterojunction,

Atoms and Molecules in Strong External Fields
Edited by Schmelcher and Schweizer, Plenum Press, New York, 1998

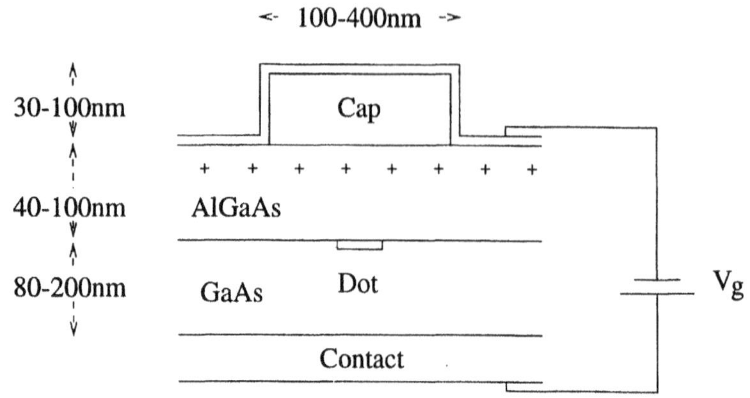

Figure 1. Schematic cross section of an electrostatic dot showing typical structure and dimensions.

although similar structures can also be made from GaAs-AlGaAs quantum wells. Electrons are introduced into the system from donors (indicated by + in figure 1) in the AlGaAs and would normally form a quasi-two dimensional sheet at the GaAs-AlGaAs interface. This occurs because the AlGaAs has a larger band gap than the GaAs so electrons released from donors in the AlGaAs transfer to lower energy states in the conduction band of the GaAs. These electrons are confined at the interface by a combination of the potential that results from the conduction band offset and the electrostatic potential that results from the electron transfer [13]. In a typical heterojunction the electron motion in the perpendicular direction is confined to within about 10nm of the interface so the electron system is effectively two dimensional. The quantum dot structure has additional components that are used to confine electrons in the lateral direction. These are a conducting bottom contact, either a heavily doped region or a delta-doped layer, and a modulated top electrode. In the dot region the gate is separated from the heterojunction by an insulating cap. When a negative voltage is applied to the gate the regions directly underneath it can be fully depleted of electrons, leaving a small number of electrons trapped underneath the cap. The scale of the lateral confinement depends on the length scale of the electrostatic potential at the heterojunction and is typically 50-100nm, which is smaller than the cap dimension. Typical single electron energies are around 2-4meV and the number of electrons in the dot ranges from zero upwards. The transport and optical properties of electrostatically confined structures have been studied extensively over the last few years and a number of authors have provided convincing evidence that single electrons can be confined in these structures [9, 14, 15, 16, 17].

Theoretical Model

Real quantum dot systems are rather complicated so it is convenient to use a simplified model that contains the key physics. This model has been used by most authors and is believed to give a good semi-quantitative description of electrostatically confined systems. The model is based on three simplifying assumptions. The first is that the electron motion is exactly two dimensional. Physically, this is quite reasonable because the length scale of the confinement in the direction perpendicular to the plane of the dot is about an order of magnitude smaller than the length scale of the confinement in the plane of the dot. The second assumption is the form of the confining potential. Typical gate voltages are around -1V. Thus the electrons are confined by a deep potential well whose depth is of order 1eV. The ground state wave function is large only near the minimum of this potential so a good first approximation is that the potential has the parabolic form associated with any minimum. Further, typical devices have a circular or square cap. In the first case the potential is circularly symmetric and in the second case circular symmetry is a good approximation because the dot region is much smaller than the square cap. The confining potential is therefore taken to be $m^*\omega_0^2 r^2/2$ in all cases, where m^* is the effective mass and $\hbar\omega_0$ is the confinement energy. The third simplifying assumption is about the form of the electron-electron interaction. It is normally assumed that a pure Coulomb interaction is appropriate although this is only a first approximation because both the dot structure and the surrounding electrodes contain carriers that can screen the interaction of the electrons inside the dot.

There is strong evidence that the theoretical model gives a good semi-quantitative description of dot behaviour. Experiments on far infra-red (FIR) absorption of dots in a perpendicular magnetic field have shown that the absorption spectrum is largely independent of the number of electrons [7, 8, 15]. According to the 2D parabolic model this is exactly what should be observed because in a 2D parabolic potential the centre of mass motion and the relative motion of the electrons decouple [18, 19]. Because it has a long wavelength the FIR radiation couples only to the centre of mass of the system so the absorption spectrum is sensitive only to the energy levels of the centre of mass. But these levels depend only on the charge to mass ratio which is independent of the number of electrons. Real absorption spectra are indeed largely independent of electron number but do have some residual features which indicate that the corrections to the 2D parabolic confinement model are not entirely negligible [20, 21]. There is also evidence that the assumption of a Coulomb interaction gives a good semi-quantitative description of dot physics. The effects of screening have been shown to be sufficiently large to have a measurable effect on electron energies [22] but modifications of the interaction potential do not change the qualitative form of the physics. This has been confirmed in recent work where the detailed forms of the confinement and interaction potentials in a real device were calculated [23, 24].

The model Hamiltonian for an N-electron electrostatic dot in a perpendicular magnetic field B has the form

$$H = \sum_{i=1}^{N}\left[\frac{1}{2m^*}\left(\mathbf{p}_i + e\mathbf{A}(\mathbf{r}_i)\right)^2 + \frac{1}{2}m^*\omega_0^2 r_i^2\right] + \frac{1}{2}\left(\frac{e^2}{4\pi\epsilon\epsilon_0}\right)\sum_{i=1}^{N}\sum_{\substack{j=1 \\ j\neq i}}^{N}\frac{1}{|\mathbf{r}_i - \mathbf{r}_j|} + g^*\mu_B B S_z, \quad (1)$$

where the first term is the one electron term, the second term is the Coulomb interaction term and the last term is the Zeeman energy. g^* is the effective g-factor and ϵ is the dielectric constant. It is convenient to use the circular gauge so the magnetic vector potential is, $\mathbf{A} = (B/2)(\hat{\mathbf{k}} \times \mathbf{r})$.

QUANTUM STATES

Single Electron States

The eigenstates of non-interacting electrons with 2D parabolic confinement are well known. They were first investigated by Fock [25] and Darwin [26] in the context of diamagnetism and in the quantum dot literature they are generally called the Fock-Darwin states. They have the form

$$\psi_{nl}(\mathbf{r}) = \frac{1}{\sqrt{2\pi\lambda^2}}\left[\frac{n!}{(n+|l|)!}\right]^{\frac{1}{2}}\left(\frac{r}{\sqrt{2}\lambda}\right)^{|l|} L_n^{|l|}(r^2/2\lambda^2)\exp(-r^2/4\lambda^2)\exp(-il\phi), \quad (2)$$

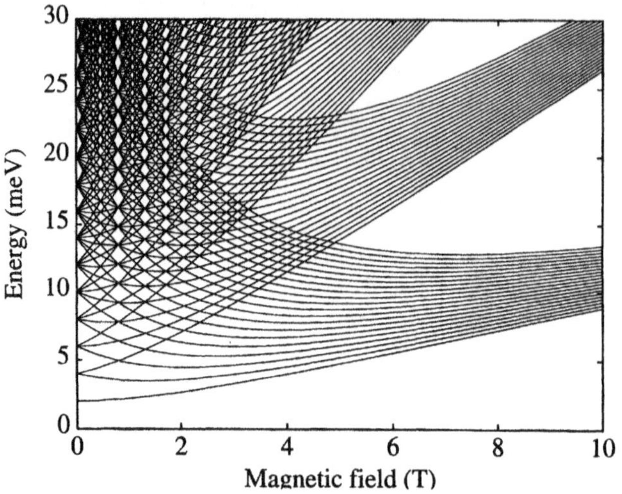

Figure 2. Single electron energy levels for electrons in a GaAs dot with $\hbar\omega_0 = 2$meV.

with energies given by

$$E_{nl} = (2n + 1 + |l|)\hbar\Omega - l\hbar\omega_c/2, \tag{3}$$

(excluding the Zeeman energy). Here l and n respectively are angular momentum and radial quantum numbers, the $L_n^{|l|}$ are associated Laguerre polynomials, $\lambda^2 = \hbar/(2m^*\Omega)$, $\Omega^2 = \omega_0^2 + \omega_c^2/4$ and ω_c is the cyclotron frequency, eB/m^*. The size of a Fock-Darwin state depends on the length parameter, λ, and on the quantum numbers n and l. A convenient measure of the size is the mean square radius

$$R^2 = \langle n, l|r^2|n, l\rangle = 2\lambda^2(2n + |l| + 1). \tag{4}$$

Clearly the radius is proportional to λ and because of the relation between λ and ω_c this means that the system size decreases with magnetic field. In addition, Eq. 4 shows that the size increases with angular momentum. The dependence of the system size on magnetic field and angular momentum is responsible for much of dot physics.

The Fock-Darwin energies are shown in figure 2. This gives the magnetic field dependence of the lowest few levels for a GaAs dot ($m^* = 0.067m_0$) with $\hbar\omega_0 = 2$meV. At zero field the levels are equally spaced and as the field increases a complicated series of crossings occurs. In the high field regime the levels coalesce into the free electron Landau levels, that is levels with constant Landau quantum number $n + (|l| - l)/2$. There is however significant broadening even at 10T. The broadening illustrates the effect of the angular momentum on the system size. For example, the states in the zeroth Landau level have $n = 0$ and $l \geq 0$. The lowest state has $l = 0$ and l increases for the higher energy states in the Landau level. Physically, this is because states with larger l have a larger size so electrons in these states experience a larger potential.

Interacting Electron States

Most calculations of quantum states of interacting electrons have been done by numerically diagonalizing the Hamiltonian [19, 27]. Typically, a Slater determinant basis

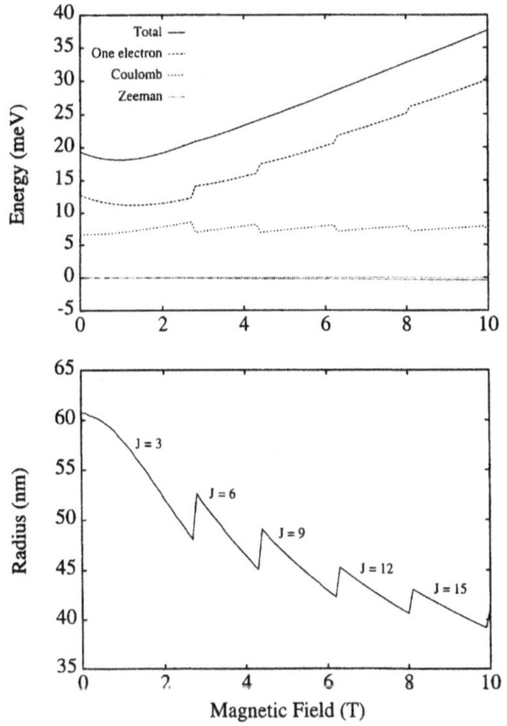

Figure 3. Ground state energy of 3 spin polarised electrons as a function of magnetic field (upper frame) and system size and orbital angular momentum (lower frame).

is constructed from Fock-Darwin states and the QR algorithm or the Lanczos method is used to find the states of the interacting system. Accurate calculations are possible for up to about 6 electrons throughout the magnetic field range. More electrons can be treated in the strong magnetic field limit, where a single Landau level gives an accurate description. Current density functional theory offers a promising alternative for treating even larger systems [20]. The magnetic field dependence of the ground state is of particular interest and is relevant to various measurable properties of dots.

Figure 3 (upper frame) shows the ground state energy as a function of magnetic field for 3 spin polarised interacting electrons in a GaAs dot ($m^* = 0.067m_0, g^* = -0.44, \epsilon = 12.4$) with $\hbar\omega_0 = 2$meV. The solid line gives the total ground state energy while the remaining lines give one electron, Coulomb and Zeeman components which were obtained by taking the ground state expectation value of the corresponding terms in Eq. 1. While the total energy is smooth, the one electron and Coulomb components change discontinuously and the discontinuities are accompanied by changes in the dot radius and total orbital angular momentum (lower frame). These changes are a consequence of the changing size of the wave function. The increasing field reduces the system size and hence increases the Coulomb energy. Because the length scale increases with angular momentum (Eq. 4) the system is able to decrease its Coulomb energy by increasing the total angular momentum quantum number, J. However this only happens at certain critical magnetic fields where the decrease in Coulomb energy can compensate for the increase of one electron energy that results from an increase in J. The cycle repeats with increasing B and this leads to oscillations in the Coulomb energy and abrupt increases of one electron energy and J value as in figure 3.

The abrupt changes in the orbital angular momentum are accompanied by abrupt changes in the system size. The system size in figure 3 was determined by calculating the radius of the circle that contains 95% of the charge, that is, R_{eff} is defined by $2\pi \int_0^{R_{eff}} n_e(r) r dr = 0.95N$, where $n_e(r)$ is the electron density [29]. Abrupt increases in R_{eff} clearly co-incide with a transition to a new ground state (figure 3, lower frame). The increase of angular momentum with magnetic field is essentially a classical effect and the transition fields can be determined quite accurately by considering the classical motion of interacting electrons in circular orbits [6, 30]. In this approach the field dependence of the total angular momentum is obtained by minimising the classical energy and then imposing the constraint that the total angular angular momentum is an integer multiple of $\hbar$ to approximate the transition fields.

The magnetic field dependence of the ground state energy in figure 3 is typical for electrostatic quantum dots containing small numbers of electrons. Extensive numerical studies have shown that angular momentum transitions always occur and in addition there are transitions in the total spin [31, 32]. These transitions generally cause oscillations in observable properties of dots such as the electronic heat capacity [19], magnetisation [31, 32] and many others – a comprehensive list can be found in a review by Maksym [6]. There is evidence that some of the transitions have been observed [16].

One of the most interesting aspects of the ground state is that only certain values of the total angular momentum are selected. For example, the angular momentum of the interacting ground state of 3 spin polarised electrons is always a multiple of 3 but for 4 spin polarised electrons the total angular momentum is $4k + 2$ where k is an integer, while for 4 electrons with total spin zero the total angular momentum is a multiple of 2. Clearly the preferred angular momentum depends on the electron number and total spin and the actual values that occur are known as the magic angular momentum numbers. The origin of the selection rules for the magic angular momenta is particularly interesting.

Origin of Magic Numbers

The earliest approach to explaining the origin of magic numbers was an argument based on exchange energy [31]. In the spin polarised case the Slater determinant that has the lowest exchange energy in the zeroth Landau level is one where all the electrons are adjacent in angular momentum space. This configuration can only occur at certain values of the total angular momentum and it turns out that they are the magic ones. For example, in the 3-electron case one way of making the l values adjacent is to choose $l = 1, 2, 3$ which gives $J = 6$, one of the magic numbers for the 3-electron system. The argument is similar to Hund's first rule in that the state of lowest total energy is the one where exchange effects are the largest. Numerical calculations show that configurations of adjacent electrons tend to occur in the ground state with high probability.

Subsequent approaches, based on symmetry, have given a much more detailed picture. It is reasonable to suppose that the ground state is localised near the minimum of the potential and in the quantum dot the form of the minimum is determined by the balance of the confining and repulsive forces. In classical mechanics, this leads to minima with well defined symmetry. For example, the minimum energy configuration of 3 classical point charges is an equilateral triangle, or a rotating equilateral triangle if the system has non-zero angular momentum. In quantum mechanics the electron wave function has to be anti-symmetric and this together with the requirement that the correlation is compatible with the symmetry of the classical minimum leads to the magic numbers. In early work on the 3-electron spin polarised system this was demonstrated by constructing a system of co-ordinates in which cyclic permutations

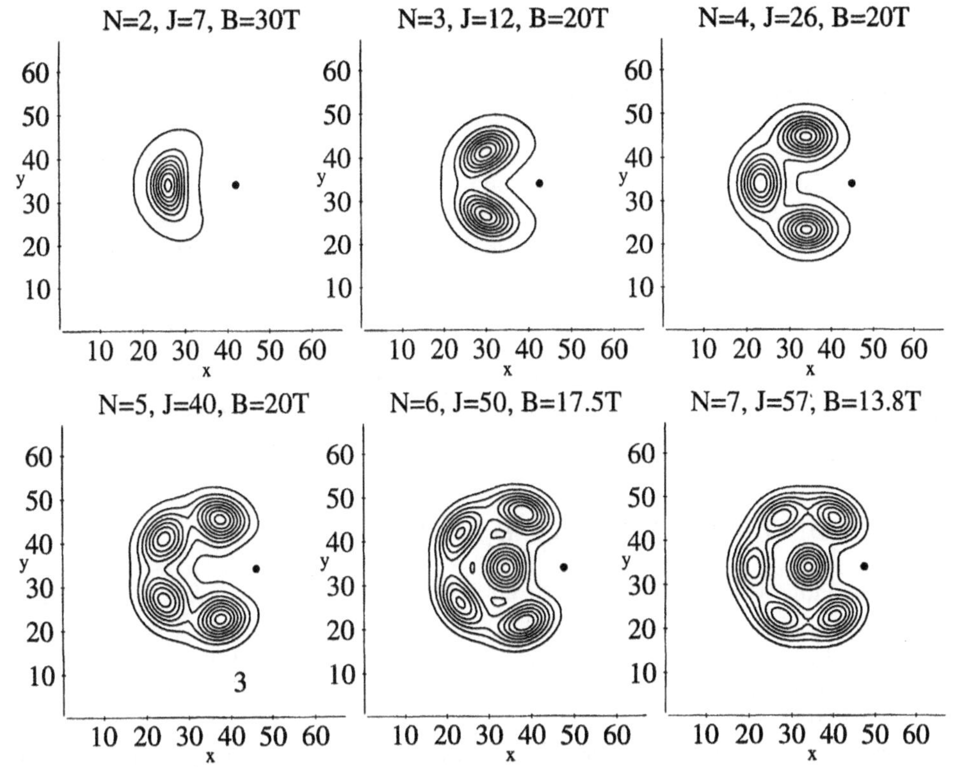

Figure 4. Pair correlation functions for magic number ground states at the indicated values of J and B in a GaAs dot with $\hbar\omega_0 = 4$meV. The length unit is 1.89nm.

and rotations are equivalent. Using these co-ordinates it is possible to show [33] that anti-symmetric ground states can only have maxima with 3-fold symmetry when the total angular momentum is multiple of 3. An alternative and simpler argument is to show that if an anti-symmetric wave function is non-zero when the electron positions have 3-fold symmetry it must have one of the magic angular momenta [34, 35, 36] and this argument can be generalised to more than 3 electrons and arbitrary spins [30, 34, 36, 37].

The link between magic numbers and symmetry has been verified by computing the pair correlation function [30, 33, 35, 36]. For spin polarised electrons this is defined as the ground state expectation value:

$$P(\mathbf{r}, \mathbf{r}_0) = \frac{(2\pi\lambda^2)^2}{N(N-1)} \langle \sum_{i \neq j} \delta(\mathbf{r}_i - \mathbf{r})\delta(\mathbf{r}_j - \mathbf{r}_0) \rangle. \tag{5}$$

The vector $\mathbf{r}_0$ is fixed while $\mathbf{r}$ is varied so the resulting function of $\mathbf{r}$ is proportional to the probability of finding an electron at $\mathbf{r}$ given that there is one at $\mathbf{r}_0$. As can be seen in figure 4, this function is highly sensitive to angular correlations. The figure shows pair correlation functions, in the strong field limit, for ground states of 2 through to 7 spin polarised interacting electrons. The magnitude of $\mathbf{r}_0$ has been chosen to coincide with the classical orbit radius and in each frame $\mathbf{r}_0$ is indicated by the black spot. In the case of 3 electrons the spot and the two peaks opposite it form the corners of an equilateral triangle. For other electron numbers other forms of symmetry occur, with ring-like symmetry for up to 5 electrons and a ring with a central peak for 6 and 7 electrons. In each case the symmetry coincides with the symmetry of the classical minimum energy configuration [38]. The idea that the magic numbers are connected with symmetry is also consistent with the argument based on exchange energy. The pair correlation function provides the link between the two approaches for it can be shown [30] that the pair correlation functions for Slater determinants composed of adjacent orbitals have ring or centred-ring symmetry as in figure 4.

In addition to the approaches based on exchange energy and symmetry the magic numbers have also been interpreted in terms of composite fermions [39]. A composite fermion consists of a magnetic flux tube bound to an electron and magic numbers are identified by considering how composite fermion Landau levels are occupied. This leads to an approximate wave function that is to used approximate the total energy. The magic numbers correspond to compact occupation of the composite fermion Landau levels. For $N \leq 5$ the composite fermion approach gives the same magic numbers as the approaches based on symmetry and agrees well with other approaches for larger numbers of electrons, although some discrepancies have been reported [36]. The good agreement of the composite fermion and symmetry approaches suggests that there may be a link between them however this is still an open question.

THE STRONG MAGNETIC FIELD LIMIT

In the limit of very strong magnetic fields the peaks in the pair correlation function become very sharp. This suggests that the wave function is strongly localised about the classical minimum and that a harmonic expansion of the Hamiltonian should give a good approximation to the ground state. Physically, the system can be pictured as an 'electron molecule' that rotates and vibrates inside the quantum dot and sometimes the term Wigner molecule is used. The equilibrium configuration of the molecule is determined by the form of the classical minimum and the lowest order approximation to the wave function is obtained by expanding the Hamiltonian about this minimum. Actually, however, there are $N!$ minima, each corresponding to a different permutation of the electrons and the main assumption of the theory based on a harmonic approximation is that quantum tunnelling between the minima can be neglected. This is reasonable when states are so strongly localised that the overlap between different minima is very small.

The classical minimum rotates and it is necessary to transform to a moving frame to develop a theory based on a harmonic expansion. The appropriate frame [30] is the Eckardt frame that is normally used for the study of molecular vibrations. This has the unique feature that the Coriolis coupling between rotational and vibrational motion is minimised. However in a strong magnetic field Coriolis coupling is not negligible and must be treated exactly to obtain accurate results. With this exception the theory is very similar to the theory of molecular vibrations.

Approximate electron states are obtained by antisymmetrizing states obtained from the harmonic expansion of the Eckardt frame Hamiltonian [30]. Each of these states is a product of the form

$$\Psi = \psi_{CM} \exp(-iJ_{RM}\chi) f_{J_{RM},n_1 \ldots n_{2N-3}}(Q_1 \ldots, Q_{2N-3}) \psi_{spin}(S_z), \tag{6}$$

where the ψ_{CM} is the centre of mass state, $\exp(-iJ_{RM}\chi)$ is a state of angular momentum relative to the centre of mass, $f_{J_{RM},n_1 \ldots n_{2N-3}}(Q_1 \ldots, Q_{2N-3})$ is a vibrational state and $\psi_{spin}(S_z)$ is a spin state. Here J_{RM} is the relative angular momentum quantum number, χ is an Euler angle, n_i is a vibrational quantum number associated with the normal mode Q_i and S_z is the z-component of the total spin. The result of antisymmetrizing this state is either zero or an antisymmetric state, and because ψ_{CM} is symmetric under all permutations the result of the antisymmetrization is determined by the remaining

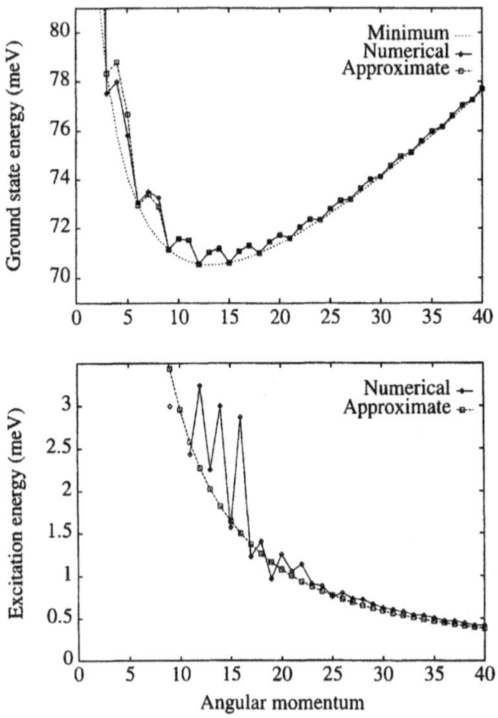

Figure 5. Approximate ground state energy (upper frame) and relative excitation energy (lower frame) for 3 spin polarised electrons with dot parameters as for figure 4 and $B = 20$T.

factors. Only certain combinations of the quantum numbers J_{RM}, n_i and S_z lead to antisymmetric states. The lowest energy states occur when $n_i = 0$ for all i and in this case non-zero states are obtained from Eq. 6 only when J_{RM} is one of the magic numbers. The allowed combinations of quantum numbers are found from group theory and this gives magic angular momenta which agree exactly with the results of the previous section. An example of the quantitative accuracy of the theory is given in figure 5. The upper frame shows a comparison of approximate and exact results for the ground state energies, together with a minimum energy defined as the sum of the classical energy and the quantum zero-point energy. The lower frame gives the lowest excitation energy relative to the centre of mass. In each case there is good agreement in the large angular momentum region which occurs in strong magnetic fields.

The approximate treatment in the strong field limit can be extended to systems with larger numbers of electrons [30, 40]. For up to 5 electrons the accuracy is similar to that shown in figure 5 but some new features occur in the 6-electron system. In this case there are competing 3-fold and 5-fold classical minima. The 5-fold minimum has the lowest energy and most ground states have 5-fold pair correlation functions. The first new feature is that although there are no 6-fold classical minima there are excited quantum states with 6-fold pair correlation functions. It turns out that quantum mixing of states localised on two degenerate 3-fold minima leads to the 6-fold correlation[40]. The second new feature is that the selection rules allow both 6-fold and 5-fold correlation at the J values 15, 45, 75... Because the J value is a measure of the size of the system it is related to the filling factor [13], ν, via the relation [41], $\nu = N(N-1)/2J$. The filling factors corresponding to 15, 45, 75... are 1, 1/3, 1/5..., that is odd denominator fractions where the fractional quantum Hall effect occurs. The pair correlation function at these J values does not have the ordered structure illustrated in figure 3; instead the quantum state is 'liquid-like' [30]. The 'liquid-like' nature of these 6-electron states is a consequence of mixing of states localised on the competing classical minima and the fact that they only occur at odd denominator fractions is a consequence the selection rule that associates particular symmetries with particular J values. Competing minima generally occur in large systems and it is an open question whether quantum mixing restricted by selection rules generally leads to 'liquid-like' states at odd denominator filling factors [44].

CONCLUSION

Interacting electrons in quantum dots in a magnetic field exhibit very rich behaviour. As the field increases a series of transitions occurs in which the orbital angular momentum and spin of the ground state change in a way characteristic of the number of electrons. In the very strong field limit 'molecular' states occur which are well described by a harmonic expansion of the Hamiltonian. In addition 'liquid-like' states occur in a 6-electron dot at odd denominator filling factors. Dot technology is advancing rapidly and there are hints that novel systems, such as double dots [42, 43] exhibit equally rich behaviour.

ACKNOWLEDGEMENTS

It is a pleasure to acknowledge contributions from T. Chakraborty, D. Pfannkuche, V. Gudmundsson, L. D. Hallam, J. Weis, N. Bruce, H. Aoki and H. Imamura. This work was supported by the UK Engineering and Physical Sciences Research Council.

REFERENCES

1. T. Chakraborty, *Comments in Condensed Matter Physics* 16:35 (1992).
2. M. Kastner, *Physics Today* 46:17 (1993).
3. D. Heitmann and J. P. Kotthaus, *Physics Today* 46:56 (1993).
4. N. F. Johnson, *J. Phys. Cond. Mat.* 7:965 (1995).
5. R. C. Ashoori, *Nature* 379:413 (1996).
6. P. A. Maksym, Springer Lecture Notes in Physics, in press.
7. T. Demel, D. Heitmann, P. Grambow and K. Ploog, *Phys. Rev. Lett.* 64:788 (1990).
8. C. Sikorski and U. Merkt, *Phys. Rev. Lett.* 62:2164 (1989).
9. S. Tarucha, D. G. Austing, T. Honda, R. J. Vanderhage and L. P. Kouvenhoven, *Phys. Rev. Lett.* 77:3613 (1996).
10. L. Solimany and B. Kramer, *Solid State Commun.* 7:471 (1995).
11. F. M. Peeters, A. Matulis and I. S. Ibrahim, *Physica B* 227:131 (1996).
12. M. Grundmann, O. Stier and D. Bimberg, *Phys. Rev. B* 52:11969 (1995).
13. T. Ando, A. B. Fowler and F. Stern, *Rev. Mod. Phys.* 54:437 (1982).
14. P. L. McEuen, E.B. Foxman, U. Meirav, M. A. Kastner, Yigdal Meir, Ned D. Wingreen and S. J. Wind, *Phys. Rev. Lett.* 66:1296 (1991).
15. B. Meurer, D. Heitmann and K. Ploog, *Phys. Rev. Lett.* 68:1371 (1992).
16. R.C. Ashoori, H.L. Störmer, J.S. Weiner, L.N. Pfeiffer, K.W. Baldwin and K.W. West, *Phys. Rev. Lett.* 71:613 (1993).
17. J. Weis, R.J. Haug, K. v. Klitzing and K. Ploog, *Phys. Rev. Lett.* 71:4022 (1993).
18. L. Brey, N. F. Johnson and B. I. Halperin, *Phys. Rev. B* 40:10647 (1989).
19. P. A. Maksym and T. Chakraborty, *Phys. Rev. Lett.* 65:108 (1990).
20. D. Pfannkuche and R. R. Gerhardts, *Phys. Rev. B* 44:13132 (1991).
21. V. Gudmundsson, A. Brataas, P. Grambow, B. Meurer, T. Knurth and D. Heitmann, *Phys. Rev. B* 51:17744 (1995).

22. L. D. Hallam, J. Weis and P. A. Maksym, *Phys. Rev. B* 53:1452 (1996).
23. L.D. Hallam, N. A. Bruce and P.A. Maksym, *Surface Science* 361/362:648 (1996).
24. N. A. Bruce and P. A. Maksym, submitted to Phys. Rev. B.
25. V. Fock, *Z. Phys.* 47:446 (1928).
26. C. G. Darwin, *Proc. Cambridge Philos. Soc.* 27:86 (1930).
27. G. W. Bryant, *Phys. Rev. Lett.* 59:1140 (1987).
28. M. Ferconi and G. Vignale *Phys. Rev. B* 50:14722 (1994); These proceedings.
29. P. A. Maksym, L. D. Hallam and J. Weis, *Physica B* 212:213 (1995).
30. P. A. Maksym, *Phys. Rev. B* 53:10871 (1996).
31. P.A. Maksym and T. Chakraborty, *Phys. Rev. B* 45:1947 (1992).
32. M. Wagner, U. Merkt and A. V. Chaplik, *Phys. Rev. B* 45:1951 (1992).
33. P.A. Maksym, *Physica B* 184:385 (1993).
34. W. Y. Ruan, Y. Y. Liu, C. G. Bao and Z. O. Zhang, *Phys. Rev. B* 51:7942 (1995).
35. P. A. Maksym, *Europhys. Lett.* 31:405 (1995).
36. T. Seki, Y. Kuramoto and T. Nishino, *J. Phys. Soc. Japan* 65:3945 (1996).
37. C. G. Bao, W. Y. Ruan and Y. Y. Liu, *Phys. Rev. B* 53:10820 (1996).
38. F. Bolton and U. Rössler, *Superlattices and Microstructures* 13:139 (1993).
39. J. K. Jain and T. Kawamura, *Europhys. Lett.* 29:321 (1995).
40. H. Imamura, P. A. Maksym and H. Aoki, submitted to Surface Science.
41. S. M. Girvin and T. Jacah, *Phys. Rev. B* 28:4506 (1984).
42. K. Bollweg, T. Kurth, D. Heitmann, E. Vasiliadou, P. Grambow and K. Eberl *Surface Science* 361/362:766 (1996).
43. H. Imamura, P. A. Maksym and H. Aoki, *Phys. Rev. B* 53:12613, (1996).
44. P. A. Maksym in "High Magnetic Fields in the Physics of Semiconductor Physics II" eds. G. Landwehr and W. Ossau, World Scientific, Singapore 1997.

DENSITY FUNCTIONAL THEORY OF QUANTUM DOTS IN A MAGNETIC FIELD

M. Ferconi[1] and G. Vignale[2]

[1] Department of Physics and Astronomy
Vanderbilt University
Nashville, Tennessee 37235

[2]Department of Physics and Astronomy
University of Missouri-Columbia
Columbia, Missouri 65211

INTRODUCTION

Two-dimensional quantum dot systems, at high magnetic fields, have been recently studied by various authors [1–3]. The basic features of the system have been described in the previous talk by Peter Maksym. Electrons in a semiconductor layer are confined to a circular region by a smooth, essentially parabolic potential extending over a region of typically $10 - 100$ Bohr radii of the host semiconductor (1 Bohr radius $\sim 100\AA$ for GaAs). The number of electrons in the dot is controlled by varying a "gate voltage". The energy scale of excitations is typically of the order of a few meV's, corresponding to wavelengths in the infrared range. Such a system has been dubbed an "artificial atom" in the literature - bringing us close to the subject of this workshop.

In contrast to what happens in real atoms, electromagnetic absorption spectroscopy is of limited value in quantum dots. The reason is that a uniform electromagnetic field couples only to the center of mass of the system which responds like a quantum harmonic oscillator of frequency ω_0 determined by the curvature of the parabolic potential [4]. Hence, infrared absorption is "blind" to excitations of the internal degrees of freedom, when a purely parabolic potential is assumed, as mentioned earlier in the paper.

All we know about the internal structure of quantum dots comes from two types of spectroscopy known as, "single electron capacitance spectroscopy" (SECS) [5] and "single electron tunneling spectroscopy" (SETS) [6]. In the first technique one measures the differential capacitance between the quantum dot and a reference electrode as a function of gate voltage V_g. In the second technique one measures the differential conductance between the quantum dot and an electrode, as a function of V_g. In both cases, the basic spectroscopic principle is that the differential capacitance /conductance is almost always zero, except when the gate voltage has such a value that the state with N electrons on the dot has the same Gibbs free energy ($G(N) = E(N) - \mu N$, where

Atoms and Molecules in Strong External Fields
Edited by Schmelcher and Schweizer, Plenum Press, New York, 1998

$E(N)$ is the ground-state energy and μ is the chemical potential) as the state with $N+1$ electrons on the dot. Only when this happens, charge can tunnel back and forth between the electrode and the dot, and one observes a sharp peak in the differential capacitance or conductance. Thus, the resonance condition is

$$\alpha V_g(N) = E(N+1) - E(N) = \mu(N), \tag{1}$$

where α is a numerical factor relating the gate voltage to the chemical potential. This gives us a sequence of peaks as the occupation number N of the dot is varied from 1 to infinity. It is easy to verify that the occupation of the dot increases sequentially in steps of 1, with increasing V_g, if and only if the ground-state energy $E(N)$ is a convex function of N. In the simplest model (known as "constant interaction model") $E(N) = e^2N^2/2C$ where C is the classical capacitance of the dot, taken to be independent of N. This model predicts a series of peaks at equally spaced values of V_g. A slightly more careful calculation, taking into account the size dependence of the capacitance leads to the conclusion that $E(N) \sim N^{5/3}$ and $\mu(N) \sim N^{2/3}$ for large N [7,8]: thus the spacing between the peaks tends to *decrease* with increasing N, but very slowly.

On top of this smooth, classical behavior, quantum mechanics introduces *irregularities* in the distribution of the peaks, caused by changes in the *symmetry* of the ground-state as a function of N. It is through a careful study of these irregularities, that one learns about the internal structure of quantum dots. For example, recent experiments by Tarucha *et al.* [9] have revealed the existence of pronounced oscillations in the spacing between peaks as a function of N for few-electron dots. These oscillations have been interpreted as a manifestation of a shell structure for an atom held together by a parabolic, rather than coulombic, potential.

The study of quantum mechanical effects is greatly facilitated by the application of a magnetic field. Following the evolution of a given capacitance or conductance peak as a function of magnetic field B one eliminates the classical background (which is independent of B) and focusses on the symmetry of the quantum mechanical ground-state. We can identify at least 4 classes of irregularities which appear with increasing magnetic field:

(1) At low magnetic field one can observe [9] a crossover from a zero-field regime, in which the atomic-like shells are filled according the usual Hund's rules, to a Zeeman dominated regime in which parallel spin occupation of adjacent shells is favored.

(2) With increasing magnetic field one sees a complicated pattern of oscillations (in the voltage of a given peak vs. B) which can be understood in terms of crossings of single-electron Fock-Darwin [10] levels with different symmetry. A particularly strong feature is observed at a value of B such that all the electrons fall in the lowest Fock-Darwin band: this is the band of levels which, in the limit $B \to \infty$ evolves into the lowest Landau level (LLL) of the uniform electron gas.

(3) At still higher B, one sees a series of oscillations related to spin flip transitions [11]: the system is gradually evolving from a spin-unpolarized (or weakly spin-polarized state) to a fully spin polarized state within the lowest Fock-Darwin band. This final state is known as the "maximum density droplet" (MDD) [12] and is characterized by a compact occupation of states with (canonical) angular momenta $0, 1 ... N-1$ within the lowest Fock-Darwin band.

(4) Finally, at the largest attainable magnetic fields, one observes a breaking up of the maximum density droplet, which becomes electrostatically unstable, due to the reduced radii of the Fock-Darwin orbitals. The "edge reconstructed " [12–14] state that arises from the optimization of the electrostatic energy, has a larger angular momentum, which continues to increase with increasing magnetic field. This phenomenon is

the large-N analogue of the transitions between states with increasing magic angular momentum values, described by Maksym in the previous talk.

THEORETICAL APPROACHES

The challenge to theory has been and remains to produce a coherent understanding of such a wealth of experimental data. Various theoretical methods have been employed to this end. Exact diagonalizations of the quantum dot Hamiltonian

$$H = \sum_{i=1}^{N} \left[\frac{1}{2m^*} \left(\mathbf{p}_i + \frac{e}{c} \mathbf{A}_i \right)^2 + \frac{1}{2} m^* \omega_0^2 r^2 \right] + \frac{e^2}{2k} \sum_{i=1}^{N} \sum_{j=1, j \neq i}^{N} \frac{1}{|\mathbf{r}_i - \mathbf{r}_j|} + g^* \mu_B B \sum_{i=1}^{N} \sigma_i, \tag{2}$$

(Here ω_0 is the frequency of the external parabolic potential, $\mathbf{A}_i$ is the external vector potential, k the dielectric constant, m^* the electron effective mass, μ_B the Bohr magneton, g^* the effective g-factor for the Zeeman splitting, and σ_i the spin component along the axis perpendicular to the plane of the electrons) have been pioneered by Maksym and Chakraborty [15], Hawrylak *et al.* [16], and Yang *et al.* [7]. This highly accurate approach is unfortunately only feasible for dots containing a relatively small number of electrons ($N \leq 10$), due to the exponentially growing dimension of the Hilbert space. Classical Monte Carlo calculations have been carried out by Bolton *et al.*[17] Mean field approximations have also been applied with considerable qualitative success. We refer, in particular to the Hartree-Fock calculations by de Chamon *et al.* [13] and Palacios *et al.* [18], and the composite Fermion mean field theories by Jain [19], Brey [20], and Chklovskii [21]. These theories neglect the correlations between their constituent particles (electrons and composite Fermions respectively).

DENSITY FUNCTIONAL THEORY

Finally, the density functional theory (DFT), originally developed by Hohenberg, Kohn and Sham [22] has proved to be a valuable tool for studying quantum dots with large N [23–26]. This approach is based on the assumption that it is possible to obtain the ground-state density of an interacting inhomogeneous electron system from the ground-state wave function of a noninteracting system (the "Kohn-Sham" system) subjected to an appropriate effective potential. Although there are some special cases (see [25], for example) in which this assumption is known to fail, the DFT is nevertheless widely used in condensed matter physics, and a finite temperature ensemble generalization is available to cope with the potentially troublesome case of degenerate, or quasi-degenerate ground-states. A serious limitation of DFT is the lack of an exact knowledge of the exchange-correlation (xc) part of the effective Kohn-Sham potential - a complicated, nonlocal functional of the density. Usually, one resorts to the so-called local density approximation in which the xc potential $V_{xc}(\vec{r})$ is written as $d\epsilon_{xc}(n)/dn|_{n=n(\vec{r})}$, where $\epsilon_{xc}(n)$ is the exchange correlation energy density of a uniform electron gas of density n. This is, in general, an uncontrolled approximation, and a major limitation on the numerical accuracy of DFT-based results.

RESULTS

In the rest of this talk we present and discuss the results obtained for the ground-state properties of quantum dots using density and current-density functional theory (CDFT), the latter being an extension of DFT designed to explicitely take into account the coupling of orbital currents to an external magnetic field. The main results have been published in ref. [24], and we refer the reader to this paper for details. Basically, it is found that the DFT and CDFT approaches produce (with minor differences between each other) ground-state energies that are typically within few percents of the exact ones, obtained by exact diagonalization for $N = 2$ and $N = 3$ quantum dots. More importantly, the symmetries of the ground state, that is, the quantum numbers of orbital angular momentum and spin, as functions of magnetic field, are qualitatively reproduced by DFT. However, the values of the magnetic field at which transitions between ground-states of different symmetry occur (for example single-triplet transitions in the $N = 2$ case) are generally *overestimated* by a fraction of a Tesla in DFT. These results are encouraging, in view of the fact that the $N = 2$ and $N = 3$ cases present a most severe test of the LDA, being very far from the uniform electron gas limit on which the LDA is based. We have also performed calculations of addition energies $E(N+1) - E(N)$ as a function of B for larger values of N within DFT: the results exhibit numerous cusps associated with crossings between ground-states of different symmetry. A detailed matching of the calculated curve to the experimental data has not been possible yet.

THE MAXIMUM DENSITY DROPLET

A question of considerable theoretical interest is that of the stability of the so-called maximum density droplet (MDD) in quantum dots. In the limit of high magnetic field, this state can be written as a Slater determinant of lowest Landau level orbitals with angular momenta $0, 1 \ldots N-1$, where N is the number of electrons. [12] In the limit of $N \to \infty$ this coincides with the incompressible state of the quantum Hall effect at filling factor $\nu = 1$. Because, within the LLL, the MDD is the *only* N-electron state of angular momentum $N(N-1)/2$, (and there is none with lower angular momentum) it follows that it must be an exact eigenstate of the Hamiltonian (2), if the small Coulomb coupling between different Landau levels is neglected.

The question is whether this exact eigenstate (or rather its continuation to finite magnetic field) can actually be the *ground-state* of the quantum dot, in some range of magnetic fields. The basic physics is simple: if the magnetic field is too large, the MDD cannot be the ground state, because the compact arrangement of the electrons costs too much electrostatic energy: the electrostatic stress is released through a rearrangement of the electrons leading to a state of higher angular momentum. If, on the other hand, the magnetic field is too weak, the confinement energy will cause the external electrons in the MDD to be transferred to the center of the quantum dot, even though, in so doing, a higher Landau level becomes populated at the center of the dot. The conclusion of these arguments is that there will exist, at most, a "window" of magnetic fields in which the MDD is stable. The "window" shrinks with increasing electron number N and it closes up completely at a critical value N_c of N of the order of 100.

We have investigated this question within DFT and refer the reader to the original paper [27] for details. The main conclusion is that the region of stability of the MDD in the $N - B$ plane, including the slopes dN/dB of the left and right boundaries are in fairly good agreement with the experiments of Klein *et al.* [28, 29], and appear to be better than the corresponding Hartree-Fock results, which do not include correlations. The critical value of N above which the MDD cannot be stable turns out to be $N_c \sim 37$. The following points are worth mentioning:

(i) The width, in magnetic field, of the region in which the MDD is stable, (typically 1 Tesla) depends crucially on the exchange-correlation energy: it is, in fact, the exchange correlation energy, that determines the difference between the slopes (dN/dB) of the left and right boundaries of the phase diagram.

(ii) Mixing of different Landau levels is quantitatively important: we would find $N_c \sim 160$ if only the lowest Landau level were included.

(iii) For $N > N_c$ the quantum dot undergoes "edge reconstruction", namely the changes on the structure of the dot occur only at the edge of the system. For large N the edge structure is controlled by classical electrostatics.

(iv) DFT is in better agreement than Hartree-Fock with the results of exact diagonalizations. In particular, when the "edge reconstruction" takes place, the DFT density profile is more compact than the Hartree-Fock one, (i.e., the reconstruction affects a smaller region around the original edge), and closer to the "exact" result.

CONCLUSIONS

To summarize the main content of this talk:

(1) We have reviewed the papers demonstrating a broad applicability of DFT to the calculation of ground-state properties and level crossings in quantum dot "atoms" in magnetic field.

(2) A specific problem – the calculation of the region of stability of the maximum density droplet – has been studied. The results are in qualitative agreement with experiment.

ACKNOWLEDGEMENTS

The authors acknowledge financial support from ONR under grant No. 00014-96-1-1042 and from NSF under grant No. DMR-9403908.

REFERENCES

1. R. Ashoori, Nature, **379**, 413 (1996), and references therein.
2. T. Chakraborty, Comm. on Cond. Matt. Phys. **16**, 35 (1992), and references therein.
3. M. A. Kastner, Rev. Mod. Phys., **64**, 849 (1992); Comm. on Cond. Matt. Phys., **17**, 349 (1996).
4. L. Brey *et al.*, Phys. Rev. B 40, 10647 (1989); ibid. **42,** 1240 (1990); S. K. Yip, Phys. Rev. B **43,** 1707 (1991).
5. R. C. Ashoori *et al.*, Phys. Rev. Lett. **68,** 3088 (1992); **71,** 613 (1993).
6. P. L. McEuen *et al.*, Phys. Rev. Lett. **66,** 1926 (1991).
7. S.-R. E. Yang, A. H. MacDonald, and M. D. Johnson, Phys. Rev. Lett. **71,** 3194 (1993).
8. M. Ferconi, Ph.D Thesis, University of Missouri-Columbia, 1995.
9. S. Tarucha *et al.*, Phys. Rev. Lett. **77,** 3613 (1996).
10. V. Fock, Z. Physik **47,** 446 (1928); C. G. Darwin, Proc. Cambridge Philos. Soc. **27,** 86 (1930).
11. P. L. McEuen *et al.*, Phys. Rev. B **45,** 11419 (1992).
12. A. H. MacDonald, S. R. Eric Yang, and M. D. Johnson, Aust. J. Phys. **46,** 345 (1993).
13. C. de Chamon and X.-G. Wen, Phys. Rev. B **49,** 8227 (1994).
14. M. Ferconi, M. R. Geller, and G. Vignale, Phys. Rev. B **52,** 16357 (1995).
15. P. A. Maksym, and T. A. Chakraborty, Phys. Rev. Lett. **65,** 108 (1990).
16. P. Hawrylak and D. Pfannkuche, Phys. Rev. Lett. **70,** 485 (1993).
17. F. Bolton and U. Rössler, Superlatt. and Microstructures, **13,** 139 (1993).

18. J. J. Palacios *et al.*, Phys. Rev. B **50,** 5760 (1994).
19. J. K. Jain, Phys. Rev. Lett. **63,** 199 (1989).
20. L. Brey, Phys. Rev. B **50,** 11861 (1994).
21. D. B. Chklovskii, Phys. Rev. B **51,** 9895 (1995).
22. P. Hohenberg and W. Kohn, Phys. Rev. **136,** B864 (1965); W. Kohn and L. Sham, *ibid.* **140,** A1133 (1965).
23. M. Macucci *et al.*, Phys. Rev. B **48,** 17354 (1993).
24. M. Ferconi and G. Vignale, Phys. Rev. B **50**, 14722 (1994).
25. O. Heinonen, M. I. Lubin, and M. D. Johnson, Phys. Rev. Lett. **75**, 4110 (1995).
26. T. H. Stoof and Gerri E. W. Bauer, Phys. Rev. B **52**, 12143 (1995).
27. M. Ferconi and G. Vignale, Phys. Rev. B (to be published).
28. O. Klein, C. de Chamon, D. Tang, D. M. Abusch-Magder, U. Meirav, X.-G. Wen, and M. Kastner, Phys. Rev. Lett. **74**, 785 (1995).
29. O. Klein, D. Goldhaber-Gordon, C. de Chamon, and M. A. Kastner, Phys. Rev. B **53**, R4221 (1996).

AN ANALYTICAL APPROACH TO THE PROBLEM OF AN IMPURITY ELECTRON IN A QUANTUM WELL IN THE PRESENCE OF ELECTRIC AND STRONG MAGNETIC FIELDS

Boris S. Monozon,[1] Colin A. Bates,[2] and Janette L. Dunn[2]

[1]Physics Department, Marine Technical University, 3 Lotsmanskaya Str., St.Petersburg, 190008, Russia
[2]Physics Department, University of Nottingham, University Park, Nottingham NG7 2RD, UK

INTRODUCTION

The problem of a Coulomb particle in the presence of external electric and magnetic fields has a common interdisciplinary character in physics. In particular, the fundamental phenomena of semiconductors and the properties of semiconductor devices are much affected by the presence of impurities formed by an electron (or hole) captured by the Coulomb field of the charged impurity ion. During the last decade, much attention has concentrated on the problem of an impurity that is confined within a single quantum well (QW) structure formed, for example, by the narrow-energy- gap semiconductor GaAs sandwiched between wide gap GaAlAs material.

In parallel with numerical calculations, analytical methods of studying impurity problems are of much interest because they enable the basic physics of the problem to be kept clearly in view. In this paper, an analytical approach to the problem of an impurity electron (hole) confined within a single quantum well (QW) subject to electric and strong magnetic fields is developed. The dependencies of the electron energies upon the magnitudes of the electric and magnetic fields, the width of the QW and the position of the impurity within the well are derived explicitly. The effect of inversion and the resonance tunnelling caused by an electric field directed parallel to the magnetic field is investigated. The results are then extended to the case of a confined Mott magneto-exciton.

GENERAL THEORY

The z-axis is chosen to lie perpendicular to heteroplanes of the QW and along the direction of both the uniform magnetic field $\boldsymbol{B}$ and electric field $\boldsymbol{E}$ (parallel to $\boldsymbol{B}$). The effect of the magnetic field is taken to be much greater than that of the Coulomb field of the impurity, and of the electric field. The QW is treated as an infinite square well of width d, and the centre of the well is taken to be the point $z = 0$. The other parameters relevant to the calculation are the impurity Bohr radius $a_0 (= 4\pi\epsilon\epsilon_0\hbar^2/\mu e^2)$, where ϵ is the dielectric constant and μ and e are the effective mass and charge of the carrier respectively), the magnetic length a_B $(= \sqrt{\hbar/eB})$ and the distance b of the impurity centre from the mid-point of the QW ($|b| < d/2$). In the strong magnetic field limit (for which $a_B/a_0 \ll 1$) the wave function Ψ of the electron at a position $\mathbf{r}(\rho, z)$ can be written in the form $\Psi(\mathbf{r}) = \chi_\perp(\rho)f(z)$ where the function $\chi_\perp(\rho)$ describes the transverse motion of the electron in the $x - y$ plane. This has an associated energy of $\mathcal{E}_\perp$. When the electric field is also directed along the negative z direction, the longitudinal function $f(z)$ satisfies the equation

$$-\frac{\hbar^2}{2\mu}\frac{d^2 f(z)}{dz^2} + (V(z) - eEz)f(z) = Wf(z) \tag{1}$$

with the boundary conditions $f(\pm d/2) = 0$ and with

$$V(z) = -\frac{e^2}{4\pi\epsilon_0\epsilon}\int \frac{|\chi_\perp(\rho)|^2}{\sqrt{\rho^2 + (z-b)^2}} d\rho. \tag{2}$$

Also, $W = \mathcal{E} - \mathcal{E}_\perp$, where $\mathcal{E}$ is the total energy of the impurity electron.

In order to simplify the calculations, we consider only the ground transverse state (although the results obtained below will be valid qualitatively for any transverse state). The relevant transverse function $\chi_\perp$ and energy $\mathcal{E}_\perp$ then have the forms

$$\chi_\perp(\rho) = \frac{1}{\sqrt{2\pi}a_B}\exp\left[-\frac{\rho^2}{4a_B^2}\right] \text{ and } \mathcal{E}_\perp = \frac{\hbar eB}{2\mu} \pm \mu_B B \tag{3}$$

where μ_B is the effective Bohr magneton for the electron. For the wide QW (for which $d \gg a_0$), the energy states for which $W < 0$ considered below have a quasi-Coulomb character.

AN IMPURITY ELECTRON IN A QW IN THE PRESENCE OF A STRONG MAGNETIC FIELD B (with $E = 0$)[1,2]

The schematic form of the one-dimensional quasi-Coulomb potential given by equation (2) is depicted in Fig 1. In the absence of the electric field and when the impurity centre is situated at the centre of the QW (or at any position in a bulk semiconductor), the electron states have a definite parity. Although the states do not strictly have a well-defined parity for arbitrary impurity positions, a classification into two groups of quasi-even and quasi-odd states can still be made. The energies of the electron states can be written in the form

$$W_\lambda = -R/2\lambda^2 \tag{4}$$

where $R = e^2/4\pi\epsilon_0\epsilon a_0$ is the impurity Rydberg constant and where $\lambda_{ng} = n + \delta_{ng}(n = 0, 1, 2, \ldots)$ and $\lambda_{nu} = n + \delta_{nu}(n = 1, 2, \ldots)$ are the quantum numbers corresponding to quasi-even (g) and quasi-odd (u) levels respectively. To simplify the explicit expressions for the energy levels, we assume that the effect of the strong magnetic field on the one-dimensional Coulomb states is larger than the confinement effect of the QW itself. The Hasegawa-Howard approach[3] is used; this is based on matching solutions to Eq.(1) in the regions $a_B \ll |z - b| \ll a_0$. In the logarithmic approximation (for which $\gamma = 2^{1/2}a_B/a_0 \ll 1$ and $|\ln\gamma| \gg 1$), the ground level ($n = 0$) has

$$\delta_0 = (-2\ln\gamma - C)^{-1}(1 + G_{\lambda_1} + G_{\lambda_2}) \tag{5}$$

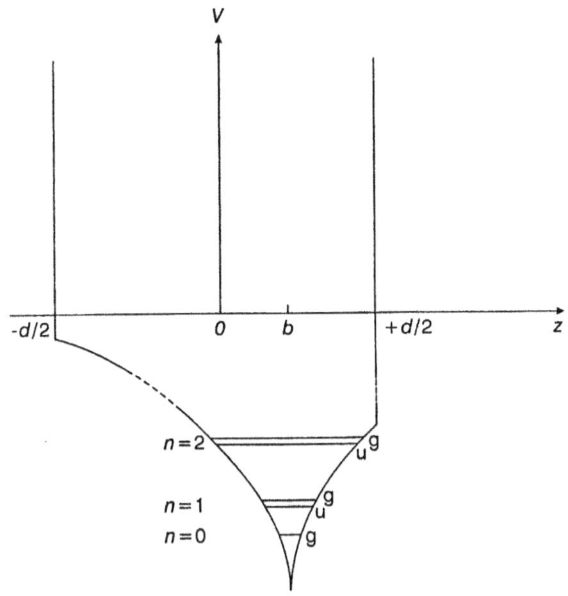

Figure 1. The schematic form of the quasi-Coulomb potential $V(z)$ from Eqn. (2). The solid lines refer to the ground ($n = 0$) and two first excited ($n = 1, 2$) quasi-even (g) and quasi-odd (u) states.

where C is the Euler constant (= 0.577) and where

$$G_{\lambda_{1,2}} = e^{-u_{1,2}+2\lambda \ln u_{1,2}} \tag{6}$$

with $u_{1,2} = (d/a_o\lambda)(1 \pm 2b/d)$. For the excited states ($n = 1, 2, \ldots$) we have

$$\delta_{n_g} = -[\ln(\gamma/n) + 3C/2 + 1/2n + \psi(n)]^{-1} + (n/2(n!)^2)(G_{n_1} + G_{n_2}) \tag{7}$$

and

$$\delta_{n_u} = (2|Q| - \psi(n) - 2C + 1 - 1/2n)^{-1} + (n/2(n!)^2)(G_{n_1} + G_{n_2}) \tag{8}$$

where

$$Q(\lambda) = \frac{\sqrt{\pi}}{2\gamma}(\ln\frac{\gamma}{\lambda} + 1 - 1/2C) \qquad \text{for} \qquad Q(\lambda) < 0 \tag{9}$$

and where $\psi(n)$ is the psi function (the logarithmic derivative of the gamma function). Thus the impurity levels consist of a non-degenerate ground level ($n = 0$) and excited levels ($n = 1, 2, \ldots$) having a doublet structure consisting of quasi-even and quasi-odd components. The energy level pattern is depicted in Fig 1.

It follows from Eqns. (4)-(9) that as the magnetic field increases in magnitude, the energy W_λ decreases. Also, the energy W_λ increases both as the width d of the QW decreases and as the impurity centre shifts from the mid-point (that is b increases). These dependencies coincide with the results obtained numerically by a variational approach[4, 5] and the method of direct integration[6].

MOTT EXCITON IN A QW IN THE PRESENCE OF A STRONG MAGNETIC FIELD B (E= 0)[7]

We develop an analytical approach to the problem of a Wannier-Mott exciton in a QW in the presence of a strong magnetic field that is directed perpendicular to the

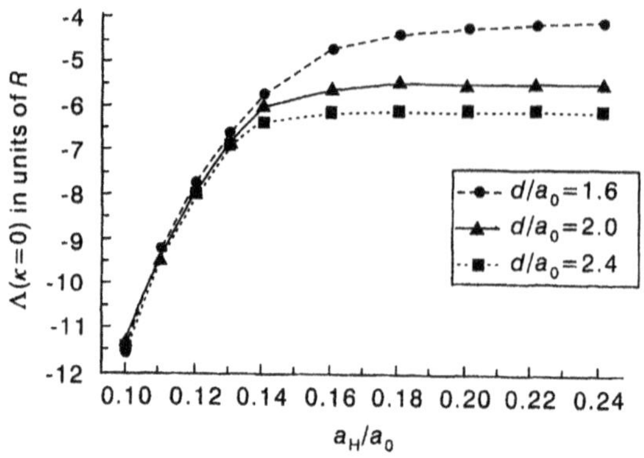

Figure 2. The energy of the exciton ground state Λ plotted as a function of scaled magnetic field for different well widths.

layers of the QW. The important approximation to be made is that the hole mass m_h is much greater than the electron mass m_e. Note that this approximation is clearly valid for the majority of III-V group semiconductors.

The potential energy of the exciton V is defined by the right hand part of the expression (2) by replacing the co-ordinate z by the co-ordinate of the electron z_e and the parameter b by the co-ordinate of the hole z_h. In general, an analytical expression for the exciton wave function $\Psi(z_e, z_h)$ can not be found as the exciton potential V depends upon the relative co-ordinate $z_e - z_h$ whereas the boundary conditions $\Psi = 0$ at $z_{e,h} = \pm d/2$ are stated in terms of the separate hole (z_h) and electron (z_e) co-ordinates. This difficulty can be resolved using an adiabatic approach based on the extreme difference in electron and hole masses. In this approximation, the exciton wave function can be written in the form

$$\Psi(z_e, z_h) = f_\lambda(z_e - z_h)\phi_\lambda(z_h) \tag{10}$$

where f is the wave function of the electron about the fixed hole position z_h and λ is a quantum number. The wave function ϕ satisfies the equation

$$\left[-\frac{\hbar^2}{2m_e}\frac{d^2}{dz_h^2} + W_\lambda(z_h)\right]\phi_\lambda(z_h) = \Lambda\phi_\lambda(z_h) \tag{11}$$

with the boundary conditions $\phi_\lambda(\pm d/2) = 0$.

In Eq. (11), Λ is the energy of the longitudinal motion of the exciton. The potential energy of the hole $W_\lambda(z_h)$ can be obtained from the expressions (4)-(9) by replacing b by z_h. On using numerical calculations at the last stage of solving equation (11), we find that the exciton energy Λ decreases both by increasing the magnetic field and widening the QW. The energy Λ of the ground exciton states as function of the scaled magnetic field is depicted in Fig. 2 for different widths of QW.

AN IMPURITY ELECTRON IN A QW IN THE PRESENCE OF PARALLEL ELECTRIC *E* AND STRONG MAGNETIC *B* FIELDS

The inversion effect of the electric field[8, 9]

In the case of the electric field directed parallel to the magnetic field, the combined potential acting on the impurity electron has the form depicted in Fig. 3. If the impurity

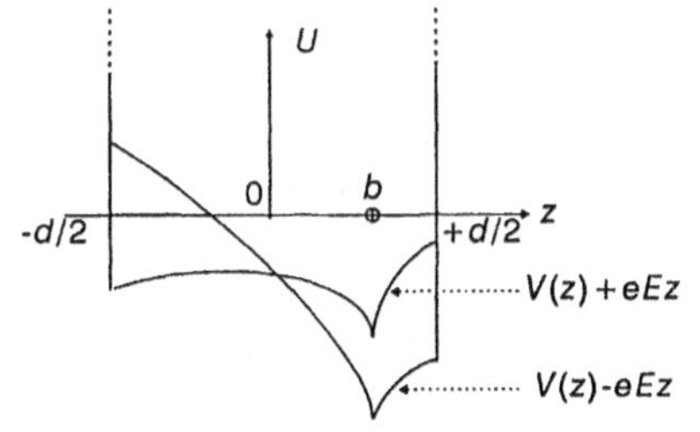

Figure 3. The combined potential $U(z) = V(z) \pm Ez$ caused by the quasi- Coulomb potential $V(z)$ (Eqn. (2)) and the electric field.

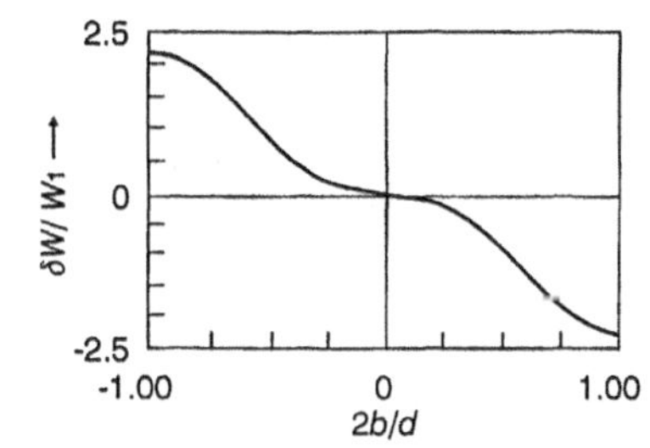

Figure 4. The energy shift of the ground impurity level $\delta W/W_1$ (Eqn.(12)) as function of the displacement of the impurity centre b from the mid-point of the QW.

centre is displaced from the mid-point of the QW ($b \neq 0$), an inversion effect arises. The energy levels change as the direction of the electric field changes.

In the case of a weak electric field, when both the Coulomb field of the impurity centre and the external electric field E are considered to be perturbations to the states of the electron in the QW, the expression for the inversion shift of the ground energy level $\delta W = W(+E) - W(-E)$ can be given in the simple form[8]

$$\delta W \sim W_1[C + \ln(\pi^2 a_B^2/2d^2)] \sin(\pi b/d) \tag{12}$$

where $W_1 = eEd(d/a_0)$. The dependence of the dimensionless inversion shift of the ground impurity level upon the dimensionless displacement of the impurity centre is depicted in Fig. 4. Estimates of the suitable values for the parameters for the GaAs QW ($\mu = 0.067m_0$, $\epsilon = 12.5$, $a_0 = 98.7$Å) for the well of width $d = 70$Å and displacement $b/d = 0.25$ are made. For a strong magnetic field ($B = 40T$, $a_B/a_0 = 0.4$) and a weak electric field $E = 15\text{kVcm}^{-1}$ the inversion shift of the ground level may be deduced from Eq. (12) and gives $\delta W \approx 1.2$meV. For the case of a strong electric field $E \sim 150\text{kVcm}^{-1}$, numerical calculations using the same parameters as above but with the displacement $b/d = 0.33$, an inversion shift $\delta W \approx 2.8$meV is obtained[9].

The coherent resonance tunnelling[10]

We consider now the case in which the impurity is positioned at the left hand boundary of the wide ($d \gg a_0$) QW. The schematic form of the combined potential is depicted in Fig. 5. Such a system may be treated as if it were a resonance structure. This is because the combined potential has a shape similar to that of a double quantum well (DQW). One of the wells is formed by the quasi-Coulomb one-dimensional impurity potential and the QW boundary close to which the impurity is located, and the other well is made up from the electric field potential and the other boundary of the QW. In the zeroth approximation, on neglecting the tunnelling of the electron through the barrier separating the effective wells, the system of the energy levels is the sum of two independent series of energies. The first series is formed by the 'electric' levels $W_k^{(0)}$ ($k = 1, 2, 3, \ldots$) associated with right hand well. The second series $W_n^{(0)}$ ($n = 1, 2, 3, \ldots$) are the quasi-Coulomb levels shifted towards low energies by the electric field E. Under the condition $W_n^{(0)} = W_k^{(0)}$, the relevant quasi- Coulomb and 'electric' levels appear to be in resonance.

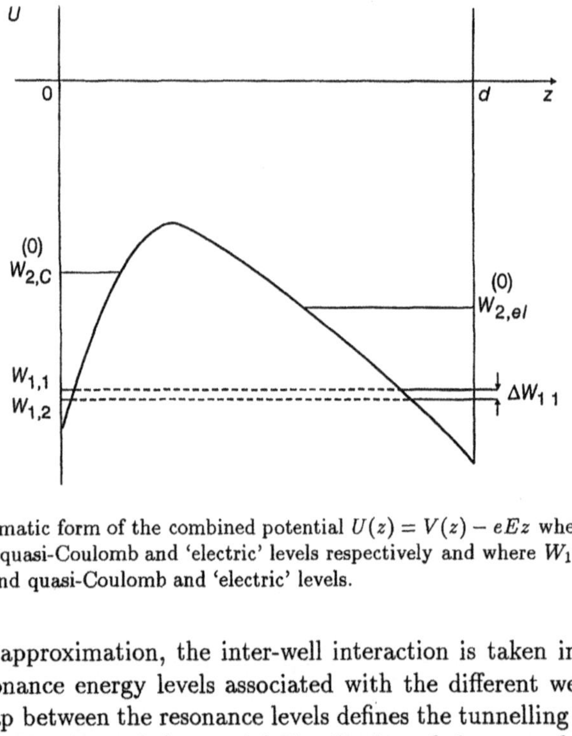

Figure 5. The schematic form of the combined potential $U(z) = V(z) - eEz$ where $W_{2,c}^{(0)}$ and $W_{2,el}^{(0)}$ are the first excited quasi-Coulomb and 'electric' levels respectively and where W_{11} is the resonance splitting of the ground quasi-Coulomb and 'electric' levels.

In the next approximation, the inter-well interaction is taken into account. As a result, the resonance energy levels associated with the different wells are found to anticross. The gap between the resonance levels defines the tunnelling time and results in an oscillatory behaviour of the spatial distribution of the wave function. This in turn defines the frequency of radiation emitted by such a structure. On equating the expressions for the quasi-Coulomb $W_n^{(0)}$ and the 'electric' $W_k^{(0)}$ levels, the scaled resonance electric fields s_{nk} has the form

$$s_{nk} = 1/n^2(a_0/d) + \alpha_k/n^{4/3}(a_0/d)^{5/3} + 1/3(a_0/d)^2 \tag{13}$$

where $s = E/(1/2)[e(4\pi\epsilon_0\epsilon a_0^2)^{-1}]$ is the electric field E scaled relative to the impurity electric field and where $\alpha_1 = 2.34; \alpha_2 = 4.09; \alpha_3 = 5.52; \ldots$ for example. The resonance fields s_{nk} define the cases in which the n-th quasi-Coulomb level and k-th 'electric' level are in resonance. The relevant resonance gap is defined by

$$\Delta W_{nk} = 2R\Delta_{nk}(s_{nk})/n^3 \tag{14}$$

In the above expression, the parameters are defined by

$$\Delta_{nk}(s) = \beta_{nk}^{1/2} s^{1/3} e^{-\phi(s)} \quad \text{and} \quad \Phi(\lambda) = 2/[3sn^3] + n\ln(sn^3/8) \tag{15}$$

and where

$$\beta_{nk} = -\frac{Bi(-\alpha_k)n^4}{4A'i(-\alpha_k)n!^2} \tag{16}$$

where Ai and Bi are the Airy functions. The above analytical expressions are valid for the weak electric fields such that $sn^4 \ll 1$.

It is clear from Eqns. (14) - (16) that the resonance gap Δn_k increases as the resonance field s_{nk} increases. Also, from the expression (13), the dependence of the resonance field s_{nk} on the 'electric' (k) and quasi-Coulomb (n) indexes and on the width of the QW d is obtained. For a fixed index for the quasi-Coulomb level n, the resonance field s_{nk} increases as a function of the index of the 'electric' level k. Meanwhile, for a fixed index k, the resonance field s_{nk} decreases with increasing index n. The wider the QW, the smaller the resonance field s_{nk}.

Taking values for the parameters for the GaAs QW applicable to the case of a strong magnetic field ($B = 40$T) and a weak electric field ($E_{11} = 3.8\text{x}10^4\text{Vm}^{-1}$, $s_{11} = 0.066$), the resonance splitting ΔW_{nk} of the ground quasi-Coulomb and 'electric' levels ($k = n = 1$, $\beta_{11} = 0.21$,) is obtained from Eqs. (14)-(16) such that $\Delta W_{11} = 1.2$meV. This gap corresponds to the frequency 0.31THz of the emitted radiation. For the value of the electric field E_{11}, the penetration through the potential barrier is relatively weak. When this value for E_{11} is exceeded, the penetration increases and the above approach becomes inappropriate. However, clearly in the presence of a stronger electric field ($s \leq 1$), the effect of the resonance splitting still holds.

REFERENCES

1. B. S. Monozon and A. G. Zhilich *Sov. Phys. JET P b* 73:1066 (1991).
2. B. S. Monozon, C. A. Bates. J. L. Dunn and M. J. Pye *Phys Rev. B* 53:12938 (1996).
3. H. Hasegawa and R. E. Howard *J. Phys. Chem. Solids* 21:179 (1961).
4. R. L. Greene and K. K. Bajaj *Phys. Rev. B* 34:951 (1986).
5. J. Cen and K. K. Bajaj *Phys. Rev. B* 48:8061 (1993).
6. T. Kuhn, G. Mahler, J. L. Dunn and C. A. Bates *J.Phys.Condens.Matter* 6:757 (1994).
7. B. S. Monozon, C. A. Bates, J. L. Dunn and M. J. Pye *Superlatt. Microstruct.* 21:151 (1997).
8. B. S. Monozon, P. Huther, P. Wies, H. Gerlach and M. Schmidt *Phys. Status Solidi (b)* 174:175 (1992).
9. B. S. Monozon and A. N. Shalaginov *Solid State Commun.* 89:167 (1994).
10. B. S. Monozon, C. A. Bates, J. L. Dunn and M. J. Pye *Phys. Rev. B* 55:9779 (1997).

LIST OF PARTICIPANTS

Dr. J. Ackermann
IMB Jena
Beutenbergstr. 11
07745 Jena, Germany
jack@imb-jena.de

Dr. P. Arnault
Commissariat à l'Energie Atomique
Centre d'Etudes de Limeil-Valenton
94195 Villeneuve-Saint-Georges, France
arnault@limeil.cea.fr

Prof. D. Baye
Physique Theorique et Mathematique
Code Postal 229
Universite Libre de Bruxelles
B-1050 Brussels, Belgium
dbaye@ulb.ac.be

W. Becken
Theoretische Chemie
Physikalisch-Chemisches Institut
Universität Heidelberg
Im Neuenheimer Feld 253
69120 Heidelberg, Germany
wolfgang@tc.pci.uni-heidelberg.de

Dr. V. Bezchastnov
Department of Theoretical Astrophysics
Ioffe Institute of Physics and Technology
Politekhnicheskaya 26
194021 St. Petersburg, Russia
vit@astro.ioffe.rssi.ru

Dr. M. Braun
Physics Department
University of South Africa (UNISA)
P.O. Box 392
0001 Pretoria, South Africa
braunm@kiaat.unisa.ac.za

Prof. P. Braun
Insitute of Physics
Theoretical Department
St. Petersburg State University
Ulianovskaja 1
SU-198904 St. Petersburg, Russia
braun@niif.spb.su

Dr. A. Buchleitner
Max-Planck Institut für Quantenoptik
Hans-Kopfermann-Str. 1
85748 Garching, Germany
adb@mpq.mpg.de

Prof. L. S. Cederbaum
Theoretische Chemie
Physikalisch-Chemisches Institut
Universität Heidelberg
Im Neuenheimer Feld 253
D-69120 Heidelberg, Germany
h35@ix.urz.uni-heidelberg.de

Prof. D. M. Ceperley
National Center for Supercomputing Applications and
Department of Physics
University of Illinois at Urbana-Champaign
1110 West Green Street
Urbana, Il 61801, USA
ceperley@uiuc.edu

Prof. J. P. Connerade
Blackett Laboratory
Imperial College
London SW 72 AZ, UK
j.connerade@ic.ac.uk

T. Detmer
Theoretische Chemie
Physikalisch-Chemisches Institut
Universität Heidelberg
Im Neuenheimer Feld 253
69120 Heidelberg, Germany
thomasd@tc.pci.uni-heidelberg.de

Dr. P. Faßbinder
Lehrstuhl für Theoretische Astrophysik
Universität Tübingen
Auf der Morgenstelle 10
D-72076 Tübingen, Germany
fassbind@tat.physik.uni-tuebingen.de

Prof. H. Friedrich
Technische Universität München
Physik Department
D-85747 Garching b. München, Germany
hfriedri@physik.tu-muenchen.de

Dr. S. Friedrich
Institut für Astronomie
Universität Tübingen
Waldhäuserstr. 64
72076 Tübingen, Germany
susanne@ait.physik.uni-tuebingen.de

Dr. S. Fritzsche
FB Physik
Universität Kassel
Heinrich-Plett-Straße 40
34132 Kassel, Germany
fritzsch@physik.uni-kassel.de

Prof. M. R. Godefroid
Laboratoire de Chimie Physique Moleculaire
Universite Libre de Bruxelles (ULB)
CP. 160/09
50, av. F.D. Roosevelt
B-1050 Bruxelles, Belgium
mrgodef@ulb.ac.be

R. Gonzalez-Ferez
Instituto Carlos I de Fisica Teorica & Computacional
Facultad de Ciencias
Universidad de Granada
Campus de Fuentenueva
E-18071 Granada, Spain
prointas@goliat.ugr.es

Dr. H. Held
Universität München
Sektion Physik
Am Coulombwall 1
85748 Garching, Germany
held@laser.physik.uni-muenchen.de

Prof. W. Hogervorst
Laser Centre Vrije Universiteit
Department of Physics and Astronomy
De Boelelaan 1081
1081 HV Amsterdam, The Netherlands
wh@nat.vu.nl

K. Hornberger
Max-Planck-Institut für Quantenoptik
Postfach 1513
85748 Garching, Germany
klh@mpq.mpg.de

Dr. W. Ihra
Department of Mathematics
Royal Holloway College (University of London)
Egham, Surrey TW20 OEX, UK
uhah010@rhbnc.ac.uk

A. Johnson
Royal Holloway College (University of London)
Department of Mathematics
Egham, Surrey TW20 0EX, UK
phah032@vms.rhbnc.ac.uk

Dr. S. Jordan
Institut für Astronomie und Astrophysik
Olshausenstr. 40
D-24098 Kiel, Germany
jordan@astrophysik.uni-kiel.de

Dr. K. Karremanns
Laser Centre Free University
Atomic Physics
De Boelelaan 1081
1081 HV Amsterdam, The Netherlands
kees@nat.vu.nl

Prof. T. O. Klaassen
Department of Applied Physics
Delft University of Technology
P.O.Box 5046
2600 GA Delft, The Netherlands
tjeerd@hfwork1.tn.tudelft.nl

Prof. D. Koester
Institut für Astronomie und Astrophysik
Olshausenstr. 40
D-24098 Kiel, Germany
koester@astrophysik.uni-kiel.de

Dr. J. Main
Institut für Theoretische Physik I
Ruhr-Universität Bochum
44780 Bochum, Germany
main@tp1.ruhr-uni-bochum.de

Dr. P. Maksym
Department of Physics
University of Leicester
University Road
Leicester LE1 7RH, UK
map@le.ac.uk

Prof. N. H. March
6 Northcroft Road
Egham, Surrey TW20 ODU, UK

Dr. V. Melezhik
Laboratory of Nuclear Problems
JINR, Dubna
Moscow Region 141980, Russia
melezhik@nusun.jinr.dubna.su

Prof. B. S. Monozon
Physics Department
Marine Technical University
3 Lotsmanskaya St.
St. Petersburg 190008, Russia
karpov@bsft.spb.su

Dr. P. O'Mahony
Royal Holloway College (University of London)
Department of Mathematics
Egham, Surrey TW20 0EX, UK
uhah005@vms.rhbnc.ac.uk

Prof. G. G. Pavlov
Pennsylvania State University
Department of Astronomy & Astrophysics
525 Davey Laboratory
University Park, PA 16802, USA
and
Department of Theoretical Astrophysics
Ioffe Institute of Physics and Technology
Politekhnicheskaya 26
St. Petersburg 194021, Russia
pavlov@astro.psu.edu

D. Pingel
Theoretische Chemie
Physikalisch-Chemisches Institut
Universität Heidelberg
Im Neuenheimer Feld 253
69120 Heidelberg, Germany

Dr. A. Y. Potekhin
Department of Theoretical Astrophysics
Ioffe Institute of Physics and Technology
Politekhnicheskaya 26
St. Petersburg 194021, Russia
palex@astro.ioffe.rssi.ru

P. Pourre
Commissariat à l'Energie Atomique
94195 Villeneuve-St-Georges, France

Dr. K. Runge
Department of Physics
PO Box 118435
University of Florida
Gainesville, FL 32611-84325, USA
runge@qtp.ufl.edu

Prof. J. R. Sabin
Department of Physics
PO Box 11843
University of Florida
Gainesville, FL 32611-8435, USA
sabin@qtp.ufl.edu

Dr. P. Schmelcher
Theoretische Chemie
Physikalisch-Chemisches Institut
Universität Heidelberg
Im Neuenheimer Feld 253
D-69120 Heidelberg, Germany
peter@tc.pci.uni-heidelberg.de

Dr. W. Schweizer
Lehrstuhl für Theoretische Astrophysik
Universität Tübingen
Auf der Morgenstelle 10
D-72076 Tübingen, Germany
and
Ruhr-Universität Bochum
Fakultät für Physik und Astronomie
Theoretische Physik I
D-44780 Bochum, Germany
schweize@tat.physik.uni-tuebingen.de

A. Scrinzi
Institut für theoretische Physik
Universität Innsbruck
Technikerstrasse 25
6020 Innsbruck, Austria
scrinzi@aimep0.imep.univie.ac.at

M. Stehle
Lehrstuhl für Theoretische Astrophysik
Universität Tübingen
Auf der Morgenstelle 10
72076 Tübingen, Germany
stehle@tat.physik.uni-tuebingen.de

Prof. K. Taylor
Department of Applied Mathematics and Theoretical Physics
The Queen's University Belfast
Belfast BT7 1NN, Northern Ireland
k.taylor@am.qub.ac.uk

J. Trost
Technische Universität München
Theoretische Physik, T30
85747 Garching, Germany
jtrost@physik.tu-muenchen.de

Prof. T. Uzer
School of Physics
Georgia Institute of Technology
Atlanta, GA 30332-0430, USA
tuzer@tamop.physics.gatech.edu

Prof. J. Ventura
Department of Physics
University of Crete and Research Center of Crete
GR-71110 Heraklion, Crete, Greece
ventura@physics.uch.gr

Prof. G. Vignale
Department of Physics
University of Missouri, Columbia
Columbia, Missouri 65211, USA
physvign@mizzou1.missouri.edu

Prof. D. Weiss
Max-Planck-Institut für Festkörperforschung
D-70569 Suttgart, Germany
dieter.weiss@physik.uni-regensburg.de

Prof. G. Wunner
Ruhr-Universität Bochum
Fakultät für Physik und Astronomie
Theoretische Physik I
D-44780 Bochum, Germany
wunner@tp1.ruhr-uni-bochum.de

INDEX

GPSR Compliance

The European Union's (EU) General Product Safety Regulation (GPSR) is a set of rules that requires consumer products to be safe and our obligations to ensure this.

If you have any concerns about our products, you can contact us on ProductSafety@springernature.com

In case Publisher is established outside the EU, the EU authorized representative is:

Springer Nature Customer Service Center GmbH
Europaplatz 3
69115 Heidelberg, Germany

Batch number: 10755805

Printed by Printforce, the Netherlands